BRYOPHYTES OF COLORADO
MOSSES, LIVERWORTS, AND HORNWORTS

BRYOPHYTES OF COLORADO
Mosses, Liverworts, and Hornworts

William A. Weber
Professor and Curator Emeritus, Herbarium COLO
University of Colorado Museum, Boulder
Fellow of the Linnean Society of London

and

Ronald C. Wittmann
Museum Associate, Herbarium COLO, Boulder

BRYOPHYTES OF COLORADO: Mosses, Liverworts and Hornworts, W. A. Weber and R. C. Wittmann

Copyright © 2007 Pilgrims Process, Inc.

Publication of this book was supported entirely by the authors

ISBN: 978-09790909-1-2

Library of Congress Control Number: 2007930118

Printed in the United States of America

0 9 8 7 6 5 4 3 2

Set in Garamond True Type

Printed on 551 PPI, white, acid-free paper

Cover design by Gary White

Dedicated to

FREDERICK J. HERMANN, 1906–1987

"His work will not end until we have traced each delicate web that ties us with all the life of the planet. His greatest worry was that we will finish the planet before we finish that work. The world and all its beauty belongs to all of us, not to politicians and mere ideologies. That our weapons could eradicate all life—plants, insects, mammals as much as man—was a terrible fear for him, and he worked hard, contributing hours and money, against that fear." Eric Hermann in Voss & Reznicek (1988).

CONTENTS

PREFACE

Bryophytes consist of a variety of small green plants including mosses, liverworts and hornworts. (Liverworts and hornworts are also called hepatics.) Although almost too small to appreciate except in mass, they are hardly simple. Unlike the higher plants, to which our naked eyes are attracted by their beautiful flowers, varied leaf forms, or stately trunks, a bryophyte has to be examined in all its parts with the aid of a hand lens or microscope. Such close examination reveals some of the most interesting, unusual, and captivating structures in the plant world.

My colleague, Ron Wittmann, a professional physicist, and I have worked together for some 30 years. Initially we studied and wrote about the flowering plants and the lichens; however, we gradually expanded our interests to the bryophytes, and now our full time has been occupied with them for more than a decade. We agree that this period has been the most exciting one of our botanical lives. (The senior author uses the first person singular).

More than thirty years ago, a preliminary publication (Weber, 1973) presented a catalog, keys, and helpful hints on recognition of the mosses. The total came to about 290 species. The liverworts were not treated. Since we began intensive field work, the number of bryophyte species recorded for Colorado has risen to over 500. Even though we covered a lot of ground, it is impossible to explore Colorado adequately in two combined lifetimes. We argue whether we have even reached the point of diminishing returns. Probably we have not. Although necessarily incomplete, this book is the most comprehensive study of Colorado mosses and liverworts to date.

Although not intended as an introductory handbook, beginners will find this book a valuable reference. The Europeans would call it an 'Excursion-Flora' since it does not have complete descriptions. Keys to the genera and species are combined with notes on recognition in the field. Much more attention is paid to habitats than one finds in other manuals, for in the Rocky Mountains habitats are not blended as they are in more humid climates.

At the present time a great collaboration of amateur and professional bryologists are cooperating in a project to produce a *Bryophyte Flora of North America*. This will be the 'Bible', encompassing three large volumes, containing keys, descriptions, references, and illustrations of the entire North American flora. So, why should we write a book on Colorado bryophytes? Because a local flora is much smaller than a regional one. Fewer species means fewer closely-related ones that require expert microscopy and access to a comprehensive herbarium in order to make identifications.

Anyone studying bryophytes must learn to rely on a variety of sources. Luckily a number of well-illustrated books are at hand, and the volume of information available on the internet is already very great. For various reasons it has been impossible for us to provide illustrations of more than a few interesting species. However, in this 'information age', the internet gives us an abundance of basic information about the life history and anatomy of bryophytes, as well as images of genera and species. The study of bryophytes is not nearly as difficult as it used to be.

We find the study of bryophytes in the field to be extremely exciting, and present this analysis as a beginning step in attracting more serious students of these wonderful little plants that are the ultimate survivors. The color green is a friendly one, and people who study bryophytes are among the most pleasant and cooperative in the world—and they walk slowly. We hope that this book will open a new, small world to the reader.

ACKNOWLEDGMENTS

We are much indebted to a number of bryologists whose help has been freely offered. Among them are: Richard Andrus (*Sphagnum*), Bruce Allen (*Fontinalis*), Hans Blom (*Schistidium*), David Cooper (fen mosses), Howard Crum(minute pleurocarps), Claudio Delgadillo *(Crossidium)*, Patricia Eckel (*Tortella*), Jan-Peter Frahm (*Campylopus, Paraleucobryum*), Henk Greven (*Grimmia*), M. T. Gallego (*Syntrichia*), Dana Griffin III (*Anacolia*), Roxanne Hastings (*Coscinodon, Grimmia*), Lars Hedenäs (*Amblystegiaceae*), Marie Hicks (thalloid liverworts), Kjeld Holmen (*Oreas*), Diana Horton (*Encalypta*), Robert Ireland (*Dicranum*), David Jamieson (*Hygrohypnum*), Timo Koponen (Mniaceae), Terry McIntosh (*Schistidium),* Brent Mishler (*Syntrichia*), Jesus Muñoz (*Grimmia*), Norton Miller (alpine mosses and hepatics), Barbara Murray (*Andreaea*), Ryszard and Halina Ochyra (*Palustriella* and *Racomitrium, sensu lato*), Ronald Pursell (*Fissidens*), Jon Shaw (*Mielichhoferia, Pohlia*), Wilfred Schofield (*Hypnum*), John Spence (*Bryum, sensu lato*), Toby Spribille (Montana taxa), Dale Vitt (*Orthotrichum*), Richard Zander (*Pottiaceae*), and William C. Steere (Arctic mosses).

We particularly appreciate the companionship of Lewis Anderson, Gay Austin, David Cooper, Dina Clark, Eilif Dahl, Alison Dibble, Vera Evenson, Dana Griffin III, Frederick G. Hermann, JoAnn Flock, David Jamieson, George Neville Jones, Jean Kekes, Tass Kelso, Vera Komarkova, Timo Koponen, Mary Lincoln, Norton Miller, Patricia Nelson, Holmes Rolston III, Mark Seaward, Bernie and Kim Smith, Toby Spribille, Ron and Jean Tidball, and Patricia Wikel, who worked in the field with us, deposited collections in the herbarium, or brought new records to our attention. Special thanks go to Paula Lehr for her enthusiasm for the project and her unflagging help and encouragement along the way.

We also thank the Colorado Native Plant Society for a John Marr grant for field expenses.

I thank my early mentors and professional colleagues whose helpful instruction over the years, in the field and laboratory, and through correspondence, have been of good counsel. Although at the age of twelve I was allowed to walk through Bronx Park with an elderly man who showed me my first prothonotary warbler, I never realized until many years later that my friend, Mr. Robert Statham Williams, was a famous bryologist. Another famous bryologist, Elizabeth Gertrude Britton, shook her parasol at me and chased me out of the park when I left the trail to pick a skunk cabbage leaf for show-and-tell. I wonder whether some of their ectoplasm wore off on me. My favorite biology teacher in Evander Childs High School in the Bronx, Grace Esternaux, took a course in bryophytes at Cornell, and showed me in 1935 how to collect and learn the bryophytes using Abel Joel Grout's *Mosses with a Hand Lens*. My first mentor in bryology was Dr. Henry S. Conard, who identified my collections from the Columbia River Gorge while I was performing work of national importance as a conscientious objector at Cascade Locks during World War II. When I was in Sweden in 1957–58, Herman Persson helped me to realize the great contributions made by Scandinavian bryologists to the knowledge of the taxa common to our two areas.

Although most of the keys are original, one can hardly improve upon a good key already available. We freely acknowledge the use of published keys and critical observations made by our colleagues and predecessors. We are most of all indebted to our sharpest critics—our students, who have encouraged and stimulated us to provide workable keys for them rather than for our own self-gratification.

INTRODUCTION

Colorado is famous for its displays of wild flowers. People come here from all parts of the world to see and photograph them, but few people realize that Colorado also has a distinctive bryophyte flora, as well. This book is the first comprehensive work treating the bryophytes known to occur in Colorado.

Wild-flower lovers might think that bryophytes are merely small versions of flowering plants. For example, on one collecting trip onto the local tundra, a passerby asked us what we were doing, and we replied, "We're studying mosses." "Oh," replied our new acquaintance, "Do you mean the little ones with the pink flowers?" She meant the vascular plant, *Silene acaulis*, commonly called Moss Campion.

About this Book

No new names or combinations are proposed in this volume.

This work treats all bryophytes known to occur in Colorado. It should also be useful in adjoining parts of neighboring states and more generally in the mountainous regions of the interior western United States.

Mosses are organized alphabetically by family, genus and species. Detailed descriptions and illustrations are not provided generally. Rather, we concentrate on distinguishing characters of similar taxa (especially 'field characters') and provide comprehensive ecological notes. Detailed descriptions and drawings may be found in the references and, in many cases, on the Internet. Novices will be well served by supplementing this guide with one of several excellent introductory texts on the science of bryology (see Source Books, page 26).

Dichotomous keys are furnished for the identification of genera and species; however, we have been unable to produce a useful key to the families. Several family keys are available (for example, on the Internet) if our readers would like to try them. For the seasoned bryologist, family identification is a matter of experience. For the beginner, there is no substitute for a mentor!

An important feature are the Catalogs of genera and species that provide a check list of the bryophytes . The Catalogs can be used to determine family assignments and also contain information on naming authorities, synonymy, and erroneous reports. We have attempted to account for all bryophyte names used in published references to the Colorado flora. In addition, an index to specific epithets is provided to help resolve generic assignments.

Liverworts and Hornworts are treated in a fashion parallel to that used for the Mosses.

Collections of relatively rare or poorly known species are cited here for the record. Specimens housed at COLO—the standard acronym for the University of Colorado Herbarium at Boulder— are cited with a running accession number prefaced by 'B-'. The authors' collections, for example, are cited as *W&W B-*. . . Specimens housed elsewhere are often cited with the herbarium acronym (see Holmgren & Holmgren, 1998).

We have relied heavily on the preliminary treatments of bryophyte taxa developed for the *Flora of North America*. The acronym BFNA is used throughout this book to acknowledge this source. In particular, we thank the BFNA project for permission to reproduce Patricia Eckel's beautiful drawings of several 'special' Colorado mosses.

Family names are often abbreviated using three-letter acronyms:

Amblystegiaceae	AMB	Calypogeiaceae	CLY	Fontinalaceae	FNT
Andreaeaceae	AND	Campyliaceae	CMP	Frullaniaceae	FRL
Aneuraceae	ANR	Catascopiaceae	CTS	Funariaceae	FNR
Anomodontaceae	ANM	Cephaloziaceae	CPH	Geocalycaceae	GEO
Antheliaceae	ANT	Cephaloziellaceae	CPL	Grimmiaceae	GRM
Anthocerotaceae	ANC	Cleveaceae	CLV	Haplomitriaceae	HPL
Aulacomniaceae	AUL	Climaciaceae	CLM	Helodiaceae	HLD
Aytoniaceae	AYT	Conocephalaceae	CNC	Hylocomiaceae	HYL
Bartramiaceae	BRT	Dicranaceae	DCR	Hypnaceae	HPN
Blasiaceae	BLS	Ditrichaceae	DTR	Jungermanniaceae	JNG
Brachytheciaceae	BRC	Encalyptaceae	ENC	Lepicoleaceae	LPC
Bryaceae	BRY	Entodontaceae	ENT	Lepidoziaceae	LPD
Bryoxiphiaceae	BRX	Fabroniaceae	FBR	Leptodontaceae	LPT
Buxbaumiaceae	BXB	Fissidentaceae	FSS	Leskeaceae	LSK

Lophoziaceae	LPH	Pelliaceae	PLL	Scapaniaceae	SCP
Lunulariaceae	LNL	Plagiochilaceae	PLC	Scouleriaceae	SCL
Marchantiaceae	MRC	Plagiotheciaceae	PGT	Seligeriaceae	SLG
Marsupellaceae	MRS	Polytrichaceae	PLT	Sphagnaceae	SPH
Meesiaceae	MEE	Porellaceae	PRL	Splachnaceae	SPL
Metzgeriaceae	MTZ	Pottiaceae	PTT	Tetraphidaceae	TTR
Mniaceae	MNI	Pterigynandraceae	PTR	Theliaceae	THL
Neckeraceae	NCK	Radulaceae	RDL	Thuidiaceae	THD
Notothyladaceae	NTT	Rebouliaceae	RBL	Timmiaceae	TMM
Odontoschismaceae	ODN	Rhytidiaceae	RHY	Trichocoleaceae	TRC
Orthotrichaceae	ORT	Ricciaceae	RCC		

Abbreviations for the Colorado counties are taken from the Smithsonian Institution River Basin Surveys with the exception of Broomfield and Montezuma counties. The following abbreviations are used in our citations of specimens:

Adams	AM	Denver	DV	Kit Carson	KC	Phillips	PL
Alamosa	AL	Dolores	DL	Lake	LK	Pitkin	PT
Arapahoe	AH	Douglas	DA	La Plata	LP	Prowers	PW
Archuleta	AA	Eagle	EA	Larimer	LR	Pueblo	PE
Baca	BA	Elbert	EL	Las Animas	LA	Rio Blanco	RB
Bent	BN	El Paso	EP	Lincoln	LN	Rio Grande	RG
Boulder	BL	Fremont	FN	Logan	LO	Routt	RT
Broomfield	BR	Garfield	GF	Mesa	ME	Saguache	SH
Chaffee	CF	Gilpin	GL	Mineral	ML	San Juan	SJ
Cheyenne	CH	Grand	GA	Moffat	MF	San Miguel	SM
Clear Creek	CC	Gunnison	GN	Montezuma	MZ	Sedgwick	SW
Conejos	CN	Hinsdale	HN	Montrose	MN	Summit	ST
Costilla	CT	Huerfano	HF	Morgan	MR	Teller	TL
Crowley	CW	Jackson	JA	Otero	OT	Washington	WN
Custer	CR	Jefferson	JF	Ouray	OR	Weld	WL
Delta	DT	Kiowa	KW	Park	PA	Yuma	YM

We ask that collectors of bryophytes in Colorado deposit critical specimens with herbarium COLO for permanent documentation. We also welcome specimens sent to us for identification. Please contact us before transmitting specimens.

The Uniqueness of Bryophytes

In a recent book review (2000), Brent Mishler presents what he calls "a radical view of bryophyte biology, or, Mosses are from Mars, Vascular Plants are from Venus." He writes: "The bryophytes are clearly a key to understanding how the embryophytes are related to each other and deciphering how they came to conquer the hostile land environment from their primitive home in fresh water—habitats still occupied by relatives of the land plants, the green algae. Yet despite their diversity, phylogenetic importance, and key roles in the ecosystems of the world, study of many aspects of the biology of bryophytes has lagged behind that of the larger land plants, perhaps because of their small size and the few scientists specializing on the group. This is unfortunate because of the intrinsic scientific interest of these plants. The bryophytes contain some of the most species-rich lineages of land plants, presenting a challenge (as well as opportunities) for understanding processes of evolutionary diversification."

In a summary, Mishler lists nine differences between bryophytes and vascular plants that are important for anyone interested in bryophytes to understand. This essay was published in a professional journal that would not be seen by most amateurs, so we are quoting these nine points in full:

1. *Haploid dominance in the alternation of generations.* The green, vegetative part of the life-cycle in bryophytes is haploid. Without the genetic benefits of dominance, genes acting in the gametophyte are presumably subject to relatively severe selection.

2. *Extensive phenotypic plasticity.* Studies have shown that bryophytes tend to have very high amounts of morphological and physiological plasticity [as opposed to genetic variation]. This may compensate for their low levels of ecotypic differentiation (perhaps due to haploidy).

3. *Poikilohydry and dessication tolerance.* Poikilohydry is the rapid equilibration of the plant's water content to that of the surrounding environment, while desiccation tolerance is the ability of a plant to recover after being air-dry at the cellular level. All bryophytes have these abilities to some extent, but . . . this was lost in the larger, more complex, and endohydric tracheophytes.

4. *Need for free water for sexual reproduction.* A residual feature of the early land plants is the constraint imposed by the swimming sperm. Swimming gametes have short dispersal distances that lead to frequent inbreeding in monoicous species and lack of sporophyte production in dioicous species.

5. *The clump as a "super-organism".* Many mosses and some liverworts are essentially social organisms. This results from the combination of clonal growth, poikilohydry, and external water conduction. The plants in a clump are subject to natural selection as a group. Intimate contact of each vegetative cell with the environment, due to poikilohydry, lends itself to inter-plant chemical communication via pheromones.

6. *Heavy reliance on asexual reproduction.* Due to the difficulty of achieving fertilization, many bryophytes have evolutionarily lost functional sexuality. Since bryophytes grow from an apical cell, somatic mutation allows genetic variation even within clones.

7. *Small stature and the occupation of microhabitats.* Small size, lack of roots, and poikilohydry means that bryophytes are in a close relationship with only their immediate micro-environment. Over geological time, they may be less influenced by climatic change, and linger in refugial habitats.

8. *Less selection pressure from the biotic component of the environment than from the physical component.* Vagility [ability to spread] and establishment abilities of bryophytes are relatively poor. Available substrates are not filled in most mesic and xeric environments (although they may be in some hydric environments). The presence of other bryophytes nearby often appears beneficial to growth.

9. *Relatively slow evolutionary rates in morphology.* The fossil record of bryophytes indicates their ancient forms are very similar to modern ones. Biogeographically, bryophytes tend to follow the same historical patterns of disjunction as tracheophytes, but at a lower taxonomic level. This may indicate that developmental constraints play an unusually important role.

The overall effect of these features on the evolutionary ecology of bryophytes makes them profoundly different. By studying bryophytes and comparing their life style to that of tracheophytes, the student can learn to observe structure closely, think critically about evolutionary inferences, and comprehend how different lineages can take different functional paths in response to the same stimuli.

Because of these features, bryophytes are resistant to the forces of natural selection, and show little variability at the species level. Mosses are an ancient group, much older than the vascular plants, and there are fewer taxa. Colorado is an example of a region with a highly relictual flora. We have no endemic species or genera. Instead we have an unusual number of widely disjunct species from the northern world.

Among Colorado bryophytes, there are a few variable species and we do not know whether the variation is genetic or environmental. *Ceratodon purpureus*, for example, shows extraordinary plasticity in the narrowness or broadness of its leaf shape. Thus, sterile specimens are very exasperating to inexperienced observers. The species of *Amblystegium* have wide ranges of variation (or is it plasticity). *Hypnum cupressiforme* also varies from robust to slender extremes. Nevertheless, the lack of genetic variation within most species is truly amazing.

Bryology in Colorado

Colorado is a bryological frontier. The eastern U. S. has had a long history of bryological exploration, professional and amateur interest, and useful publications. Field and laboratory studies have flourished. Contacts between regional scientists and those dealing with the similar floras of continental Europe are well-established. The eastern U. S. is no longer a frontier. It is entering a stage of maturity, in which regional and state floras are fairly common.

During the 1800's there were two great bryologists in America, who rightly may be called the 'fathers' of American bryology—William Starling Sullivant and Leo Lesquereux (Rodgers, 1940). These men and their collaborators, Thomas P. James and Coe Finch Austin, published the first 'comprehensive' moss flora of North America (Lesquereux & James 1884).

Very few mosses were collected in Colorado on the U. S. Geological Surveys (see Rothrock, 1878) and the Hayden Surveys (see Foster, 1994). Most collectors were medical men doing double duty. It is no secret that botanists were last in line for positions on general exploring expeditions. They were usually held hostage by the demands of the more important members. Their collecting spots were usually accidental, such as rest stops along the way. When the geographers, surveyors, military, geologists, and zoologists wanted to move their camp, the botanists had to go along with them. Even those who have been identified as botanists collected only vascular plants.

The interior western United States (Montana to New Mexico) is still a bryological frontier. Only Utah has a comprehensive bryophyte flora. Some states do not even have a published check list, and for states that do, these may be undocumented by specimen citations. Serious field work in some has not progressed far beyond the historical governmental surveys of the nineteenth century. Herbaria do not routinely have collections of bryophytes.

Colorado bryophytes have never been fully treated. The earliest paper attempting to list them is that of Leo Lesquereux (1874). A. J. Grout visited Colorado once and published a short paper (Grout 1916) on his collections near Tolland, in Gilpin County. Fred J. Hermann published a few papers, one on Rocky Mountain National Park (1987) and another on additions to the flora (1976). Hermann's life and career are reviewed by Voss and Reznicek (1988). J. H. Craft and I. H. Craft who taught at Adams State College at Alamosa, published a three-page paper (1952) listing by county some 40 mosses they had collected in southern Colorado, but cited no specimens. H. S. Conard, Frederick McAllister, and others collected mosses sporadically but no publications are based on their collections. The ground breaking, but somewhat premature, account of Weber (1973) is the precursor of this present guide.

The origin of the bryophyte collection at the University of Colorado can be traced to Geneva Sayre. She came from Menlo, Iowa, and went to Grinnell College, where she came under the wing of Henry S. Conard. When she graduated, during the Great Depression, she could not afford to go to graduate school, so Dr. Conard invited her to stay another year at Grinnell so that he could continue to be her mentor. In 1934 Conard arranged for her to spend the summer with Dr. Grout in Newfane, Vermont. With Grout she worked on the preparation of the text for the Splachnaceae, Timmiaceae, and Aulacomniaceae for his *Moss Flora of North America* (1928–1939). In 1935 she received the Master's Degree at the University of Wyoming, her thesis being the work just mentioned. In 1935 she went on to get a doctorate at the University of Colorado under Joseph Ewan.

Geneva had no mentors in the region so she kept in touch with Conard and Grout. The difficulties of writing a moss flora for Colorado (her thesis subject) were immense. There was no moss herbarium here; the library was poor. Grout's flora was only two-thirds published. Nevertheless, she did write a thesis. She was never very proud of it. Through no fault of hers, the information was just not ready for her, and field work must have been extremely difficult without a car. During this time she also studied the vascular plants. It is said that she wrote a spring flora of the Laramie area and a mountain flora of Boulder County; these were probably never really published and evidently no longer exist. But she did leave a small collection of her bryophytes, some 2000 specimens, which I discovered in a cardboard box soon after I arrived on campus in1946. This was the beginning of the University of Colorado bryophyte collection that now comprises over 116,000 specimens.

Geneva held an instructorship in the Biology Department and taught at the University Camp near Ward (now the Mountain Research Station) from 1938–1941 before taking an Assistant Professorship at Russell Sage College in Troy, New York. There she had to teach almost everything in the Biology Department (except bryophytes!) and spent her entire career there, becoming head of the department in 1946. She took early retirement in 1972 and held a part time position at Harvard in the Farlow Library and Cryptogamic Herbarium.

Because of the pressures of her various academic duties and the lack of an herbarium, she was never able to pursue a career in the taxonomy of bryophytes. Instead, she embarked on several bibliographic projects that are extraordinarily important basic tools for research in bryology (see Sayre, 1959–1975).

I got to know Geneva at national meetings and always looked forward to finding her there. She was a bright young lady full of ideas and conversation topics in bryology. In her biography there is nothing said about her having anything to do with the formation of the bryological herbarium at Boulder, but I feel we owe her a great deal for all that she achieved.

Fred Hermann (see Voss & Reznicek, 1988) came to Fort Collins in 1970 from the U. S. Forest Service herbarium in Washington, D.C. From then until his death in 1987 he devoted his spare time to collecting bryophytes, mostly in Rocky Mountain National Park. Fred and I frequently went into the field together and the herbarium has duplicates of most of his Colorado collections. His private herbarium went to the University of Michigan, where he started his botanical career as a student in the 1920s. It was a great privilege to have had his companionship throughout the years.

Fred was fortunate to find a compatible colleague, Holmes Rolston III, a philosophy professor at Colorado State University at Fort Collins. We are fortunate that these two men spent many days collecting bryophytes in and near Rocky Mountain National Park. Holmes was just the companion Fred needed, for it is very difficult to be a bryologist and hike alone in safety in the wilderness during one's declining years. Holmes proved to have not only stamina, but an excellent eye, and collected a number of exciting specimens. Holmes is a philosopher who applies religious teachings and ethics to environmental conservation. He has published many significant papers and has traveled world-wide on the lecture circuit. In 2003 he earned the Templeton Prize for Progress toward Research about Spiritual Realities, joining Mother Theresa and Alexander Solzhenitsyn in the list of notable recipients. We bryologists can be very proud that such an individual has worked in our midst.

In Colorado liverworts (technically called Hepatics) are much less well known than mosses. Alexander Evans (1915a) published a list of Colorado liverworts, but never visited Colorado to see them in the field. T. C. Frye & Lois Clark (1937, 1945) published a work on the North American liverworts and cited specimens from Colorado, but likewise did not visit the state. Their work at least permitted us to compile a check list from the literature. While Rudolph Schuster published a giant six-volume work (1966–1992) on the liverworts of part of eastern North America, he did not include observations on the western American flora. At the present time Won Shic Hong (1979–2002c), College of Great Falls, Montana, has published on the western American leafy liverworts. He has examined our collections and provided identifications which have proved very helpful to us. Nevertheless, the American West needs more researchers in the field of hepatics.

The following is a sampling of some other important historical collectors of Colorado bryophytes. This list is not exhaustive and excludes contemporary investigators.

Brandegee, Townshend Stith (1843–1925). Civil engineer with the Atchison, Topeka and Santa Fe Railroad, assigned in 1871 to the Canyon City area, became the county surveyor of Fremont County, and is said to have 'laid out' the town of Florence. He served as botanist on the Hayden Survey in 1875 and provided an early and very useful description of the flora of southwestern Colorado. His bryophytes came from 'S. W. Colorado' or 'within 100 miles of Cañon City'. He was the only resident botanist who collected any mosses here. A few of the fragmentary specimens on which new species were based have not been located again. Brandegee later went on to have a distinguished career in California, where he was a botanist at the University of California, Berkeley.

Conard, Henry Shoemaker (1874–1971). Important American bryologist, author of two guides for amateurs, teacher of bryologists, and professor at Grinnell College. He collected some mosses near Buena Vista while on vacation in the summer of 1941. His herbarium is at Iowa State University.

Downie, Timothy Campbell (1830–1875). Nicknamed 'Major Downie', he collected some mosses in Colorado in 1868, at Twin Lakes, with the Hayden Survey (Ewan & Ewan, 1981). According to Hayden (1869), the expedition came into Colorado first from Cheyenne, visiting points in the outer Front Range to Colorado Springs and Raton Pass. It returned to Colorado from Santa Fé, up the Colorado River, through the San Luis Valley, over Poncha Pass, through South Park, and into the upper Arkansas River Valley. Presumably Twin Lakes was their last camp before returning to Denver. The collections were to be turned over to the Smithsonian Institution, but we have been unable to find them.

Flowers, Seville (1900–1968). For many years Bill Flowers was the only active bryologist in the interior West. He spent his entire life in Utah, and had a full career at the University of Utah, where he was a revered teacher, a talented artist, and a bryophyte authority who collected extensively and published many papers. His field notes, original drawings, and specimens came to the Herbarium COLO, where they have been invaluable to us. Flowers accomplished, in his comparative isolation, an extraordinary service to bryology. See Behle (1984).

Grout, Abel Joel, (1867–1947). Grout made a short visit to Colorado in 1916, and collected in the vicinity of the University of Colorado's summer camp at Tolland in Gilpin County. He privately published the first comprehensive moss flora of North America (1928–1939). See Steere (1948).

Haines, Mrs. Mary Parry (1826–1884). A naturalist of eclectic tastes, custodian of the Paleontology Department, Joseph Moore Museum, Earlham College, Richmond, Indiana, and an amateur horticulturist. Although she never visited Colorado she was responsible for funneling several odd specimens [cf. *Orthotrichum hainesiae*) collected by her friends, Mrs. E. J. Spence and T. S. Brandegee, to bryological specialists (Flowers, 1942).

Holzinger, J. M. (1853–1929) was German-born Minnesotan who collected bryophytes in Colorado in Arapaho and Pike National Forests in 1896. His Colorado collections contributed to his published exsiccati, *Musci Acrocarpi Boreali-Americani* (see Sayre 1971, 1975).

Kiener, Walter B. (1894–1959) was a Swiss-born Colorado mountaineer who collected on Long's Peak in 1938, where he was a climber and guide. His bryophyte and lichen collections were acquired by the University of Nebraska in 1960. Duplicate specimens are at COLO.

Komarkova, Vera (1942–2005) a Czechoslovakian-born botanist and mountaineer whose team made the first successful climb of Annapurna by women. She was the pre-eminent plant ecologist of the Rocky Mountains, with experience in the European Tatra and Antarctica, a field botanist par excellence, with a keen eye for rare bryophytes. She published a two-volume work on the ecology of the Indian Peaks area (Komarkova, 1979). Her collections are at COLO.

Porter, Thomas Conrad (1822–1901), a botanist and clergyman who was attached to the Hayden Surveys from1869–1874 (see Foster, 1994). He was a professor at Lafayette College, Lancaster, Pa. Porter and Coulter (1874) published the first account of Colorado mosses by Leo Lesquereux.

Bryophyte Habitats

Cliffs, talus slopes, screes, and boulders. This is where the Rocky Mountains got their well-deserved name. North-facing exposures are often very rich in bryophytes. Massive granite surfaces are commonly dominated by pure carpets of *Hypnum revolutum*. In the foothills and montane zone other very common genera are *Grimmia, Hypnum, Orthotrichum*, and *Schistidium*. Protected alcoves usually support shade-loving genera such as *Neckera*. The shaded spaces on the underside of talus blocks also occasionally provide similar protection. Less common or rare cliff-dwelling genera include *Amphidium, Andreaea, Anoectangium, Barbula*, and *Didymodon*.

Forested areas. The largest number of bryophytes. including the most common and abundant species, occur in forested areas. However, forests are very diverse. Forests on steep slopes contain a mixture of dry forest species and cliff species. Steep drainages with seasonal run-off water support a few characteristic species, such as *Atrichum* and *Bryoerythrophyllum*. South-facing slopes, except those watered by seepage, are usually devoid of mosses.

Dry lodgepole pine forests are usually quite barren of bryophytes; so, except for scattered boulders, are dry aspen groves. Forest trails mostly traverse relatively dry forests, but these are very good places to discover and become acquainted with the more common boreal bryophytes of second-growth forests and the common mosses of granite boulders. The most interesting areas for unusual and rare species are the old growth and subalpine forests that are close enough to snow-melt streamlets to remain quite moist throughout the year. In the transition zone between the subalpine forests and the alpine tundra, interesting bryophytes can be found at the downslope peaty ledges of solifluction lobes at the lower ends of stunted tree islands.

Wet meadows, carrs and fens. Wet meadows on the eastern plains are very poor in mosses. They usually support *Drepanocladus aduncus*, but little else. Carrs are willow-dominated wetlands common on mountain flood-plains. These are not very productive of bryophytes because of erosive scouring by water, and density of *Carex* stands. However, at the bases of willow shrubs, a characteristic cluster of moss species includes *Climacium dendroides, Helodium blandowii*, and *Tomenthypnum nitens*.

Colorado has no true bogs. What have been called bogs in Colorado are really fens. Bogs are wetlands that are very poor in nutrients because most of the minerals are obtained from precipitation rather than from ground water. They are also referred to as ombrotrophic peatlands, meaning simply that all nutrients come strictly from precipitation. They are usually characterized by a pH of 4 or less. *Sphagnum* is the dominant vegetation of a bog.

Fens are similar to bogs, but they receive their nutrients from the surrounding ground water. Many of the same plants that are found in bogs are also found in fens. Fens are often referred to as minerotrophic peatlands because of the greater influx of nutrients from ground water. The term fen may be somewhat misleading, because the category includes some diverse habitats. For example, the word fen is used to describe some alkaline wetlands that are dominated by sedges and grasses and very few peat mosses.

So-called poor fens generally obtain more water from precipitation than from run-off from elevated places, and their pH is higher (5 to 6.5). High Creek Fen, in South Park, may be so characterized. Rich fens occur in level areas into which water drains from higher altitudes and flows through and out although at a rather slow rate, thus allowing nutrients to slowly accumulate. Chattanooga Fen, in the San Juan Mountains, is an example. See Cooper (1991–2002).

Sites that are dominated by running water are among the most varied habitats in the mountains. Streams vary from slowly flowing or intermittent streamlets to rushing cataracts. Some very common species, such as *Brachythecium rivulare* and *Hypnum cupressiforme* are widespread, while certain other species require some subtle characteristics of the stream that are not easily pin-pointed.

Walls, lawns, sidewalks, and greenhouses. A great deal has been written about the moss vegetation of walls in Europe, much less in the United States. In Colorado our buildings are not very old, and we lack the stone walls separating agricultural fields. Our dry climate does not encourage very much wall vegetation. Nevertheless, wall mosses can be found here, for example, on the buildings of the University of Colorado, where native stone, particularly the Lyons sandstone, has been used for facings. Mosses occur particularly on calcareous mortar in the cracks between the flagstones on east-facing walls. Old retaining walls also support mosses. When these occur in areas that are shaded, the number of species increases.

Some of the species that are common on mortar in horizontal cracks are: *Bryum argenteum, Barbula convoluta, Ceratodon purpureus, Coscinodon calyptratus, Didymodon rigidulus, Grimmia anodon, G. pulvinata, Orthotrichum hallii, Ptychostomum imbricatulum, Schistidium confertum, Syntrichia virescens*, and *Tortula muralis*. Shaded retaining walls, in addition to supporting these species, have some species of more mesic preferences, such as *Brachythecium erythrorrhizon, Rosulabryum flaccidum, Tortula atrovirens*, and *T. mucronifolia*.

There are a few weed mosses in Colorado. Lawns that are over-irrigated or poorly drained support mostly *Brachythecium erythrorrhizon* and *Ceratodon purpureus*. Sidewalk cracks, as they become eroded, are commonly filled with *Bryum argenteum*. For many years the University of Colorado had a system of shallow concrete ditches connected to a major irrigation canal. These ditches, which were once used for flood irrigation, are now dry but receive enough seepage moisture that their vertical sides are often plastered with mats of *Amblystegium varium*. A drain-pipe on the corner of Denison Building watered the surface at its base enough to support a nice colony of *Gemmabryum subapiculatum* until the Facilities Management people decided to cleanse the building of asbestos and built a temporary shack over the site. Most of the moss disappeared but is slowly coming back! Another site for the species is in a shaded garden along Broadway opposite the National Institute of Standards and Technology. The species has never been found in a natural situation in Colorado.

Golf courses along the Front Range have problems with weed mosses on their greens. Some of their troubles have to do with over-watering, but probably the mosses are encouraged by the heavy doses of fertilizers. This is evidently a problem of some local economic importance.

Greenhouses are undoubtedly the source of some of our weed mosses. Flower pots in greenhouses and nurseries usually contain masses of *Funaria hygrometrica* and often *Lunularia cruciata*. In a greenhouse on the University of Colorado campus we recently found *Amblystegium serpens* var. *juratzkanum* in many pots. Other species were more infrequent contaminants of vascular plants that might have come from anywhere in America. *Rosulabryum flaccidum* is a native species that commonly behaves like a weed in town. *Thuidium delicatulum* was found in one pot but has never been found in the wild in Colorado.

Other habitats. Alkaline flats are seasonally wet areas that sometimes support mosses. On the Western Slope, particularly around Delta, these are commonly filled with greasewood (*Sarcobatus vermiculatus*). In the shade of these shrubs a few species of mosses, such as *Crossidium* and *Syntrichia* are frequent.

Gypsum/salt domes retain water longer than the flat desert pavements. The most notable one in Colorado is in Paradox Valley. The gypsum retains rainwater longer than the adjacent sandy flats, and is densely covered by a variety of crustose lichens. The moss flora is poor but especially interesting. *Didymodon nevadensis* occupies bare spaces between the lichens, and *Syntrichia caninervis* occurs in the shade of low *Atriplex* shrubs.

Around mineral springs travertine rock is formed by the interaction of mosses and water containing dissolved salts. Some characteristic species are *Didymodon tophaceus, Gymnostomum aeruginosum,* and *Hymenostylium.* So-called 'copper mosses' belonging to *Mielichhoferia* are locally abundant on mine tailings in very moist sites, especially in the San Juan Mountains around Red Mountain Pass.

Animals Associated with Mosses

Colorado bryologists are waiting for some field botanist/zoologist to combine the exciting study of some minute animals that live in intimate contact with mosses. The tardigrades or 'water bears' are interesting enough because of their extremely small size (hardly a millimeter long), their complicated anatomy (they have a mouth, gullet, and anus), and their remarkable shapes, colors, and hooked appendages. They wander around in moss tufts and have no apparent importance to us, but they have habitat preferences and geography. The study of tardigrades elicits money grants in exotic places such as Antarctica, or long-term studies in Italy. Here in Colorado, where virtually nothing is known about tardigrades, it would be an ideal, inexpensive, and rewarding hobby for someone willing to team up with an aficionado of mosses.

Maggie W. Ray (reference not available), zoologist at North Carolina State University, writes: "Among the meiofauna that live in the film of water on most moss plants (including *Sphagnum*) are three taxa which can enter into an ametabolic state when faced with dehydration or other extreme environmental challenges: Tardigrades, free-living nematodes, and rotifers. Animals which have undergone the process known as cryptobiosis are virtually indestructible. They are able to withstand normally lethal environmental extremes: High and low temperatures, high and low pH, very high pressure, and very low vacuum. Once the environmental stress is relieved, they absorb water, expand, and revive with apparently no ill effects. Their harmless nature may explain in part why they have not been very well-studied. Their ability to survive has been of interest lately in some areas of cellular research. Apparently while they are 'asleep' they do not age, and they can remain in this cryptobiotic state for decades. A few researchers suggest that this trait would be excellent for space travel."

Recently it was found that spring-tails and mites, which commonly roam in moss tufts, are able to mediate fertilization between separated male and female moss plants. They are attracted to fertile shoots and are able to carry sperm when a continuous water film is lacking. "Mosses, spring-tails, and mites are extant representatives of taxa that originated after the early phase of land colonization (circa 440–470 million years ago). Animal-mediated fertilization in mosses therefore potentially antedates similar syndromes in other plant groups" (Cronberg et al., 2006).

Mosses, Wet and Dry

In a semi-arid state like Colorado, mosses tend to be more inconspicuous than elsewhere because, in the absence of moisture, the foliage dries up and photosynthesis ceases, often for very long periods. It is well known that a large percentage of desert-steppe mosses tend to soak up water immediately, while aquatic mosses, when dried, resist re-wetting. Every amateur bryologist has observed the dramatic change in the attitude of the leaves of *Hedwigia ciliata* when wetted. When dry, the leaves are appressed to the stem, but suddenly, upon encountering a single drop of water the leaves spread widely at a 45 degree angle. This phenomenon is our common experience with moss leaves; just wet the stem and the leaves will assume a new position which is advantageous for exposing photosynthetic surfaces. We tend to regard this as a simple feature of hydration.

However, we cannot assume that this is a simple hydrodynamic movement. During our examination of a large number of species of *Bucklandiella* from all over the world, we discovered that hydration is by no means a simple matter. Most of us are accustomed to seeing mosses completely dry, and we douse them with water to freshen them up. I believe that few of us actually watch the process from beginning to end.

In examining dry material and wetting up single stems for microscopic study we watched, under the stereo microscope, the entire sequence of events, from completely dry to completely wet, and observed a remarkable phenomenon. We see two distinctly different patterns of leaf movements: in some species the leaves simply spread out, either stiffly or gracefully curved; in others the introduction of water causes the leaves to suddenly spread out away from the stem and form a complete circle, what one might call a complete recoil or a revolution! However, the process does not stop there, as one would expect. The leaves then more slowly reverse the process and straighten out again, assuming a straight or gently curved, erect form. Unless one sees the first movement one might remain oblivious of it.

Proctor (2000) describes the various ways in which moss stems and leaves receive moisture and the subsequent turgor pressure that results. He writes: "Many bryophytes, especially those of intermittently dry habitats, are ectohydric. Water conduction is predominantly external, in an interconnecting network of capillary spaces on the outer surface of the plant. These include the spaces between sheathing leaf bases, in the concavities of overlapping imbricate leaves. . . within felts of rhizoids or paraphyllia. . . in the interstices between the papillae that cover the leaf surfaces . . . and between tightly packed shoots or between shoots and the substrate."

Our observations demonstrate that, beyond these general means of conduction there are important cellular structures that, in the end, determine the ultimate appearance and orientation of the leaves of wetted mosses. The alar cells are of different sizes and wall thicknesses; the basal cells often differ in size and wall thickness from the median and distal cells. The structure of the costa can be very simple or very complex, comprised of very specialized cells, with different tissue features abaxially and adaxially. All of theses things combine to create the aspect of the plant in the field.

Proctor also states that "the division between apoplast water in the cell walls, symplast water within the cells, and external capillary water, and especially the latter two, is important for several reasons." He goes on to develop the matter in relationship to the survival of mosses in nature. For our purposes we are interested more in the possible taxonomic value of the complex interaction of the structural mechanics of imbibition by moss leaves.

We believe that this aspect of moss hydrodynamics needs to be studied, for it has been too superficially observed and has unquestionable taxonomic importance. It may be a D'Arcy Thompson who will be required to develop a comprehensive theory, but we have been impressed with the obvious utility in distinguishing simpler species of *Bucklandiella*.

Winter Activity

We used to think that the first snow would bring an end to our work in the field, because, we thought, once the mosses become covered with snow, there would be no use looking for them until May or June. At Hall Ranch and on the out-wash colluvium around Boulder, we found that this is not necessarily the case.

In 2005 we were very successful in finding mosses at low altitudes in what ordinarily would be considered extremely dry habitats. In the winter, however, snow cover is often spotty, and ground mosses, such as *Tortula acaulon*, can be locally abundant in grasslands, with their sporophytes perfectly formed and, although green, ready to pop when spring arrives.

This situation surely obtains in the canyons and plateaus of the Western Slope, but unfortunately we have not had the opportunity to visit these areas in the winter. A resident bryologist needs to explore the gypsum flats for *Crossidium* and *Aloina*, and the rim-rock for *Entosthodon tucsonii*, *Didymodon convolutus*, *Funaria muhlenbergii*, and other minute ephemeral species. There is much work to be done also on the high plains, where there is always a possibility of discovering *Aschisma kansanum* on the underside of quartz pebbles in the easternmost counties.

Geography of Colorado Bryophytes

As far as we know, Colorado has no endemic species of bryophytes. On the other hand, the state is rich in species with great gaps in their ranges, both in America and world-wide. This makes Colorado an exciting place to study bryophytes. The ruggedness and relative inaccessibility of the high mountain areas and the vagaries of the weather help explain why the moss flora is still poorly known.

It is a boon to scientists that the Lord did not distribute bryophytes at random. If He or She did, we scientists would have a terrible time, because we would have to scour every inch of the planet. Bryophytes are tied very closely to their microhabitats, which are discrete and can be recognized by an astute observer.

Bryological collectors move at a snail's pace compared to wild flower people. We might walk a few hundred yards in a day—longer distances if we close our eyes until we reach the destination where we know we will find a particular micro-habitat. Usually we tend to find an interesting spot at the roadside or only a few hundred yards away, and spend the day right there!

We should be very thankful that man has built roads and trails, despite the fact that we tend to resent these intrusions into pristine habitats. But for these we might never find some of the rarer species. When looking at a dot distribution map of a bryophyte we must remember that this is where collections have been made. A dot map will give us a general idea of other localities, but few areas are so well collected that one can make detailed dot maps covering the smaller geographical divisions such as counties, townships, ranges, and sections. Thus it is not very meaningful to declare a bryophyte rare because few collections have been made. Few bryologists collect the same common moss in more than a few counties in a state. At the same time, because we have found a rare micro-habitat does not mean that we have exhausted the possibility that other such places may still be found farther from the road.

Bryophytes often have much larger or more widely disjunct distribution patterns than do flowering plants. The Southern Rocky Mountains draws its flora from the obvious migration pathway afforded by the immense north-south extent of the Western American Cordillera. Every drainage system that radiates from the core of the system also serves as a highway of migration for riparian and lowland or desert species. An apt analogy is that of a great wheel, whose hub, the Southern Rockies, preserves the most ancient survivors, and whose axle and spokes—the Cordillera and the drainage systems, provide the pathways along which migration has taken place through time under the pressures of climatic shifts and orogenic movements.

The Colorado bryophyte flora is predominantly boreal-montane, as testified to by the fact that all but a very small number of species are common to both Colorado and Scandinavia. A bryologist trained in Scandinavia or Siberia will be very comfortable in Colorado.

To summarize our findings, we have filled in many gaps in the distribution of the common widespread Rocky Mountain and desert-steppe bryoflora of the region. We have stated elsewhere that the Colorado Flora is somewhat analogous to an old house in which the walls are papered with flaking layers representing different tenants. The flora is essentially a relictual one, characterized by a total lack of local endemism. The distribution patterns that have emerged are indicative of the complicated history and antiquity of the flora. Many people ask us, "How did these species get here?" Our answer is, "They probably were here in the first place, having once occupied much larger territories dating back to the times when the continents were joined. In this respect, we feel somewhat akin to Wegener, who proposed continental drift before geologists found direct evidence for continental motion. We recognize, by the following list, several albeit imprecise examples of the local distribution patterns of our flora and their connection to the broader world flora.

1. Desert-steppe taxa disjunct from Middle and Southwestern Eurasia: *Anoectangium handelii, Bryum veronense, Didymodon anserinocapitatus, Orthotrichum hallii, Jaffueliobryum* (the genus), *Schistidium atrofuscum, Syntrichia sinensis.*

2. Eastern American woodland species disjunct along the eastern mountain front as a result of the desiccation of the Great Plains: *Dicranum brevifolium, D. flagellare, D. montanum, Mnium hornum, Platygyrium repens.*

3. Relicts of ancient relictual floras with disjunct worldwide distributions: *Bryoxiphium norvegicum, Anacolia laevisphaera, A. menziesii, Entodon concinnus, Grimmia teretinervis, Hedwigia nivalis, Hylocomiastrum pyrenaicum, Imbribryum alpinum, Leptodon smithii, Leptopterigynandrum (the genus), Oreas martiana, Plagiobryum demissum, P. zierii, Rhabdoweisia crispata,*

4. American Andean cordilleran species ranging from southern Colorado to southern South America: *Bartramia potosica, Brachythecium stereopoma, Homomallium mexicanum, Pseudocrossidium replicatum.*

5. Pacific Northwest disjunct woodland species: *Brachythecium hylotapetum, Meiotrichum lyallii, Rhytidiopsis robusta.*

6. Arctic-alpine and circumboreal disjunct species: *Andreaea heinemannii, Bartramia subulata, Campylopus schimperi, Catoscopium nigritum, Cirriphyllum cirrosum, Dicranum groenlandicum, Funaria microstoma, Hydrogrimmia mollis, Isopterygiopsis*

alpicola, Mnium spinosum, Paludella squarrosa, Pleurozium schreberi, Rhytidiadelphus triquetrus, Ptilium crista-castrensis, Ptychostomum cryophilum, P. cyclophyllum, Schistidium boreale, Sphagnum balticum, Timmia norvegica, Voitia nivalis.

7. Disjunct species of "copper-mosses": *Mielichhoferia mielichhoferiana, M. elongata.*

Conservation

In a large state with few active collectors, it may premature to describe any species as rare or threatened. It seems best to simply state that a species is poorly known, from only one or a few collections. Rarity can be more safely claimed when the habitat itself can be demonstrated as rare. In Colorado, the rare habitats include South Park with High Creek Fen, (PA), Blue Lake in the Ten Mile Range (ST), the gypsum/salt domes in Paradox Valley (MN), the Front Range foothill canyons (from LR to PE), and Chattanooga Fen (SJ). These are truly endangered habitats, and if destroyed, their botanical components likely will be lost forever.

The occurrence of bryophyte species is much more microhabitat- than geographically-controlled. We feel that concentrating on saving rare and endangered ones is an expensive and futile exercise; too little is known about their autecology. Although bryophytes have survived for millennia, they cannot withstand the deliberate destruction and elimination of their habitats by a growing human population.

Most crucial is the loss of our wetlands. South Park, one of the great high altitude inter-mountain areas of central Colorado was until recently an extremely rich and varied wetland. The human population, mostly ranchers, was small. The meadows were rich in grasses, sedges, and herbaceous perennials and were interspersed with many calcareous fens. As their populations increased, Denver and its suburbs demanded new sources of water. Soon South Park water was needed, the cattlemen sold their water rights, and within a few years the rich wet meadows had become deserts dominated by the weedy *Artemisia frigida*. The Nature Conservancy has succeeded in protecting High Creek fen, in the middle of the Park, but we will never know what other botanically rich areas have been lost. The few thin layers of mountain peat in South Park have been scraped for the gardens of the Denver area. An entire mountain ecosystem thus has been ruined.

Summit Lake, on Mount Evans, was established in 1964 as the first National Natural History Landmark (see Weber 1991) because of its rich assemblage of rare alpine and arctic plants, including many bryophytes. It is the crown jewel of the Colorado Rocky Mountains. A ceremony drew international scientists from as far away as Japan, Scandinavia, India, and Iceland. A permanent plaque was placed at Summit Lake, the stocking of fish was ended, and boulders were placed in order to prevent fishermen from driving their vehicles to the shore. The access road is now a toll road, the mountain is protected by the Forest Service, and plant collecting is restricted. However, the area is still vulnerable and under siege. The lake itself was, and possibly still is, potentially subject to draining in the event of a great emergency in the Denver area. This may not happen for many years.

Illogically, we have on Mount Evans a management mechanism that with one hand makes a genuine and laudable effort to protect rare plants, but with the other hand introduces voracious Rocky Mountain Goat, *Oreamnos americanus*, to make the area more attractive to the public. This self-destructive combination should be ended immediately. Despite its name, the Rocky Mountain Goat is not indigenous to Colorado and is native only to the northern Rockies. The early reports of this animal were misidentifications of the Mountain Sheep. Like domestic goats and sheep, it is a ravenous vegetarian. Following its recent introduction into the area, these goats have become very numerous on Mount Evans and Grays Peak. Rocky Mountain National park will not allow any to enter. On Mount Evans in summer they are fed by tourists, raise families, and now have started to congregate en masse on the mossy wet gravelly tundra of the alpine saddle just below the summit, between Mount Evans and Epaulet Mountain. It is inconceivable that the disturbance of an ecologically-sensitive area should be tolerated by a public that claims to care about the native vegetation.

For many years Rocky Mountain National Park has had its own problem of its becoming an outdoor zoo. The American Elk, or Wapiti, has become so abundant that it has invaded the adjacent town of Estes Park. In the National Park, the elk have decimated the willows in the wetland meadows and stream-sides, but public pressure to provide entertainment has so far defeated any proposals for controlling the size of the herds. Moving part of the herds is not an option because of the chronic wasting disease that would infect herds elsewhere, and there is now the possibility of controlling by killing animals. The destruction of the subalpine wetlands by elk is a strong threat to the survival of the bryophyte flora.

There are other threats, admittedly minor, that affect the survival of our bryophytes. Again, it is a matter of water. On the plains, the muddy shores of drying ponds were once the habitat of many species of thalloid liverworts. To enhance the water-holding capacity of the ponds, their margins have been dredged. What mud-flats remain are being fouled by the droppings of the Canada Geese, which now abound in greater numbers than before.

To paraphrase President Clinton, we can say, "It's the ecosystem, stupid!" Instead of focusing on endangered species or conspicuous megafauna, we should realize that every organism on earth is an integral part of the whole, and we pick and choose at our peril.

The Role of Amateurs

The junior author of this book, Ron Wittmann, is not a professional botanist but rather a physicist specializing in esoteric studies in radar. His hobbies are climbing, skiing, and classical music. Ron came to me with a germinal interest in flowering plants. Within the space of a few decades he became a highly competent field man and joint author of my books on the vascular plants of Colorado. As my specializations extended to lichenology and finally bryology, he developed in parallel, and became my computer guru and companion in the field. As my motor skills began to fade with age, Ron became my legs and navigated rocky screes and cliffs, and his 'magnifier' eyesight has been responsible for discovering many of our recent additions to the flora. Investing in the necessary microscopes and technical volumes on bryophytes, he has become a true amateur.

Amateur botanists have never really received their due, because their productivity has largely been in collecting specimens and depositing them in herbaria, where the 'real' botanists have published books and papers based on amateurs' gleanings. The word amateur comes from that most important Latin verb, amare—to love. In the old days every schoolboy or girl learned to conjugate it—amo, amas, amat. 'Amateur' is often considered synonymous with 'tyro' or 'dilettante', but it literally means 'lover of one's subject'.

In Victorian times in England, there were few professional botanists. Amateurs all had their 'cabinets of natural history'. Darwin and Wallace were amateurs. Darwin, except for a fairly brief exposure to the field, achieved fame through his lifetime of studies mostly at his estate in the countryside. No one considers Darwin to be an amateur. Wallace spent most of his life in the field in southeast Asia, collecting and describing organisms from his experience in wild nature. Wallace came to the same conclusions about natural selection as Darwin had, but Wallace produced the vast collections that others took credit for. The amateur field naturalist has the opportunity to make invaluable observations concerning habitat, occurrence, behavior, and population size, that the indoor scientist can only speculate about.

As a high school student, I was invited to the meetings of the New York Linnean Society, which was composed partly of scientists working at the American Museum of Natural History. It was in the early days of bird-watching. Bird-watching, as opposed to the more scientific 'ornithology', was considered to be a pleasant, harmless and unimportant hobby, and a great many young people were drawn to this 'sport'. Roger Tory Peterson had not yet published his bird guide, and most of the other birders were 'week-end warriors'. At the Linnean meetings, a professional held the stage on some phase of museum natural history, be it flatfish, snakes, or fish evolution, and we youngsters and amateurs reported our recent spotting of rare birds in the region. The *New Yorker Magazine* ran satirical cartoons dealing with birders who in great numbers surrounded one little warbler in a tree in the Ramble in Central Park.

One meeting, I now realize, marked a great turning-point in the status of the amateur ornithologist. Margaret Morse Nice came over from New Jersey to give a talk on the life history of the Song Sparrow! She was a bird watcher of a different breed. Mrs. Norse watched families of song sparrows from her kitchen window and patiently recorded all of their movements, over a period of years, following the individuals and their offspring, their movements, habits, and copulations, and she finally wrote a book about it. It was then that field ornithology rose from being a pleasant but useless avocation to a scientific study that could be done by a rank amateur. She showed that amateurs could make tremendous contributions to ornithology through field observation from a very homespun base. She practically founded the science of animal behavior in America.

Another amateur who had a similar effect on another phase of ornithology was Mrs. Rosalie Edge, who came to these meetings and acted like a real political rabble-rouser. The axe she was grinding was the need for conservation of the environment, and the protection of avian wildlife, of all things! But Rosalie Edge was largely responsible for the protection and establishment of the Hawk Mountain Preserve in Pennsylvania over which thousands of hawks and eagles passed during migration. This was the first of innumerable bird preserves.

Amateur science is a characteristic of civilized society. England, of course, was the mother lode that yielded so many great scientists who began as dedicated amateurs and never left off being amateurs. The centers of amateur bryology in America are the Eastern states, where the science took off originally, and more recently California and the states of the Pacific northwest. These are centers of population, higher education, great museums and libraries, and have easy contact with the rest of the literate world.

The Rocky Mountain region has only a short history of botanical exploration, most of it involving the vascular plants. The bryophyte flora has been largely neglected. Most of the amateur interest in plants in the Rocky Mountains lies in photography, preservation of rare and endangered vascular plants, and wetland studies that give a nodding appreciation of the fact that mosses are part of the ecosystem. Lichens also have gained popularity, although many do not know what lichens are (there is a booming business in facing buildings and fireplaces with 'moss rock'). However, there is virtually no serious amateur interest in bryophytes. Driving through Colorado, one hardly realizes that there are bryophytes, and the mountain trails are usually built on the dry side of the slope where bryophytes are not common. What we have done during the past ten years is only a beginning. We have just begun to scratch the surface.

In the twenty-first century the laboratory aspects of botanical science, particularly those pertaining to phylogenetic studies that employ the methods of cladistics and DNA analysis, are popular and lucrative. It is no secret that these parts of the science, being 'in', tend to not only dominate the departments of universities but also to ostracize what remains of the so-called classical disciplines. These 'high-tech' developments fragment the discipline of biology into disjointed specialties. This tendency is not likely to change. The future for field-oriented studies of the less popular groups of plants like the bryophytes, we feel, will be up to amateurs. They will require mentors. We hope that books like this, based on field as well as laboratory study, may provide amateur bryologists with mentors for the future. The study of bryophytes is not nearly so difficult as it used to be.

Source Books

The following references should be in every bryologist's library for use in conjunction with this book.

Crum, H. 1991. *Liverworts and Hornworts of southern Michigan.*

Crum, H. 2004. *Mosses of the Great Lakes Forest*, 4th edition. University of Michigan Herbarium.

Crum, H., & L. E. Anderson. 1981. *Mosses of Eastern North America.* Two volumes. Columbia University Press, NY.

Damsholt, K. 2002. *Illustrated Flora of Nordic Liverworts and Hornworts.*

Doyle, W. T., & R. E. Stotler. 2006. *Contributions toward a bryoflora of California. III. Keys and annotated species catalogue for liverworts and hornworts.* Madroño 53(2):1–109.
  Available from the California Botanical Society.

Flowers, S. 1961. *The Hepaticae of Utah.* Univ. of Utah Biol. Ser. 12(2):1–89. 18 line drawings.
  Has an excellent introduction; the treatment is still quite useful and applicable to Colorado.

Flowers, S. 1973. *Mosses: Utah and the West.* Brigham Young University Press, Provo, Utah.
  Indispensable for its introductory chapters on morphology and life history and the excellent line illustrations of many Colorado species.

Frye, T. C., & L. Clark. 1937, 1945. *Hepaticae of North America.*

Ireland, R. R. 1982. *Moss Flora of the Maritime Provinces.* National Museums of Canada Publications in Botany No. 13. Ottawa.

Lawton, E. 1971. *Moss Flora of the Pacific Northwest.* Hattori Botanical Laboratory, Nichinan, Japan.

Malcolm, Bill and Nancy. 2006. *Mosses and other bryophytes: An illustrated glossary*, second edition. 336 pp. Color illustrations. Micro-Optics Press. Nelson, N.Z.

Norris, D. H., & J, R, Shevock. 2004. *Contributions toward a bryoflora of California. I. A specimen-based catalogue of mosses. II. A key to the mosses.* Madroño 51: 1– 269. Available from the California Botanical Society.

Schofield, W. B. 1985. *Introduction to Bryology.* New York. Macmillan.

Schofield, W. B. 2002. *Field Guide to Liverwort Genera of Pacific North America.*

Schuster, R. 1953. *Boreal Hepaticae: A manual of the liverworts of Minnesota and adjacent regions.* 684 pp. Univ. of Notre Dame Press (originally published in American Midland Naturalist 49: 257–694.)

Sharp, A. J., H. Crum, & P. Eckel, eds. 1994. *The Moss Flora of Mexico.* Two volumes. Memoirs of the New York Botanical Garden 69.

Shaw, A. J., & B. Goffinet (eds.). 2000. *Bryophyte Biology.* Cambridge University Press. Cambridge.

Vitt, D. H., & W. R. Buck. 1992. *Key to the moss genera of North America north of Mexico.* Contr. Univ. Michigan Herb. 18: 43–71.

Zander, R. H. 1993. *Genera of the Pottiaceae: Mosses of Harsh Environments.* Bulletin of the Buffalo Society of Natural Sciences 32:1–378.

PART 1. MOSSES

Taxonomic Treatments by Family

AMBLYSTEGIACEAE (AMB)

The pleurocarpous mosses occurring in fens present a difficult problem, not only to ecologists analyzing wetlands in Colorado, but to professional taxonomists as well. In their introduction to the Amblystegiaceae, Crum & Anderson (1981) say, "This is one of many families belonging to the great complexes characterized by 'perfect, hypnaceous' peristomes." This is small comfort to us, because most of our species rarely or never produce sporophytes. The authors continue: "It is most representative of wet habitats subject to changing water levels. The result of fluctuating conditions of growth, seasonal and otherwise, is a considerable variability that has led to overclassification and taxonomic defeat." Suffice to say that in Colorado, a wetland pleurocarp that has long, often pinnately branched stems, and lanceolate to ovate, straight or falcate leaves, is likely to belong to the Amblystegiaceae. At the present time, the family is crumbling under the increased attention being given by specialists and, like the old Hypnaceae, many smaller families are like the crumbs falling from the cookie.

The following key is provided for working field ecologists who need to know the genera of what have usually been included in the Amblystegiaceae. To use the key effectively, there is no aid better than the superb illustrations in Flowers (1993) and in Crum and Anderson!

Field key to the pleurocarpous fen mosses (several families)

1a. Leaves broadly ovate, ovate or rounded-triangular, not falcate, in upper part suddenly rounded, narrowed or apiculate. The inflexed upper leaf margins may give the apex a more mucronate appearance . . . . (2)

1b. Leaves straight or falcate, from an ovate or triangular basal portion gradually narrowed or somewhat abruptly narrowed and tubular towards leaf apex . . . . (5)

2a. Costa short, double, or single, not or hardly visible with a hand lens; shoots turgid, slightly and irregularly branched; alar cells not hyaline, in an indistinctly delimited group. **Pseudocalliergon turgescens**

2b. Costa long, reaching leaf middle or further, single or branched, usually easily seen with a hand lens; alar cells various . . . . (3)

3a. Leaves usually broadly ovate or broadly rounded-triangular; green or brownish-green species, sometimes with a pale pinkish hue. **Calliergon**

3b. Leaves usually ovate or narrowly oblong-ovate; plants of alpine pools . . . . (4)

4a. Red colors common, otherwise green to dark green; usually strongly branched; stem leaf apex usually distinctly apiculate, at least in young leaves; leaf point often bent inwards; rarely found with leaf-borne rhizoids. **Warnstorfia sarmentosa**

4b. Pale- or yellow-green, usually sparsely branched; leaf apex rounded or rounded-obtuse; rhizoid initial cells or leaf-borne rhizoids common near the leaf tips. **Straminergon stramineum**

5a. Costa of stem leaves short, usually double, not or hardly visible with a hand lens . . . . (6)

5b. Costa long, reaching middle of leaf or further, normally single, usually easily seen with a hand lens . . . (8)

6a. Large or medium-sized species; shoots often turgid, not flattened; green or often with brown, yellow-brown, red, or blackish colors. **Scorpidium scorpioides**

6b. Medium-sized or small species; shoots often somewhat flattened; foliage green or pale to yellow-green
. . . . (7)

7a. Leaves with large and distinct groups of alar cells; costa short and forked, or lacking; leaves strongly falcate **Calliergonella** (see Hypnaceae)

7b. Leaves with small and indistinct groups of alar cells; costa short and inconspicuous; leaves weakly falcate. **Breidleria** (see Hypnaceae)

8a.   Leaves straight, lanceolate, strongly plicate . . . . (9)
8b.   Not as above . . . . (10)

9a.   Underside of the straight stems densely clothed with brown tomentum; leaves narrowly lanceolate. **Tomentypnum**
9b.   Underside of stems not tomentose; leaves ovate-lanceolate. **Brachythecium turgidum**

10a.   Stem leaves usually more or less broadly triangular to broadly cordate, rather quickly narrowed to the base and acumen; branch leaves much smaller, falcate; alar groups large, triangular and well-delimited, reaching from margin to costa; plants usually densely pinnate; leaves not plicate. **Cratoneuron**
10b.   Leaves rounded-triangular, ovate or broadly ovate to linear, more gradually narrowed toward the apex; alar groups usually less distinct; branching rarely densely pinnate (stem tomentum may occur in *Palustriella falcata* and *Tomentypnum*) . . . . (11)

11a.   Paraphyllia present (tear off a few leaves; the paraphyllia are usually distinctly visible with a hand lens); large species with falcate, distinctly plicate leaves; costa strong; tomentum frequently present. **Palustriella**
11b.   Paraphyllia absent . . . . (12)

12a.   Leaves conspicuously plicate; commonly fruiting, the capsules normally horizontal. Not common in fens, but common in wet spruce forests. **Sanionia**
12b.   Leaves not plicate or indistinctly so; capsules infrequent except in *Drepanocladus* . . . . (13)

13a.   Leaves from a more or less straight and rather erect basal portion with rather strongly curved upper part; alar groups undifferentiated or small and not visible in the field. . . . . (14)
13b.   Leaves curved more or less along their entire length, or nearly straight; alar groups distinctly differentiated and usually visible on torn off leaves (indistinctly differentiated in *Pseudocalliergon angustifolium*) . . . . . . (15)

14a.   Shoots relatively short, stiff, and more or less regularly branched; leaves green or yellow-green with brown-red costa (often also leaf base), somewhat dull (due to relatively short leaf cells with squarish ends). **Scorpidium cossonii**
14b.   Shoots longer and less branched; red, blackish-red (brown-red) or green, glossy (due to long cells with gradually narrowed ends). **Scorpidium revolvens**

15a.   Costa long-excurrent. **Drepanocladus longifolius**
15b.   Costa not excurrent . . . . (16)

16a.   Yellow-brown, brownish-yellow, or green species, typically with a golden gloss when dry; alar cells indistinctly differentiated; plants of strongly calcareous habitats. **Pseudocalliergon angustifolium**
16b.   Color varying, not with golden gloss when dry; alar groups usually large and more or less distinctly differentiated; not restricted to calcareous sites . . . . (17)

17a.   Shoots more or less radially branched, often red; leaves commonly with rhizoid initials; in less nutrient-rich situations. **Warnstorfia**
17b.   Shoots distichously branched, never red; rhizoidal initials absent; mostly in nutrient-rich habitats . . . (18)

18a.   Plant habit *Drepanocladus*-like; that is, with leaves falcate-secund, rarely straight and erect; leaf acumina in straight-leaved plants plane or at most slightly furrowed. **Drepanocladus aduncus**
18b.   Plant habit *Campylium*-like; that is, with leaves from straight and erect bases; usually with leaf acumina more or less spreading or squarrose, furrowed. **Drepanocladus polygamus**

Hints to recognition of genera of flowing water

Aquatic pleurocarpous mosses that grow over seeping rocks or attached to rocks in small streams may belong to genera of several families. *Brachythecium rivulare* may be mistaken for *Amblystegium*, but the leaves are more broadly ovate and finely serrulate distally, and the alar cells are conspicuously inflated.

1a. Large and elongate plants almost or entirely submerged with ecostate leaves in three more or less distinct ranks (see Fontinalaceae)

1b. Not as above . . . . (2)

2a. Alar cells inflated, sharply differentiated; paraphyllia usually present; leaf cells more or less prorulate (with protruding cell ends) . . . . (3)

2b. Alar cells various; paraphyllia absent; leaf cells smooth or nearly so . . . . (4)

3a. Paraphyllia abundant, linear, branched. **Palustriella**

3b. Paraphyllia absent or few, lanceolate, unbranched. **Cratoneuron**

4a. Leaves usually squarrose. **Campylium** and relatives **(see Campyliaceae)**

4b. Leaves never squarrose . . . . (5)

5a. Leaves with a single costa extending at least to mid-leaf . . . . (6)

5b. Leaves lacking a costa, or the costa short and double . . . . (10)

6a. Leaves cordate-ovate, oblong-ovate, or oblong, never long-acuminate; margins entire. **Calliergon**

6b. Leaves always distinctly acuminate; margins either entire or serrulate . . . . (7)

7a. Leaves usually falcate (except in some submerged forms) or plicate, or both. **Drepanocladus**

7b. Leaves neither distinctly falcate or plicate . . . . (8)

8a. Leaves larger; plants rather coarse; plants of wetlands and running streams. **Amblystegium**

8b. Leaves small, 2 mm or less long; plants small and delicate; plants of crevices or bases of shrubs, or streamsides . . . . (9)

9a. Leaves and branches small and slender; leaves up to 0.45 mm long, ovate-lanceolate, never squarrose, entire or often serrulate at the base; 2–3-celled gemmae present in the leaf axils. **Platydictya** (see Hypnaceae)

9b. Leaves larger, otherwise not as above. **Amblystegium serpens**

10a. Leaves small or large, less than twice as long as wide or, if longer, then somewhat falcate; basal cells rarely with pitted walls; on wet rocks in or near streams. **Hygrohypnum**

10b. Leaves usually 2 mm long or longer, commonly twice as long as wide or more, deeply concave, never falcate; walls of basal cells often pitted; not attached to rocks, but either submerged in fen ponds or in loose gravel of snow-melt rills . . . . (11)

11a. Leaves more or less falcate-secund, at least those of the tips of the stems and branches; alar cells hyaline, thin-walled and inflated, in small, inconspicuous groups. **Scorpidium**

11b. Leaves loosely imbricate to spreading, not at all or at most slightly secund; alar cells shortly oblong and sub-quadrate, neither hyaline, thin-walled, nor inflated. **Pseudocalliergon**

Amblystegium [blunt operculum]

We do not claim infallibility for this or any other published keys to the genus. Except for **A. serpens**, the species are extremely variable or plastic.

1a.   Costa extending into the acumen, sometimes filling it, occasionally excurrent. **A. varium**
1b.   Costa less than 3/4 the leaf length .... (2)

2a.   Leaves less than 1 mm long, generally serrulate for at least part of their length; basal marginal cells and adjoining alar cells subquadrate or short-rectangular to transversely elongate, firm-walled. **A. serpens**
2b.   Leaves more than 1 mm long, entire; basal marginal cells short- to rather long-rectangular, the adjoining alar cells quadrate to rectangular and rather lax. **A. riparium**

A. riparium. Common in wet places, particularly in areas where livestock or other pastoral activity is strongly evident. It is abundant in drinking troughs, on check-dam sluices and in relatively still water through the middle altitudes. In our distribution of specimens, we made the mistake of thinking a large, robust, submerged population in the bed of an intermittent stream was *Fontinalis hypnoides*. Evidently we were not alone. Crum & Anderson (1981), p. 994, write: "*Amblystegium laxirete* represents an extreme development of a *Fontinalis*-like habit. It sometimes occurs in long, streaming masses in swiftly flowing waters . . . In this habitat, the plants are striking, but in less vigorously flowing waters they are less distinctive." A number of other named forms evidently are variants of this extraordinary modification. Moral: Don't assume that everything that looks like *Fontinalis*, is *Fontinalis!*

A. serpens subsp. **juratzkanum** [for J. Juratzka]. A small, nondescript species, usually fruiting. The species is generally distributed on wet boulders, soil, bases of saplings or shrubs, exposed roots, rotten logs, etc., from the foothills up to the subalpine. The leaves are wide-spreading, about 0.6 mm long and gracefully acuminate, with costa to about mid-leaf. The median leaf cells are about 6:1 or 35μm x 5–7μm, smooth and moderately thick-walled. The leaf margin is slightly denticulate from projecting distal ends of the cells. The alar region is clearly differentiated, consisting of quadrate and broader cells forming a more or less triangular patch at the basal angles. The capsule is curved, the operculum conic, and the urn strongly constricted below the mouth. We are inclined to follow Nyholm (1954–1969) and others in treating *A. juratzkanum* as a variety or subspecies of *A. serpens*. The variety is distinguished by having the leaves more widely spreading and with the marginal basal cells rectangular instead of quadrate. Nevertheless, in Colorado, the plants are relatively constant in displaying the very widely divergent leaves illustrated by Figure 452 in Crum and Anderson (1981). Subsp. *serpens* has been found only once to date. BL: University of Colorado campus, *W&W B-116471*.

A. varium. An exceedingly diverse species of the lower altitudes, having had many forms recognized as separate although indistinct species, placed either in *Amblystegium* or *Hygroamblystegium*. Crum and Anderson (1981) amply show the frustrations inherent upon separating them. *Hygroamblystegium* has been a vexing problem because of its variability. Recently, molecular biology has come to the rescue (see Vanderpoorten 2004), in which the author demonstrates that *H. fluviatile, H. humile, H. noterophilum, and H. tenax* are justifiably synonymized under *H. varium*. In fact, *A. riparium* seems to be separated only by the variable length of the costa. We remain skeptical about lumping *A. (Hygroamblystegium) tenax* since it is easily distinguished by the stiff black stems with the stumps of costae left by the erosion of the lower leaves.

Calliergon [beautiful form]

Calliergon is generally not a plant of the edges of swiftly flowing brooks. *C. richardsonii* grows submerged in still pools. *C. cordifolium* occurs on saturated, swampy ground in forest clearings filled with tall willows. *C. giganteum* was found on sloping rock faces beside a quiet backwater on a level bench of a mountain stream. A few species formerly included in *Calliergon* are common in alpine pools. These include *Warnstorfia sarmentosa* and *Straminergon stramineum*.

1a.   Costa usually ending well below the leaf apex, with short branches or forked at the apex; shoots with rather long, thick branches, more sparsely branched and with branch leaves more erect or imbricate than in *C. giganteum* (see below). **C. richardsonii**
1b.   Costa ending almost in the leaf apex (appearing to reach the apex as seen with a hand lens) .... (2)

2a. Costa strong; alar groups of stem leaves large, triangular and sharply delimited from surrounding cells, extending from leaf margin to or almost to the costa; leaves broadly triangular (shoots when well-developed are densely branched like a spruce tree, with more or less spreading branch leaves except near branch apices). **C. giganteum**

2b. Costa weaker; alar groups similar but diffusely limited from surrounding cells. **C. cordifolium**

C. cordifolium [with heart-shaped leaves]. Pools and lakeside swamps, upper montane and subalpine. Our records are from the plateaus of western Colorado and fens in the Front Range.

C. giganteum [huge]. We have two collections: LR: Cirque Meadows trail, Pingree Park, in a streamlet tributary of Fall Creek, 9600 ft, *Hermann & Rolston 80114, W&W B-112731*.

C. richardsonii [for Sir John Richardson]. Uncommon in subalpine pools and fens. On wet tundra this can be a small, compact plant on merely wet ground. Submerged in pools it can be extremely long-stemmed and large-leaved, resembling *C. megalophyllum*. BL: Moraine pool, Niwot Ridge between University Camp and Silver Lake, Boulder Watershed, 9600 ft, *Weber & LaFarge B-43728*; CC: Pool in willow fen, Guanella Pass, 11200 ft, *Hermann 27286 (B-57409)*; LR: West of Chambers Lake, *Hermann 27592 (B-57273)*.

Cratoneuron [strong costa]

C. filicinum [fern-like]. This species is abundant in wet sites in the subalpine forests, willow carrs, and fens. The plants are small, pinnately-branched, and the stem leaves are conspicuously larger than the branch leaves. The branch leaves are usually falcate. A characteristic feature are the paraphyllia, of various shapes and sizes, minutely leaf-like, which vary in abundance and are sometimes almost lacking. The stem leaves are broadly ovate, with a strong costa. The median laminal cells are narrowly rhomboid, and the alar cells are enlarged, in a conspicuous triangular group. The species is said to be strongly calciphilous.

Drepanocladus [sickle-like branches]

Crum and Anderson (1981) accept *Drepanocladus* and *Scorpidium* in the traditional sense, which for field botanists seems reasonable, and they reject the transfer of several other species into the genera *Limprichtia*, *Warnstorfia*, and *Scorpidium*. However, we follow the recent work of Hedenäs (1970–2002). Hedenäs suggests one way to distinguish *Drepanocladus* from *Warnstorfia*: "It is always good to look for rhizoidal initials near the leaf apices. If you study 10–15 leaves (or a shoot apex with several leaves left) in the microscope, there are almost always at least some leaves with such initials in *Warnstorfia* species, but never in *Drepanocladus s. str.*" *Drepanocladus* has distichous branching (in one plane) and is usually green (except occasionally brownish in *D. longifolius*). *Warnstorfia* has radial branching and commonly has reddish leaves.

1a. Leaves from more or less straight and erect bases, usually with the acumen more or less spreading or squarrose; leaf acumen furrowed; plants more golden-brown than green. **D. polygamus**

1b. Leaves strongly falcate-secund; leaf acumen in less falcate-leaved plants plane or, at most, slightly furrowed; plants green with no other tints . . . . (2)

2a. Costa of stem leaves strongly excurrent; one or both leaf margins usually partly finely denticulate. **D. longifolius**

2b. Costa of stem leaves ending well below the leaf apex; leaf margin entire, or only occasionally finely denticulate . . . . (3)

3a. Alar groups transversely triangular, reaching the costa or almost so, the outermost cells longer and broader; stem leaves variously straight or falcate-secund, the latter especially so in small plants growing under relatively dry conditions. **D. aduncus**

3b. Alar groups quadrate or rectangular, not reaching the costa; stem leaves mostly falcate-secund to strongly so, rarely weakly so or almost straight, the latter especially in submerged plants. *D. sordidus*

D. aduncus. A common and variable species, wide-ranging in altitude from the plains up to the alpine. The row of elongate, inflated alar cells, that decrease in size from margin to costa, is diagnostic. The plant is weak, yellowish-green, never with reddish or brown colors, little branched and not at all pinnate; the costa is slender and elongate. However, some collectors have confused this with *Calliergonella (Hypnum) lindbergii*. The latter stands stiffly erect in close order. The leaves are broad at the base (triangular-ovate) and have large alar cells at the basal angles just where the leaf becomes decurrent. The costa is absent or short and double, and faint, and the leaf margins tend to curve inward near the apex.

D. longifolius. In subalpine pools. The long-exserted costa distinguishes this from other *Drepanocladus* species. Some collections have been misidentified as *Warnstorfia trichophylla*, which might be expected in Colorado; however, in *W. trichophylla*, the shoots are radially branched and the apices of stems and branches are stiffly involute, resembling sharp pencil tips, frequently becoming red when emergent.

D. sordidus. This species has only recently been recognized in the Americas, but it has been found in California, Yellowstone National Park, and the Peruvian Andes. Although there are no Colorado records, it may be expected in fens in North Park.

Hygrohypnum

Hygrohypnum is attached firmly to rocks which are inundated by water of flowing streams or at least wet periodically by spray. Plants are often coppery in color. Species with falcate leaves might be mistaken for *Drepanocladus* but lack the prominent percurrent costa. *Drepanocladus* does not occur firmly attached to rocks. We do not understand the subtleties of the microhabitats for *Hygrohypnum* species or we might be able to separate them quite nicely on their preferences. They probably do not occur together in the field. Taxonomically they are difficult if one considers relationships of the species over a broader range. Our keys conform with Jamieson's unpublished thesis (1986a) on the genus, together with some of his *in voce* hints on recognition of the species. Ignatov & Ignatova (2004) separated out *H. alpestre*, *H. cochlearifolium* and *H. smithii* into a new segregate genus, *Ochyraea*, but we defer judgment until the rest of the North American species have been treated.

1a. Stem cross-section with an epidermis of enlarged, fragile, thin-walled cells, the inner cortical cells in 3–4 layers, thick-walled, the core cells large and thin-walled. **H. ochraceum**
1b. Stem cross-section with an epidermis of several layers of small, thick-walled cells, the outer layer not fragile or thin-walled . . . . (2)

2a. Leaves broadly ovate to orbicular . . . . (3)
2b. Leaves ovate to oblong-ovate or ovate-lanceolate . . . . (6)

3a. Median marginal leaf cells 60 μm long or more. **H. bestii**
3b. Median marginal leaf cells rarely longer than 55 μm . . . . (4)

4a. Alar cells clearly differentiated. **H. duriusculum**
4b. Alar cells undifferentiated or formed of a few quadrate or short-rectangular cells . . . . (5)

5a. Costa usually single, stout, extending to mid-leaf or slightly beyond, sometimes forked or short and double; plants coarse and rigid. **H. smithii**
5b. Costa almost always short and double; if single, the costa slender and the plants soft and pliable; leaves deeply cochleariform, usually 0.8–1.2 mm long, the apex tapering to an obtuse or broadly rounded tip. **H. cochlearifolium**

6a. Alar cells clearly differentiated, either inflated and mostly thin-walled, or smaller, incrassate and quadrate to short-rectangular. **H. luridum**
6b. Alar cells undifferentiated or with but a few quadrate to short-rectangular cells which do not form a recognizable group . . . . (7)

7a. Leaf apex abruptly acuminate, tapering to a slender tip. **H. styriacum**

7b.   Leaf apex obtuse or acute, with or without a blunt tip . . . . (8)

8a.   Costa predominantly single to mid-leaf or beyond, sometimes short and double (leaf apex acute, plants coarse); leaves green, spreading, not highly cochleariform, lacking a squarrose apiculus. **H. smithii**

8b.   Costa double, reaching mid-leaf; plants green or yellow, strongly julaceous; the leaves closely overlapping, highly cochleariform except at the apex, blunt or rounded, usually with a tiny squarrose apiculus (best seen in dry material). **H. alpestre**

H. alpestre. The species is known by its julaceous, stout stems clothed below with remnants of old leaves clogged with silt; the convex, yellow-brown broadly oblong, concave, appressed leaves; faint double costa; and minute recurved-squarrose apiculus. The apiculus is best seen when dry, on leaves nearest the summit of the stem. It is poorly known, but possibly frequent in the high San Juan Mountains. Jamieson made our single collection after writing his thesis. This is evidently the only known occurrence in the contiguous United States. (Elsewhere in North America it is known from northern Canada and Alaska.) SJ: S side of Mountaineer Creek, N slope of Sugarloaf Mt., 12000 ft, on seepy gneissic rock on a tundra terrace, *Jamieson 10154 (B-88608)*.

H. bestii [for G. N. Best]. "Of the broad-leaved species, *H. bestii* is distinguished by the long marginal leaf cells, which range from 60 to 250μm, the large leaves which reach 3 mm long x 2 mm wide, and the dioicous sexuality. In many instances, discoloration in the basal leaf cells imparts the appearance of a radiating 'sunburst' in the leaf base which is typical of the species" (*Jamieson*, thesis).

H. cochlearifolium [spoon-leaved]. "A soft species; leaves small, deeply cochleariform, often with margins explanate or more or less recurved. Often gets saturated with mud; much smaller than *H. smithii*, not at all falcate, alar cells not inflated, poorly developed" (Jamieson, thesis). This seems to be the most common and widely distributed of the broad-leaved species.

H. duriusculum [dim. of hard, tough]. A common mid-altitude species commonly having stiff, black, defoliated stems, on which the stumps of the costae remain. The leaves are broadly ovate and blunt-pointed, and spreading. "Under the microscope, the species can be recognized best by its usually oblong-elliptic to broadly ovate leaves and the well-defined group of thick-walled, usually discolored, quadrate, short-rectangular or slightly irregular alar cells." (Jamieson, thesis).

H. luridum [pale yellow]. A common montane species.

H. luridum and *H. ochraceum* tend to have narrower leaves than the other common species and are difficult to distinguish without making stem sections. Both species have green or reddish, straight or falcate-leaved, imbricate or spreading-leaved forms, and there is a wide range in leaf size. It is suggested that the outer cortical cells may be showing differences in response to environmental changes, in which case the species may not be separable. "An exceedingly variable species . . . The species varies virtually continuously in nine or ten features, which have been used singly or in various combinations as criteria for the recognition for numerous sub-specific taxa." (Jamieson, thesis).

H. ochraceum. A common moss on wet rocks in subalpine and alpine rivulets. " . . . a polymorphic, yet distinctive species, which may be distinguished . . . by the dioicous sexuality, the outer layer of inflated cortical stem cells, the variable costa, and the nature of the alar cells." (Jamieson, thesis).

H. smithii [for J. E. Smith]. "May be recognized by its coarse, rigid habit and the usually broadly ovate to orbicular, loosely imbricated to spreading leaves and the stout, generally single costa." (Jamieson, thesis). A rare species occurring in high, wet tundra and in chasms with waterfalls. BL: Diamond Lake Trail, Eldora, *Hermann 26660 (B-49830)*. In a letter to Fred Hermann about this collection, 2 Sept. 1975, Jamieson wrote: "Of any species of *Hygrohypnum* that has ever been described as stiff or coarse, *H. smithii* most consistently expresses the feature. Note in your duplicate that the leaves tend to be held rather stiffly spreading. I think the coarse nature of the costa of *H. smithii* has a great deal to do with this. The singly costate leaves of *H. cochlearifolium* bear a far more slender costa. The apical leaf cells of *H. smithii* are quite regularly short-rhombic."

H. styriacum [of Styria, a province of Austria]. "*H. styriacum* looks like a 'weird' *H. luridum*; the leaf tip is flexed sideways, the leaf base is bowed out more than in other species, the leaves are triangular ovate, widest just above the base. There are few alar cells. They are enlarged, 'blistered', more well-developed than in *H. luridum*, the stems are usually more or less julaceous and the leaves may be falcate. The capsule of *H. luridum* has a persistent annulus; in *H. styriacum* it is deciduous." (Jamieson, voce). CC: Summit of Loveland Pass, at inlet of Pass Lake, 3600 m, *Weber B-111133* (!Jamieson); ST: Monte Cristo Creek, 11900 ft, *Miller & Wittmann B-115617.*

Excluded Taxa

H. molle (Hedwig) Loeske. "*H. molle* may be distinguished from other species in the genus by the broadly ovate, cochleariform leaves, which generally taper into an acute but blunt point and often denticulate apex, the undifferentiated alar cells, and the inner perichaetial leaves in which the cells on the abaxial surface of the leaf apex are prorulate" (Jamieson, thesis). This species, soft in texture compared to the others, is doubtfully reported for Colorado. Jamieson (*in litt.*) wrote that "virtually all but one of the specimens of *H. molle* from North America come from well west of the Rockies. The one troublesome specimen is a Kiener collection (no. 4113) from Longs Peak . . . Should you get in the Longs Peak area again I would appreciate your carefully looking around."

Palustriella [little swamp plant]

Palustriella falcata. A common moss of cold running water of streamlets in the alpine and subalpine. Stems pinnately branched; leaves falcate-secund, plicate, broadly triangular-ovate, abruptly narrowed at the base, the cells linear; paraphyllia abundant (strip leaves from the stem), linear, 2–3-cells wide; rhizoids red, branched, with thick cell walls; alar cells not differentiated but a few rows of basal cells inflated or much enlarged.

Pseudocalliergon

1a.   Stem leaves more or less straight, broadly ovate, not falcate, suddenly and abruptly apiculate. **P. turgescens**
1a.   Stem leaves weakly or strongly falcate, from basal leaf portion gradually or suddenly narrowed to a short or long acuminate apex. **P. angustifolium**

P. angustifolium. This is a *Drepanocladus*-like plant with essentially straight leaves, little branched and lax, the leaves gradually long-attenuate with slender apex; costa ending below the apex. The alar cells are not inflated but rectangular, with thickened walls which turn yellow; a band of short, almost quadrate cells run across the leaf base. Our specimens are from acidic iron fens. It is commonly confused with *Drepanocladus aduncus,* which is always clear green, never brownish, reddish, or yellow. BL, CC, GL, SM.

We have found that it is not easy to distinguish this from *Drepanocladus polygamus,* so we asked Lars Hedenäs, who has always been patient and helpful, for his advice. He writes: "Actually, I originally collected this species as a strange *Warnstorfia fluitans* that grew in mineral-rich ('incorrect' habitat for *W. fluitans*) early-thawing late snow-beds in the Swedish mountains. *W. fluitans* almost invariably has rhizoid initials (sometimes even rhizoids) in the upper leaf lamina (these are not present in *P. angustifolium*), is autoicous (versus dioicous), and has a smooth lamina (the distal cells in *P. angustifolium* often but not always are distally and dorsally prorulate). There are also general differences in the organization and ontogeny of the alar cells, but there are instances where mature alar cells are rather similar in the species. *D. polygamus* is also autoicous, but lacks rhizoidal initials in the lamina and usually grows in relatively mineral-rich places. On the other hand, the leaves tend to be spreading like in '*Campyliums*' in at least parts of the shoots, whereas *P. angustifolium* looks like a *Drepanocladus* as far as leaf orientation is concerned. *D. polygamus* has a smooth lamina, lacking prorulate cells, and the leaf margin is entire as in *P. angustifolium*. *P. angustifolium* usually has a distinct golden metallic gloss in part when dry and observed with a dissecting scope. This is lacking in the other two species. Finally, the axillary hairs tend to have a more elongate apical cell than in the other two species."

P. turgescens. A large, sparingly branched, flaccid, brown moss, slightly falcate and with broad, usually obtuse, cochleariform leaves. The terminal leaves of the year are yellow-green. Commonly found floating in calcium-rich willow-peat fens and forming short, dense mats on rocks in tundra snow-melt rills, South Park and vicinity. The species is sterile here.

Sanionia [for C. G. Sanio]

S. uncinata. A common moss resembling *Hypnum*, found in relatively dry sites in the montane and subalpine forests, at the bases of trees and shrubs, and up to the tundra, but not in the wetter fen sites. The clearly plicate leaves suggest *Hypnum revolutum*, and care must be exercised not to mistake the leaf folds of that species for a costa.

Scorpidium [a small scorpion]

1a.   Large species (stem leaves 0.7–2.4 mm wide); stem leaves strongly cochleariform, almost orbicular and obtuse to acuminate, or from a broadly ovate-lanceolate base narrowed to an apiculate, acute, or acuminate point, falcate or (rarely) straight; costa usually double, more rarely single or lacking, rarely reaching above mid-leaf; hyalodermis of stem often incomplete. **S. scorpioides**

1b.   Small species (stem leaves 0.45–1.1 mm wide); stem leaves cochleariform, from an ovate to ovate-lanceolate base narrowed to a short or long acuminate apex, falcate; costa single, ending in upper half of leaf; hyalodermis of stem complete. .... (2)

2a.   Median cells (of stem leaves) 14–95(–120) μm long, with squared to shortly fusiform ends; plant distinctly reddish-brown. **S. cossonii**

2b.   Median cells (of stem leaves) 61–140(–178) μm long, with shortly to long fusiform ends; plant dark brownish-green. **S. revolvens**

S. cossonii [for E. St. Charles Cosson]. A *Drepanocladus*-like pleurocarp with distinct pinnate branching, closely ranked falcate-circinate leaves forming prostrate mats. The color alone is enough to separate it from *D. aduncus*, a clear green moss for which it has been repeatedly mistaken. The *Scorpidium* is not green, but has a variety of brown and reddish tints. The stem is stout and brown here, but weak and slender in the *Drepanocladus*.

The species is not confined to fens but evidently is frequent enough in streamside situations. It seems amazing that this common plant has only recently been recognized in America. With a little practice it can be distinguished easily in the field without a lens.

One caution: When the leaves are stripped from the stem, often some of the stem epidermis comes along with them. The cortical stem cells are inflated-rectangular, and often red-brown. They may be mistaken for alar cells except for the fact that they begin where the leaf base (as measured by the costa base) ends. The leaf base is also slightly decurrent alongside these stem cells.

S. revolvens [for the circinate leaves]. A beautiful dark blackish species of alpine fens and tundra pools. It is not abundant. The leaves commonly are quite circinate, much more so than in *S. cossonii*, with the narrow tip curved back again. The alar cells are few and small. The terminal branches show the leaves completely curved into a circle. With experience, the shape and size of the median leaf cells is critical. In *S. revolvens* the leaf cells are so long and so narrow that they are difficult to measure, and the ends of the cells are acute. In *S. cossonii* the leaf cells are elongate but their borders are easily discerned, and the end walls are more blunt. *S. cossonii* is definitely a smaller plant and more pinnate. See Hedenäs (1989), for discussion of this and *Scorpidium cossonii* (dioicous). *S. revolvens* is said to be autoicous, but neither fruits here. ST: Blue Lake, 3000m, *W&W B-111205*.

S. scorpioides [curved like a scorpion's tail]. Floating in pools in calcareous subalpine fens, South Park and Guanella Pass. A large water moss, with soft and pliant, densely foliate stems with short pinnate branches. No other aquatic moss looks anything like this. The plant is black except for the apices of the leaves exposed above the water level. The leaves are short, convex, and falcate-secund. The stems reach up to more than a dm long. CC: Guanella Pass, in fen pools, 3550 m, *W&W B-110848*.

Straminergon [straw-like shoots]

S. stramineum [straw-colored]. Common in wet ground along the edges of tundra pools and in shallow water, upper subalpine and alpine. It frequently occurs with *Warnstorfia sarmentosa*, but is light green in color and has fairly narrow oblong leaves with rounded or broadly acute apices. The leaf tips often contain one or more colorless cells (rhizoidal initials) that can give rise to brown rhizoids.

Warnstorfia [for C. F. Warnstorf]

1a. Leaves ovate to narrowly ovate, not falcate, the upper part suddenly narrowed to a rounded but minutely apiculate apex, often bent inwards over the leaf. **W. sarmentosa**
1b. Leaves gradually narrowed to the apex, usually falcate . . . . (2)

2a. Shoots distichously branched, never red; rhizoids never growing from the leaves; mostly in nutrient-rich habitats. See **Drepanocladus**
2b. Shoots more or less radially branched, most distinct when growing with stem in vertical position; sometimes with red coloration, and sometimes with rhizoids growing from the leaves; in less nutrient-rich situations . . . . (3)

3a. Alar groups mostly large, triangular and well-delimited; costa rather strong; green or partly to entirely red to blackish-red plants. **W. exannulata**
3b. Alar groups either large, triangular and well-delimited, or indistinct and more or less ovate; costa weak or strong; plants usually green to brownish, hardly ever red, but sometimes reddish brown. **W. fluitans**

W. exannulata [lacking an annulus]. Common in still water of subalpine and tundra pools. The plants are usually brown, purplish or blackish. The curvature of the leaves is usually pronounced, but may vary. The leaves do not have an excurrent costa, and are distinctly serrulate especially near the apex. The alar cells are inflated, rectangular, and form a triangular patch that reaches the costa.

This species is variable. Specimens with long, narrow leaf apices have been misidentified as *W. trichophylla*. However, in *W. trichophylla* the costa clearly is long-excurrent, while the leaf apices in *W. exannulata*, although long and narrow, are twisted so as to demonstrate that the laminal cells are present. An excurrent costa would be solid, without marginal laminal cells.

W. fluitans [floating]. Common in iron fens, especially in the San Juan Mts. This species infrequently has a really red coloration, and has straight or gently curved leaves (only rarely as falcate as in *W. exannulata*) and a rather faint costa that ends far below the leaf apex. *W. exannulata* has a strong costa almost (or quite) reaching the leaf apex, and is usually red-pigmented. We believe the alar region is not different enough to serve as a diagnostic character. Perhaps they have markedly different ecologies and would rarely occur together.

W. sarmentosa [with crowded branches]. Common in tundra pools, where it forms floating or submerged purple-red masses along with *Straminergon*. However, the two are readily distinguished in the field. *Straminergon* never has reddish coloration, and the leaves are more or less cucullate at the apex, with an incurved, merely acute or obtuse tip several cells wide. *Warnstorfia sarmentosa* is almost always purple-red, and the leaves, while blunt at the apex, have a slender, short, needle-like apiculus.

ANDREAEACEAE (AND)

This family is characterized by the peculiar dehiscence of the capsule, which splits into four valves with the valves remaining attached at the top and bottom, elastically constricting the capsule in the manner of a paper-lantern. The sporophyte parts are not strictly homologous to those in other mosses, so that this family forms a separate class of mosses. *Andreaea* is a slender-stemmed moss growing in dense tufts on granitic rocks, often above timberline. The stems are more slender than most species of *Grimmia*, and usually there are at least a few of the characteristic capsules present. *Didymodon subandreaeoides*, a similar moss that forms dense, wide 'turfs' on limestone terraces and is never fertile, can be easily mistaken for *Andreaea*.

Andreaea [for J.G.R. Andreae]

1a.   Costa lacking; leaves lanceolate. **A. rupestris**
1b.   Costa present; leaves subulate. **A. heinemannii**

A. heinemannii [for F. Heinemann]. Rare, on granite boulders on the highest peaks. Easily distinguished in the field by the loosely spreading, acuminate leaves. The costa is weak, flattened above, 4-cell-layered and often absent in the base of the leaf. Murray (1987) states: "*Andreaea heinemannii* is easily identified by its small size, untidy look due to divergent leaf tips, obtuse leaf apices, the more or less flattened subula and the costa often weak or lacking basally." CC: Summit Lake, 13500 ft, *Weber, Porsild, & Holmen, B-4475*. ST: Quandary Peak, 4150 m, *Komarkova B-14141*.

A. rupestris [on rocks]. On granite, subalpine and alpine forests. More frequent than the last, but always a welcome discovery. GA: Rocky Mountain National Park, Long Meadow trail, 10400 ft, *Weber & Blackwell B-112852*.

ANOMODONTACEAE (ANM)

Anomodon [unusual teeth]

1a.   Leaves ending in a short or long, hyaline hair-point; leaf margins revolute. **A. rostratus**
1b.   Leaves not ending in a hair-point; leaves acute or merely apiculate; leaf margins plane. **A. attenuatus**

A. attenuatus. In addition to the merely acute or just apiculate leaf apices, this differs from the next by having coarser, conical papillae We have a very few collections, all from the outer foothills, where the plant forms extensive carpets over the downslope edges of granite boulders. BL: Ravine, N-facing slope of Boulder Canyon, 7000 ft, *Weber B-10547*.

A. rostratus. The stems when dry have the leaves appressed, more or less catenulate; when wet, they are widely spreading, and the stems are very densely foliate. The leaves are multi-papillose with very fine, sharp papillae, easily seen along the recurved margins of the leaves. The costa is very conspicuous, and appears sunken; the cells are elongate and lack papillae. This species occurs on cliffs in the southern and western plateau and canyon country at low altitudes. LA: Purgatoire River, at spring on cool, N-facing slope, with *Populus tremuloides, Cooper B-85560*; MN: Foothills of La Sal Mts., near Buckeye Reservoir, west end of Paradox Valley, *Weber B-5547*.

AULACOMNIACEAE (AUL)

Aulacomnium [ribbed or channeled capsule]

1a. Leaves small, mostly less than 1.5 mm long; basal leaf cells green, not distinctly different from the upper ones, unistratose; stems characteristically with a terminal pseudopodium bearing a spherical cluster of few-celled, fusiform gemmae; habitat on rotting, often charred, wood. **A. androgynum**

1b. Leaves larger, 2–4 mm long; basal leaf cells often enlarged, bistratose; pseudopodia, when present, naked or bearing relatively few green, leaf-like, many-celled gemmae in an erect flabelliform cluster; terrestrial . (2)

2a. Leaves often yellow-green, variously spreading, contorted when dry. Stems matted together by conspicuous masses of red-brown rhizoids. Gemmae usually present. **A. palustre**

2b. Leaves usually green, erect and imbricate, not contorted when dry. Stems separating easily, the rhizoids largely hidden by the leaves. Gemmae absent. **A. palustre** var. **imbricatum**

A. androgynum [hermaphrodite, based on a misconception of the gemmae]. An infrequent species of a specific habitat: rotten, often charred, wood, on the ground, on slopes, from the foothills canyons, under *Pseudotsuga*, to the subalpine forests.

A. palustre [of swamps]. One of the most abundant mosses of wet areas, occurring in willow carrs and fens in the upper montane and subalpine. The plants typically have narrow, contorted or twisted yellow-green leaves and abundant tomentum on the stems.

A. palustre var. **imbricatum**. Restricted to the highest wet alpine areas. This is a plant that strongly suggests the Arctic *A. turgidum*. In fact, the report in Lawton (1971, p. 202), attributing the species to Colorado, is probably based on this variety. It has been considered to be merely a high altitude modification of no taxonomic significance. However, this form is turgid, never produces gemmae and does not expose its rhizoids, hence the stems do not cling to each other. The tufts do not grow intermixed with other mosses and tend to be darker green rather than yellowish. Vitt (*in litt.*) has grown this in the greenhouse and reports that under those conditions it reverts to type, but we have not seen vouchers. We believe that no other mosses in our region exhibit such a pronounced ecological modification.

The most recent, and really the only description of var. *imbricatum* is given in Pedrotti (2001, p. 686): "Cushions dense. Stems scarcely tomentose; leaves erect-imbricate and acute, linear-lanceolate, apex obtuse, margin entire. Vegetative reproduction absent. Ecology: Environment similar to that of the species, subalpine and alpine flats. Because of the erect-imbricate leaves in the dry state it may be confused with *A. turgidum*, a species not yet reported for Italy which differs in having ovate leaves with obtuse or rounded apices somewhat cucullate-cochleariform, the cells not or scarcely papillose and with a distinctly sinuose costa" (our translation).

If this were an environmental modification there would be intermediate forms along an ecological gradient. This is such a strikingly different plant in the field from the abundant lower altitude *A. palustre* that we feel it should be accorded specific status, especially in view of the fact that it occurs in isolated tundra sites in Austria (the type locality), Norway, Italy, and Russia (Lake Baical). CC: Saddle between Mount Evans and Mount Epaulet, 13500 ft, *Weber, Corbridge, & Wittmann B-104083.*

BARTRAMIACEAE (BRT)

1a. Leaves from a more or less clasping or sheathing, differentiated and greatly enlarged basal portion, the lamina subulate. **Bartramia**

1b. Leaves from a non-clasping, non-sheathing, usually ovate base, the lamina subulate or broader . . . . . . (2)

2a. Lamina lanceolate or ovate; with terminal branches frequently in whorls, with the spreading, short branches subtending antheridial buds; plants of springs, seeps, streamsides and fens. **Philonotis**

2b. Lamina subulate or linear; tufted or matted plants of cliffsides, not associated with seeps, springs, or running water . . . . (3)

3a. Rhizoids with high, sharp-pointed papillae; leaves broadly linear; stem 3-angled in cross-section; usually fruiting. **Plagiopus**

3b. Rhizoids smooth or with low rounded papillae; leaves with subulate lamina; sterile in our area. **Anacolia**

Anacolia [short neck]

According to Dana Griffin (*in litt.*), sterile material of *Anacolia* can be distinguished from *Bartramia* by the cross-section of the stem. In *Anacolia* the cortical cells are highly mammillose, and there is no hyalodermis. In *Bartramia* a hyalodermis is present and thus the cortical layer is smooth. The axillary hairs (difficult to make out) are of two cells, the basal one more or less quadrate and brown, the terminal one globose and hyaline. In *Bartramia* they are of 3 or more cells, the terminal cell elongate, thick- or thin-walled, the intercalary cells hyaline or with pigmented cross-walls, the basal cells sometimes brown (Griffin 1998).

1a. Distal cells of leaf papillose at the ends on both surfaces; inner basal cells linear; upper lamina 2–3-stratose. **A. laevisphaera**

1b. Distal cells smooth or only a few with low papillae at the ends on the abaxial surface; inner basal cells quadrate or short-rectangular; upper lamina 1–2-stratose. **A. menziesii**

A. laevisphaera [smooth sphere]. One record (verified by Flowers and by Griffin) from eroding soil over sloping granite outcrop on steep slope just below abandoned mine working, west side of Boulder Falls, *Weber B-14905*. The site has been disturbed but climbing is now prohibited. Possibly plants still occur in protected adjacent areas. The species otherwise barely gets into the United States in New Mexico and southern Arizona. We have one collection from northern New Mexico.

A. menziesii [for A. Menzies]. A beautiful moss, forming sprawling mats. Frequent on north-facing granite cliffs in the Front Range near Boulder. *A. menziesii* roughly follows the distribution of the redwood forests of California, with outliers in northeastern New Mexico and northwestern Wyoming.

Bartramia [for John Bartram]

1a. Plants larger, forming loose tufts; usually abundantly fruiting; leaves lax to more or less contorted; leaf sheaths prominent, whitish; generally in forested mountains, often on cliffs. **B. ithyphylla**

1b. Plants smaller, 1–2 cm high; leaves erect, appressed; leaf sheaths not so prominent . . . . (2)

2a. Inner leaf base fenestrate; rare plants of north facing cliffs in the foothills. **B. potosica**

2b. Not as above; locally abundant on wet tundra. **B. subulata**

B. ithyphylla [stiff-leaved]. A fairly common species on cliffs and boulders from the foothills to the subalpine—the only *Bartramia* that will be generally met with. It is frequently fruiting. The plants are characterized by a peculiar blue-green color that is difficult to describe, but easy to recognize in the field.

B. potosica [for Potosí, Bolivia]. A rare species that is nowhere abundant. In Colorado this can be found on rock ledges in the foothills canyons, on north-facing cliffs, from Larimer to El Paso counties. Occurrence in the Front Range reinforces the pattern of ancient Tertiary disjunct species. *B. potosica* ranges from South Dakota to Argentina (Fransén 2004). It is always sterile here. The remarkable sheath, with its margin of thin-walled cells, at first suggests a member of the tropical family Calymperaceae! TL: Between Divide and Cripple Creeks, 9500 ft, *Weber B-16060*.

B. subulata subsp. **americana**. Usually occurring on wet tundra. Known in Colorado from several locations (mostly alpine) including Summit Lake on Mt. Evans

(CC) and Blue Lake (ST); however, some of our best material comes from the montane-subalpine ecotone along Lost Creek, in PA. The type specimen of subsp. *americana* was collected in the Indian Peaks Wilderness Area (*Komarkova B-42055*). This subspecies is restricted to the Rocky Mountains from Colorado to Alaska. The typical subspecies is found in Eurasia "on higher mountains of dry temperate areas" in the Alps, Tien-Shan, Himalaya, and Altai-Sayan (Fransén, 2004). The short and stout seta, erect, symmetrical capsule which becomes black and loses its peristome at maturity, and shorter, straight leaves distinguish it from *B. ithyphylla.* Sterile plants resemble *Campylopus schimperi*, but lack the propagulate branch tips. CC: Saddle between Mount Evans and Mount Epaulet, 13500 ft, *W&W B-110976.*

Philonotis [moisture-loving]

1a.   Leaf cells with a single papilla projecting from the lower (proximal) end; perigonia (male buds) terminal on the stems, when fresh open wide, when dry appearing nut-like; some marginal leaf cells with a projection at each end, suggesting a double tooth. **P. fontana**

1b.   Leaf cells with a papilla projecting from the upper end or from both ends; perigonia similar; marginal leaf cells with a single projection or tooth. **P. marchica**

P. fontana [of springs]. Common in wet places throughout, especially abundant on seepage slopes along highways, where its yellow-green color makes it easily recognizable from a moving car. The subspecies *pumila* (*P. tomentella*) is characterized by its dense, compact growth form, the copious production of tomentum, the non-plicate leaves that are slenderly acuminate and with a long-excurrent costa. This expression of *P. fontana* tends to be most frequent at high latitudes in the Northern Hemisphere and at high elevations in the temperate mountain ranges.

P. marchica [for the principality, Mark Brandenburg]. Evidently relatively uncommon. Our records are from the San Juan Mountains. It possibly is restricted to areas of high mineralization.

Plagiopus [curved seta]

P. oederianus [for G. C. von Oeder]. This calciphile forms dense and extensive sods over seeping rocks in the subalpine and lower alpine. GN: South of Emerald Lake on quartzite cliffs, *Weber B-23762.* SJ: Near S Mineral Campground, *Douglas 436 (B-23763).* Nyholm (1960) writes, "On dry alpine rocks forma *alpina*, with smaller capsules, grows in small tufts." This variety is locally abundant on seeping limestone terraces. ST: Monte Cristo Creek, Blue Lake, 11700 ft, *Weber & Anderson B-34242.*

BRACHYTHECIACEAE (BRC)

In order to understand some of the problems involved with the taxonomy of this family, read Robinson (1962). This is an important paper and deserves more attention. Recent studies (Ignatov & Huttonen 2002, and Vanderpoorten et al. 2005) suggest that the small species of *Brachythecium* (*B. collinum, B. fendleri, B. leibergii,* and *B. velutinum*) deserve separate generic status under the name *Brachytheciastrum*, and *B. nelsonii, B. oedipodium,* and *B. plumosum* under *Sciurohypnum*. This is not surprising. However, in their discussion, Vanderpoorten et al. state: "Although these species compose a well-supported monophyletic group, the genus *Brachytheciastrum* lacks a clear, unambiguous morphological definition." While we agree on the reclassification, we feel that, for our purposes, it is not useful to employ it in the local flora.

1a.   Stems stiffly erect, densely brown-tomentose; leaves appressed, linear-lanceolate, and multi-plicate; costa slender, difficult to distinguish from the folds; rhizoids also arising from its abaxial side. **Tomentypnum**

1b   Not as above . . . . (2)

2a. Leaves cochleariform, cucullate, rounded at the apex but then abruptly extended into a long filiform point; stems often julaceous; plants of wet tundra. **Cirriphyllum**

2b. Leaves without the above combination of characters; habitat otherwise .... (3)

3a. Leaves blunt-pointed or rounded at the apex, erect-spreading when moist, serrulate from base to apex; costa ending below the leaf apex, the terminal cell protruding from the dorsal side of the lamina as a small spine. **Eurhynchium**

3b. Leaves definitely acute or acuminate, variously smooth or denticulate along the margins; costa not protruding as a dorsal spine .... (4)

4a. Stems with the branches up-curved when dry, of a shining golden-green color; leaves linear-lanceolate; plants of vertical cliff faces in the foothills. **Homalothecium**

4b. Stems not up-curved; leaves not linear-lanceolate; plants of various habitats .... (5)

5a. Stems complanate, delicate; leaves concave, more or less distant, never plicate, the attenuate apex twisted; alar cells not differentiated. **Steerecleus**

5b. Stems not complanate; leaves usually not distant, often imbricate, plane or plicate, the apex not twisted; alar cells usually differentiated .... (6)

6a. Small plants in deep compact cushions on moist calcareous sandstone cliffs or ledges; leaves never plicate; margin serrate, the teeth at the leaf base often double and recurved; short-cylindric gemmae commonly produced along the costa near the leaf apex. **Conardia**

6b. Medium-sized to large plants on the forest floor, soil of canyonsides, fens, or tundra; leaves plane or plicate; teeth of leaf base not as above; gemmae absent. **Brachythecium**

Brachythecium [short capsules]

The species in this genus may be sorted out into size groups (large vs. small plants and leaves), serrulate versus entire leaves, auriculate triangular versus gradually broadened or narrowed leaf bases. Luckily, in Colorado we do not have a large number of species, and they are fairly well-marked in their appearance and ecology. Nevertheless, Crum & Anderson (1981) call it one of the most difficult of all moss genera: "The differences are often difficult to describe, and no keys work well."

1a. Leaves strongly falcate-secund, (resembling a small *Hypnum*), biplicate; branches distinctly pinnate and also curved; forming extensive, tightly adhering mats on bark; commonly fruiting. **B. leibergii**

1b. Leaves either straight or sometimes somewhat falcate at the branch tips, but never falcate-secund throughout .... (2)

2a. Plants minute, leafy stems 2 mm or less wide, not extensively branched; leaves less than 2 mm long .. (3)

2b. Plants larger, the stems wider; leaves over 2 mm long .... (6)

3a. Plants green; stems not julaceous; leaves spreading .... (4)

3b. Plants yellow-green; stems julaceous; leaves appressed .... (5)

4a. Plants in soft, dark green, silky mats; leafy stems less than 1 mm wide; branch leaves serrulate throughout; leaves somewhat falcate, long-tapering to a slender point; alar cells in a short group, not nearly reaching the costa. **B. velutinum**

4b. Plants bright green, not in silky mats; leafy stems broader; leaves cochleariform, abruptly narrowed to a slender apex, faintly serrulate near the apex; alar cells more numerous, reaching close to the costa. **B. plumosum**

5a. Capsules (if present) straight and erect; stem leaves strongly plicate. **B. fendleri**

5b. Capsules (usually present) inclined to horizontal; stem leaves not plicate or weakly so at the base. **B. collinum**

6a.   Leaves broadly triangular-ovate, constricted abruptly at the base (auriculate) .... (7)
6b.   Leaves from narrow to broadly ovate but not auriculate .... (9)

7a.   Leaf apices acute; leaf margin entire or nearly so; alar cells numerous, inflated; primary stems creeping and secondary stems erect and dendroid; coarse plants in running streamlets. **B. rivulare**
7b.   Leaf apices attenuate; leaf margin finely denticulate to serrate; alar cells various; stems all prostrate; habitat otherwise .... (8)

8a.   Leaf apex twisted; one or two rows of narrow marginal basal cells merging with the inflated alar cells, which overlap to give the impression of bistratose tissue but not decurrent; plants of forest floors. **B. hylotapetum**
8b.   Leaf apex plane; alar cells inflated, decurrent in a broad band; plants of shallow still water in willow carrs. **B. nelsonii**

9a.   Plants robust, in loose, somewhat shiny green or golden-brown tufts; stems subjulaceous, erect-ascending, sparsely branched; leaves appressed; strictly alpine. **B. turgidum**
9b.   Not as above .... (10)

10a.   Leaves plicate .... (11)
10b.   Leaves not plicate; small plants with leaves not much larger than those of *B. collinum*, but the branching is looser and the leaves more spreading and never julaceous. **B. oedipodium**

11a.   Yellow-green or stramineous plants, common and abundant, forming loose mats in relatively dry forests, the branch tips often curved; leaves narrowly ovate–lanceolate, gradually acuminate, not convex; small tufts of red rhizoids arising from parts of the main stem. **B. erythrorrhizon**
11b.   Bright green plants, evidently infrequent and restricted to the Front Range foothills; leaves very broadly orbicular-ovate and abruptly acuminate, convex, the shoots julaceous when moist; rhizoids sometimes present but infrequent. **B. stereopoma**

B. collinum [of foothills]. Our most abundant species on slopes in the forested outer foothills where it inhabits sites that are dry except in springtime. It also occurs in optimum moss tundra and probably throughout the middle altitudes. Along with *B. fendleri*, it is our smallest species. The plants are yellow-green, with short and commonly julaceous branches; the leaves are closely spaced along the stem, non-plicate, convex, appressed, and are sharply serrate from apex to near the base. The alar cells are numerous and quadrate. Fortunately, it fruits in the lower foothills, where it occurs with *B. fendleri* and is distinguished by the inclined rather than erect capsules. *B. collinum, B. fendleri, B. curtum, B. leibergii, B. oedipodium,* and *B. velutinum* form a group of species that have much smaller leaves (1–1.5 mm long) than the group that includes *B. erythrorrhizon, B. nelsonii,* and *B. rivulare*.

B. erythrorrhizon [with red rhizoids]. Our most abundant species in the forested regions. Our records are scattered through the middle altitudes and upper subalpine, where it occurs on the forest floor and over rotting wood. In suburban areas it commonly occurs mixed with *Ceratodon purpureus* and *Hypnum cupressiforme* in poorly drained areas of lawns. The plant forms loose, spreading patches, with light green or straw-colored foliage, gradually long-acuminate, slightly falcate, plicate leaves, so that the branches are somewhat curved. At intervals along the stem, there are almost always a few clusters of long, reddish brown rhizoids. Eventually these may become extremely thin and hair-like. The only other species having these is *B. stereopoma*, which occurs only rarely in southeastern Colorado. In Colorado *B. erythrorrhizon* has been consistently misidentified, mostly as *B. salebrosum*, but also as *B. oxycladon, asperrimum,* and *albicans*.

B. fendleri [for August Fendler]. Frequent in rocky canyon-sides in the outer foothills, especially in the southern counties. The erect capsule and plicate stem leaves serve to distinguish this from the very closely related *B. collinum*. Evidently *B. fendleri* is a plant of the southern counties or in low elevations elsewhere.

B. hylotapetum [forest carpet]. A coarse, bright-green plant that forms wide, loose, prostrate, irregularly branching mats on the forest floor. Leaves plicate with serrate to serrulate distal margins; basal and alar cells hyaline, short, and more or less inflated; stem leaves deltoid-ovate, the apex acuminate and often twisted;

branch leaves narrower, acute; branches tending to be complanate. Although often regarded as a Pacific Northwest endemic, it is disjunct in the Front Range along with *Rhytidiopsis robusta, Rhytidiadelphus triquetrus, Ptilium crista-castrensis,* and *Pleurozium schreberi.* BL: Ceran St. Vrain trail. *Wittmann & Lehr B-115628.*

B. leibergii [for John Leiberg]. A species of the Pacific Northwest, known in Colorado only by a few collections from humus and rotting logs in dry spruce forests. With its small, strongly falcate-secund leaves, this is like no other *Brachythecium,* and except for the costa, it might be mistaken for a tiny *Hypnum.* It is a very much smaller plant than *Sanionia uncinata,* which also combines a costa with falcate leaves, and its habitat separates it from *Drepanocladus.* The seta is rough. RT: Park Range, near beaver pond 1–2 mi above Slavonia, E of Clark, *Weber & Nelson B-49372*; BL: N Fork Middle Boulder Creek 8 mi N of Nederland, *Flowers 9819 (B-54933).*

B. nelsonii [for Elias Nelson]. A beautiful species of wet ground in willow fens and wet tundra. It does not occur along flowing streamlets and is a plant of high altitudes. It usually does not form dense clumps. The branching is pinnate, with the lateral branches more or less at right angles with the stem. The stem leaves are very translucent, rather remotely spaced, and are broadly triangular ovate, with distinctly auriculate alar cell groups. The branch leaves are distinctly smaller.

B. oedipodium [swollen foot]. This relatively small species is fairly common in dry forests in the montane and subalpine. The leaves are small, and serrulate or serrate, so one might think it to be a very loosely branched form of *B. collinum.* However, the branches are not julaceous, the leaves are weakly serrate, the quadrate alar cells are larger (about 20 µm) and there are some oblong cells along the basal margin. The seta is rough.

B. plumosum. An uncommon species, of which we have only a few collections. Some specimens of *Eurhynchium pulchellum* have been called this by Lawton (1971), but in that species the leaves are strongly serrate and with a blunt or even rounded apex. Flowers does not appear to have understood this species, either. His Colorado collections belong to *B. oedipodium. B. plumosum* does not appear to have very slender points except under the microscope. We find the rather short, not filiform, and abruptly narrowed apex, combined with the concave leaves, good for recognition. The leaves are somewhat spreading and often turned to one

side. Very few treatments do a good job of finding distinctiveness, but it is one of a few species that have small bright green leaves. LR: Moist face of granite boulder along trail near summit of Greyrock Mt., 7400 ft, 10.5 mi WNW of LaPorte, *Hermann 23652 (B-38552)*; Rocky Mountain National Park, trail to Mills lake, *Hermann 26957 (B-111799).*

B. rivulare. Abundant wherever there are springs and moss-lined forest seeps and brooks. This is really an easy species to recognize. Most other forest species of *Brachythecium* are soft, forming flat mats on the ground. Unlike them, *B. rivulare* is stiff and somewhat wiry, with somewhat arching stems with erect, not overlapping, branches (some call these dendroid!). The stem leaves are strongly decurrent, and are difficult to remove with forceps. The stem leaves are larger than the branch leaves. The leaves are broadly ovate, the apex acute but not drawn out, with the costa not entering the acumen. Very large patches of inflated alar cells continue down the decurrent margin of the leaf. The leaves are entire or very faintly denticulate in the uppermost part.

B. stereopoma [thick operculum]. A most unusual species in the group of large types inhabiting the forest floors. The leaves are bright green, highly concave, swelling when moistened to create julaceous shoots. The leaves are broadly orbicular-ovate, suddenly narrowed to a subula, not at all decurrent, sharply serrate distally, with quadrate alar cells. Save for the subula and the single, rather stout, costa ending in mid-leaf, there is some resemblance to species of *Entodon.*

B. stereopoma is a species of the dry subtropics ranging south into Argentina, and evidently reaches its northern limit of distribution in eastern Colorado. BL: Saddle between Green and Flagstaff Mts., 7000 ft, on N-facing cliffs and ledges, *W&W B-112049*; PE: Devils Canyon, Beulah, outer foothills of Wet Mountains, under *Abies concolor, Weber, Wittmann, & Kelso B-112071, B-112073.*

B. turgidum. A robust species, generally growing in almost erect clumps, and it is the only large *Brachythecium* occurring in moist tundra. The stems are more or less erect, very sparsely branched, and sub-julaceous. The leaves are densely packed, plicate, 2.5–3.0 mm long, abruptly narrowed to an acuminate tip, and strongly appressed to the stem. It is said to be a calciphile, but it does not seem to be so in Colorado. BL, CC, EA, GA, LK, PA, ST.

B. velutinum [velvety]. This species has been found in the outer foothills in the *Pseudotsuga* zone near Boulder.

It is not common, and occurs in extensive mats on vertical faces of shaded boulders near small streams. The leaves are small, as in *B. collinum*, but they are widely spreading, falcate, and not at all julaceous, and the foliage is dark green, not straw-colored as in *B.*

oedipodium. It is usually fruiting and the seta is rough throughout. BL: Upper Gregory Canyon between Green Mt. and Flagstaff Mt., on large boulders beside stream in deep shade of *Corylus* shrubbery, *W&W B-111354*.

Cirriphyllum [hair-pointed leaves]

C. cirrosum [curled]. This is frequent in the upper subalpine and alpine on wet tussocks in tundra. Often it occurs as scattered solitary stems mixed with other mosses, but occasionally pure stands occur as loose mats. It is the only tundra pleurocarp with oblong

leaves suddenly extended into a long hair-point. Recently, we have found this in a compensating environment at 8800 ft, protected in moist crevices under an overhanging granite outcrop. LR: Bridal Veil Falls, *Wittmann B-116464*.

Conardia [for Henry S. Conard]

C. compacta. This slightly resembles *Amblystegium serpens*, but the leaves are distinctly serrulate from base to apex, the median cells are longer and narrower (sometimes up to 10:1), the costa going well into the apex, often expanding there, and stouter than in *A. serpens*. A good separating characteristic is the conspicuous serrations, often recurved, along the basal margin. Also, the long, narrow median leaf cells separates it from *Amblystegium serpens*, and the well-developed costa from *Isopterygiopsis*. Fertile specimens

may be distinguished from *Amblystegium* because the operculum has a narrowed beak in *C. compacta* but is conical in *Amblystegium*. When the characteristic gemmae are present there is no problem. *C. compacta* appeared to be restricted to calcareous sandstone cliffs and ledges at low altitudes, but we find it especially abundant at High Creek Fen! The plant is light green, often rather densely matted and mixed with calcareous deposits, which the moss evidently is instrumental in accumulating a travertine deposit. BL, GN, MZ, ST.

Eurhynchium [alluding to the beaked operculum]

1a. Robust plants with triangular-ovate, somewhat vertically plicate stem leaves; main stems and branches well differentiated, not appressed to the substrate but somewhat arching and rigid. **E. striatum**
1b Small, weak, often julaceous plants; with little differentiation between main stems and branches, forming low mats appressed to the substrate. **E. pulchellum**

E. pulchellum [pretty]. In this species the main stems are short, the branchlets short and numerous, and the leaves are generally appressed to the stem when dry (suggesting *Brachythecium collinum* in size), and the whole plant is pressed close to the surface of the ground. It is thus inconspicuous, but nevertheless abundant, in the Front Range foothills. Once recognized, it easy to know. The branch leaves are short triangular-ovate, either pointed or rounded at the apex, and serrulate from base to apex. Even under a hand-lens the neatly rounded, serrate apices on the glossy leaves spreading at only a 45 degree angle have a characteristic appearance shared by no other species. Macroscopically it might be mistaken for a small species of *Brachythecium*. Most Colorado collections of *Eurhynchium* from relatively dry forested slopes belong to this taxon.

E. striatum [striped = plicate]. We find no evidence of

intergradation between the Lilliputian *E. pulchellum* and this large, handsome, and conspicuous plant characteristic of mesic old growth *Pseudotsuga* forests of the foothill canyons. It resembles a large *Brachythecium*, with wiry elongate main stems quite free or arching from the substrate, and strongly pinnate branching, with the stem leaves broader and more attenuate than the branch leaves, and widely divergent from the stem. In similar mesic canyon-bottoms we have found *Rhytidiadelphus, Rhytidiopsis, Pleurozium,* and *Ptilium*. These assemblages indicate a relictual habitat. In the European literature these two growth types are always separated as species, but in western North America the nomenclature has been confusing. Flowers (1973) admitted confusion but settled on calling this *E. substrigosum* Bescherelle in Cardot. Norris & Shevock (2004a, b) were the first to report *E. striatum* from California. The stem leaves in European specimens are more strongly plicate. LR: Crystal Mt. road 12 mi NW of Masonville, *Mazurek & Wittmann B-112837*; Twin Cabin Gulch, N of Buckhorn Road, *W&W B-114382*, *Weber & Baker B-34140, B-35464*.

Homalothecium [straight capsule]

H. aureum. One of the most beautiful mosses of our area. When moist, its large size, more or less regularly pinnate branching and burnished golden-green color is attractive on cliffsides at relatively low altitudes. When dry the species has a totally different aspect, with the short lateral branches having their leaves strongly appressed and the axes curved upwards, of a distinctly golden color. It is always sterile here. Our collections are mostly from the outer foothills of the Front Range, but we have one collection from MF: Yampa River Canyon, *Lehr 2020 (B-113301)*.

In his classic paper on the Brachytheciaceae, Robinson (1962) we believe justifiably synonymizes *H. nevadense* and a number of other northwestern species under *H. aureum* (Lagasca) H. Robinson, based on *Hypnum aureum* Lagasca, 1802. Hofmann (1998) recognizes two subspecies under *H. nevadense*: subsp. *nevadense* and subsp. *aeneum* (Mitten) Hofmann, and cited specimens of each subspecies from Boulder Canyon, which is unlikely. In subsp. *aeneum* the alar cells are 2–7 along the margin of detached leaves, more clearly visible, and mainly quadrate. In subsp. *nevadense*, the alar cells are 1–4 along the leaf margin, less clearly visible, and mainly irregular in shape. The capsule is curved to rarely nearly straight in subsp. *aeneum*, and straight or rarely somewhat curved in subsp. *nevadense*. Our plants have never been found in fruit. Flowers' collections from the Wasatch Range in Utah are richly fruiting.

Steerecleus [for W. C. Steere, + *kleos*, honor]

S. serrulatus. This resembles a rather laxly branched and complanate *Brachythecium*. The leaves are light green, loosely complanate, serrulate, ending in a slender acumen twisted at the tip. The leaves are not decurrent and have few if any differentiated alar cells. One fragmentary collection extracted from a mat of *Timmia*, ME: Mesa Verde National Park, Wetherill Mesa, at the west pour-off of Bobcat Canyon, where it was growing under the lip of a cliff over which water seeps intermittently, 6900 ft, *Erdman (B-3688)*.

Tomentypnum

T. nitens [shining]. A handsome moss with evenly pinnate branching in one plane. The stiffly erect, straight stems, the multi-plicate, linear-lanceolate, and narrowly acuminate leaves with a slender costa (hard to see because of the folds), and the lower stems heavily covered by brown rhizoids, are diagnostic. This is one of the most common and constant components of subalpine fens and willow carrs, commonly occurring with *Climacium* and *Helodium blandowii*. In unusually cool, moist sites, the species may occur in ravines along streams in the foothills, down to 7000 ft, but this is a rare situation. Its status within this family is unclear. Treatments place it variously in the Hypnaceae, Amblystegiaceae, and Campyliaceae. This genus is unique in any of these families.

Excluded Taxa

Isothecium (Pseudisothecium) stoloniferum (Hooker) Bridel, a species of the Pacific coast, was reported by Grout (1928–1939) from a collection by Shockley. Likely an error in transcription of the label (Cal, for California, read as Col, for Colorado). Shockley did not collect in Colorado. Such transcription errors have occurred frequently in the vascular plant collections.

Scleropodium obtusifolium (Mitten) Kindberg was reported by Sayre (1938) from a fish stomach, but has not been collected since. It might be expected in the north-central tier of counties.

BRYACEAE (BRY)

The Bryaceae are roughly characterized as being acrocarpous mosses with a complete peristome having both an exostome and endostome, and rhombic distal laminal cells without ornamentation. Most commonly the sporophyte, when present, is pendent, with a dome-like operculum. Sterile plants have been a real problem, causing even specialists to throw up their hands and give up on them, assigning them to such trash-heaps as '*Bryum caespiticium*'. The principal genus in this family has been a bugaboo, at least to American bryologists, for over a century. Keys to the species of *Bryum* have proved to be frustrating, since they have relied so much on having mature sporophytes.

Unfortunately, in Colorado most species of '*Bryum*' in the broad sense are almost always sterile. A great number of the collections have been misidentified or left without names. However, during the past few years John Spence has examined the group and discovered that vegetative characters of the leaves, stems, and rhizoids are as useful as the sporocarps in classification. He has divided this monster genus up into a number of smaller ones based largely on their vegetative features. This has been a blessing for Colorado, for now with his insight we can feel confident in our identifications and, with a little practice, we can recognize most of them in the field.

The family evidently is polyphyletic. *Pohlia* is now believed to belong to the Mniaceae; *Mielichhoferia* and *Haplodontium* are genera in a new family, Mielichhoferiaceae.

Here are some characters that need to be observed, since they may have taxonomic value:

Vegetative features

The stem and its leaf arrangements

a. The stem may be red or green, simple or branched; the stem and branches are erect and tend to make for a dense tuft. Dense turfs of slender tightly packed stems are said to be *confertate*. Stem color is often an important feature.

b. The leaves may be distributed uniformly the length of the stem (most species). They may also be clustered at the end of the stem (*comose*), forming a bud-like tip, or they may be *rosulate*, forming a spreading terminal rosette. In some species the stem is *julaceous*.

c. Stem length may be characteristic of a species. Stems may be short, not more than a few mm. Medium length stems range from 5 mm to 30 mm. Long stems run to more than 30 mm.

The leaf

a. The leaf shape is usually oblong-ovate or obovate. It may be widest at the base or contracted to a narrower base. Many species, especially of *Ptychostomum*, have ovate-lanceolate leaves. The only linear-leaved species, *Leptobryum pyriforme*, now belongs to the Meesiaceae.

b. The leaf margin may have a border (called a *limbidium*) of narrower cells; this may be inconspicuous (hardly differentiated in length and width from the laminal cells), conspicuous, or even bistratose.

c. The basal margin of the leaf sometimes is decurrent.

d. The leaf may have a distinctive form when dry. It may be essentially flat and appressed to the stem, or it may be convex (cochleariform) and curving toward the stem distally, or it may be variously contorted, sometimes distinctly spirally arranged around the stem. In some species the leaves are cucullate.

e. Some species, especially those occurring in fens or in running water, may resist wetting. This is a common adaptation occurring in wetland mosses. Dry-land mosses usually imbibe water immediately.

f. The costa may be red at the base or not. It may not reach the leaf apex or it may be excurrent to form a slender or stout awn. A few species have an excurrent costa so slender and colorless to be called a hair-point. An excurrent costa is usually stout and best referred to as an arista.

g. The leaf color may be bleached, with almost no chlorophyll in the distal cells. Constant shades of green, turquoise, red, or brown may be characteristic of species.

h. Leaves may be larger at the stem apex and reduced in size lower down on the stem.

Cell detail

The leaf lamina usually has areas of distinctly different cells: distal, median, basal, and alar. Spence (2004) describes four areolation types that are basic to his reclassification:

a. *Bryum* type: Cells heterogeneous; the distal and median cells elongate, mostly 4:1 or greater, long-rhomboidal or linear; the proximal cells wider and shorter than the median cells, quadrate or short-rectangular, the transition often abrupt.

b. *Plagiobryum* type: Cells homogeneous; all those except the alar cells elongate, mostly 4:1–8:1; no strong distinction between the cells of the upper 2/3 and lower 1/3 of leaf.

c. *Rhodobryum* type: Heterogeneous; the distal and median lamina cells short rhomboidal to hexagonal, 2:1–4:1; proximal cells the same width or narrower, and the same length or longer, more regularly rectangular with squared-off ends; the transition rather gradual, but still giving the areolation a heterogeneous appearance.

d. *Pohlia* type: Homogeneous; all lamina cells except those at the leaf base long and narrow, mostly linear to hexagonal, often thick-walled.

Rhizoids

The stem may or may not be clothed with brown rhizoids. Rhizoids have no chlorophyll, and the end walls of the cells are slanting. Rhizoids may be completely smooth or densely papillose. This feature is doubtless of taxonomic importance, but descriptions too often make no mention of this character.

Sexual structures

Even when they lack sporophytes, the male and female sex organs are often present and can aid in identification. See Glossary for Sexuality.

Asexual reproductive structures

Mosses commonly reproduce by simple fragmentation when they are dry and brittle. But several species have specialized reproductive structures that are usually present:

a. Bulbils. Short, plump structures that have a narrow opaque base and usually a few green, short, leaf primordia appressed to the sides. They occur in the axils of the upper leaves and may be released when teased.

b. Filiform axillary gemmae. These are found in the axils of the leaves, sometimes abundant, looking rather like fat rhizoids that are filled with chlorophyll and have perpendicular, rather than the diagonal cross-walls of rhizoids. They also fall free when the leaves are dissected out. They are green when young but brown or red when mature, and full of granules and chloroplasts, neither brown-walled nor clear inside.

c. Rhizoidal gemmae or tubers. These are spherical or irregular multicellular bodies ranging from 120 μm to 250 μm. They are loosely attached to the rhizoids, in our species orange- or red-brown.

Sporophytic features

The seta and capsule

a. The seta may be long and slender or short and thick, and twisted to the left or right. While the nature of the seta may sometimes be diagnostic, for the most part it is not.

b. The operculum. This may be wider or smaller than the diameter of the urn. flat, convex, or rostrate.

c. The capsule may be cylindric and slightly wider at the mouth. It may or may not have a distinguishable neck that becomes shriveled. The mouth of the capsule may have a prominent ring of red cells. The cells of the capsule wall may be thick or thin, long or short.

d. Most capsules are pendent. A few are erect or horizontal.

e. The outer peristome. The exostome is usually stout, yellow or brownish. In only one species *(Ptychostomum pendulum)* does the exostome exhibit any especially remarkable feature. Instead of the ladder-like sequence of cells from top to bottom, the adaxial side of these teeth have an additional layer that resembles circular or variously shaped chambers (this is the endostome that is adherent to the exostome).

f. The inner peristome (endostome) has a hyaline basal half that consists of united segments, which separate upward to form teeth. Between two teeth there is usually one (rarely more) slender filament called a cilium, which may have one or more short lateral perpendicular branches, which makes it 'appendiculate'. A peristome may be examined by softening the capsule by boiling momentarily, and cutting the upper part of the capsule free with a piece of a razor-blade and separating the resulting ring with fine needles.

1a.   Leaves large and translucent (*Mnium*-like), 5–6 x 2–3 mm, in a rosette at the stem apex, not much changed or scarcely contorted in drying, papery in texture; leaf cells thin-walled and large (150 x 50 μm). **Roellia**
1b.   Leaves smaller, usually somewhat altered in drying; leaf cells never so large . . . . (2)

2a.   Plants whitish (when dry) with chlorophyll-deficient distal cells . . . . (3)
2b.   Not as above . . . . (4)

3a.   Forming dense, extensive tufts, the stems tightly packed; leaves minute (up to 0.3 mm in length), ovate-lanceolate to ovate, acuminate, the costa not conspicuous. **Bryum**
3b.   Occurring as scattered stems in tufts of other alpine mosses; leaf length greater than 0.5 mm (the older stem leaves up to 2 mm long), broadly ovate, suddenly recurved-apiculate; costa conspicuous to the apex; all cells of leaf pale and often the basal cells more or less collapsed. **Plagiobryum zieri**

4a.   Capsule horizontal, the seta stout, the urn curved, bloated distally with the operculum pushed to the side. (This tundra species is not easily recognizable without capsules, but these are frequently present). **Plagiobryum demissum.**
4b.   Capsules not as above, or plants without capsules . . . . (5)

5a.   Stems slender, with minute leaves (0.5 mm long) . . . . (6)
5b.   Stems not very slender; leaves larger (most common genera with capsules will key here) . . . . (8)

6a.   Stems slender, ropy-julaceous, creeping; leaves minute, short and broad, blunt or rounded at apex; costa not excurrent. The stems are usually solitary and intermingled with other mosses. **Bryum julaceum**
6b.   Stems filiform, leaves less than 0.5 mm, lanceolate, costa present, sometimes excurrent. The stems are confertate, forming tight erect blocks . . . . (7)

7a.   Plants restricted to heavy-mineral rocks, often on mine tailings; leaves turquoise-green when fresh; leaf cells narrow and elongate (like a minute species of *Pohlia*). **Mielichhoferia**
7b.   Plants generally distributed on soil and various rock types; leaf cells broad and short. **Bryum**

8a.   Inflorescences appearing lateral; leaves soft, with costa not reaching apex or percurrent, rarely weakly excurrent; laminal areolation *Plagiobryum*-like; alar cells somewhat differentiated from juxtacostal cells, quadrate; capsule shortly pyriform, lacking a peristome. (The likelihood of finding this is extremely slight.) **Haplodontium**
8b.   Not as above . . . . (9)

9a.   Median leaf cells 5:1 or longer; leaves rarely bordered, often serrulate distally; costa ending below the apex or percurrent, never excurrent. **Pohlia**
9b.   Median leaf cells 4:1 or often shorter and broader, entire; if longer, then with at least one of the following characters: Leaves broadly ovate, oblong, or oblong-ovate, usually not serrulate; apex obtuse to rounded; costa excurrent; leaf margins bordered (sometimes weakly) by longer cells. . . . . (10)

10a. Lamina areolation open, transparent (as stained glass windows), the cells usually broad; limbidium often narrow or absent; gemmae often present, either as tubers on the rhizoids or axillary bulbils . . . . (11)
10b. Lamina areolation various, limbidium distinct; vegetative propagula lacking or filamentose . . . . (12)

11a. Large mosses with elongate leaves, on seeping granite rocks; leaves often reddish; rhizoidal gemmae present in some species; capsules slender-necked and pyriform but rarely present. **Imbribryum**

11b.   Minute mosses with short, broad leaves; terricolous, often weedy species; rhizoidal gemmae present; commonly fruiting capsule short and almost neckless. **Gemmabryum**, but see discussion under *Imbribryum gemmiparum*.

12a.   Leaves ovate to ovate-lanceolate or lanceolate, widest at or below the middle, not distinctly rosulate; lamina margins smooth to finely serrulate but not strongly serrate; limbidium uni- or bistratose; tubers lacking (most of the common densely tufted mosses belong here). **Ptychostomum**

12b.   Leaves usually obovate or oblong to spatulate, widest above the middle usually rosulate; margins serrulate to serrate, unistratose; filiform gemmae commonly produced in the leaf axils; tubers often present. **Rosulabryum**

Bryum [*Bryon*, a moss]

Note: *Bryum* is said to have only axillary gemmae, but a few species have rhizoidal tubers and would belong to *Gemmabryum*. However, Spence (in litt.) suggests that those taxa with *Bryum* vegetative anatomy might best be placed in this genus until the taxonomy of *Gemmabryum* is more thoroughly studied. In our treatment, only the two weedy species that have scarlet rhizoidal gemmae are retained in *Gemmabryum*.

1a.   Leaves, at least the upper parts, white and lacking chlorophyll .... (2)
1b.   Leaves yellow-green, brown, or variable in color .... (3)

2a.   Shoots fusiform, pointed at apex and base; leaves narrowly ovate, tapering to a slender acuminate, but not widely spreading, apiculus; costa not reaching apex. **B. argenteum**
2b.   Shoots club-shaped, the apex blunt; leaves broadly ovate; costa excurrent; stems frequently brittle, the apical part breaking off, leaving a cup. **B. lanatum**

3a.   Leaves appressed, narrow; leaf cells narrower than in most species; stems slender, ropy, tightly julaceous, often growing solitary or a few stems mixed with other mosses. **B. julaceum**
3b.   Not as above .... (4)

4a.   Plants not forming tufts, the stems short, gemmiform, in a terminal cluster; sporophytes always present, with short seta and broad ovoid purple capsule; rare subalpine species. **B. blindii**
4b.   Plants forming dense tufts, the stems sometimes elongate; sporophytes unknown; leaves cochleariform, often distinctly julaceous .... (5)

5a.   Plants lively green; the stems 1 mm wide including the leaves, julaceous; leaves strongly cochleariform, broad and overlapping, the terminal ones crowded into cabbage-like heads, broad and short, circular, obtuse or rounded, cucullate and with a short, often recurved apiculus. Common from foothills to alpine. **B. calobryoides**
5b.   Plants pale or dull green; stems only 0.3 mm wide, not julaceous; leaves plane, the terminal leaves not crowded into heads .... (6)

6a.   Leaves pale green, lance-ovate, appressed and imbricate; costa strong, excurrent; plants forming a hard confertate turf. **B. species v–c**
6b.   Leaves dark green, leaves broadly acute, narrowly cochleariform when dry, not overlapping; costa not excurrent. **B. veronense**

B. argenteum [silvery]. A common weed in sidewalk cracks, lawns and gardens, but also occasionally in natural situations. It rarely fruits in our region. The leaves are acuminate, with the hair-points not strongly spreading. The capsules, when present, are on relatively short setae, are ovoid not cylindrical, and are bright red.

The shoots are narrowed above and below, in contrast to the club-shaped shoots of *B. lanatum* which has broadly rounded leaves and strong, spreading hair-points.

B. blindii [for J. J. Blind]. *B. blindii* is unique, easily recognized with only a hand lens. The stems are minute, only a few mm long, julaceous, with pale green or straw-colored leaves. It has small, rounded leaves and a short and broad ovoid purple capsule on a short red seta. We have only two collections, from GA: Fraser River Valley, north base of Berthoud Pass, 3100 m. On steep gravelly slope just above roadside, *Weber & Dahnke B-91852*. The moss was discovered when a group of us sat down to eat our lunches! The site has recently been scoured by bulldozers and is now barren, but other populations possibly might be found up slope. A second collection is from ST: Blue Lake, 3000 m, *Weber, Wittmann, & Spribille B-111126*. Evidently these are the only records in the contiguous United States. Evidently, *Bryum blindii* favors disturbed gravel habitats at high altitudes. At Blue Lake the gravels were deposited below the dam during construction.

B. calobryoides [resembling *Calobryum = Haplomitrium hookeri*]. This species cannot be confused with any other. The populations form compact, low, bright green tufts, with cabbage-like terminal stems, which may sprout slender microphyllous stems. The terminal leaves are only up to 1 mm long and broadly obovate or circular, cucullate, often with a small slender recurved apiculus. The costa is strong, brown or reddish. The median leaf cells are hexagonal, almost isodiametric, nearly uniform on the apical leaves. The lower leaves are narrower and up to 1.5 mm long, less cucullate and lacking an apiculus. In the lower leaves the median cells are similar, but more of the laminal cells are rectangular, especially near the margin. A leaf border is indistinct in the short leaves and well-developed on the longer leaves. BL, CC, FN, GA, LR. Spence (*in litt.*, 2007) suggests that this might be synonymous with *B. gerwigii* (Müller Halle) Limpricht.

B. julaceum. Easily distinguished by the slender ropy, julaceous stems. Our early collections were made at Summit Lake, Mount Evans, but recently we have found it at low altitudes (6000 ft) in lower Boulder Canyon on stream boulders, mixed with *Bryum pseudotriquetrum*. Jamieson has collected it often in the San Juan Mts. It probably is not a rare moss, but it occurs in small patches mixed with other mosses of seasonally wet places. The stems are bright green or reddish and, according to Crum & Anderson (1981), they often produce red, ovoid gemmae in the leaf axils. Shaw (1982) discusses the reasons which Ochi (1981) used to place this in *Bryum* rather than *Pohlia*. "It is the long [leaf] cells that might indicate a relationship to *Pohlia*, but this trait is to be found rather frequently elsewhere in *Bryum*. In addition, the costa is shortly excurrent, and the leaf margins are entire (or vaguely serrulate), features uncommon in *Pohlia*." In our collections, the median leaf cells are about 5 x 30–40 µm, not vermiform. Spence (*in litt.*) suggests that we might possibly have *B. concinnatum* (Spruce) Lindberg in Colorado, which differs from *B. julaceum* in having shorter leaf cells

B. lanatum [woolly]. A common species, often occurring with *B. argenteum* in weedy situations, but also in more natural habitats and probably is native here. In Colorado it has a wide range of sizes. A specimen from CT: Among *Ericameria* bushes, San Luis, *Maslin B-16565* demonstrates the lower limit of size. However, it is usually larger than *B. argenteum*, and the club-shaped stems with broad almost circular leaves, with abruptly spreading hair-points are distinctive. BL, CC, CT, GN, LR. This was described from France as a variety of *B. argenteum* and authentic material hardly appears to differ from that. We suspect that what is here being called *B. lanatum* is an undescribed species.

B. veronense [from Verona, Italy]. A tiny species on arid slopes in the outer foothills. The stems are only 0.3 mm wide and no more than 4–5 mm long. The leaves are dark green, cochleariform and slightly catenulate and only shortly straight-apiculate so that the one stem leaf does not cover and obscure the one above. The leaf apices do become colorless, but they lack the silvery color of those of the other 'white bryums'. *B. calobryoides* is much larger, and has convex-cucullate light green leaves, and the apex has a well-developed, often recurved, apiculus. It was described from the mountains of Italy, and it occurs in Central Europe, southern Scandinavia, and Greenland. See Spence (2007b). BL: West slopes of Steamboat Mountain above the Stone Mountain Lodge, Lyons, 5600 ft, *Wittmann & Lehr B-114757, W&W B-114835*.

B. species v–c. (undescribed). Plants minute, the active portion up to 5 mm, slender-stemmed, forming a tightly packed (confertate) turf of erect stems held together by fine rhizoids; leaves pale green, plane, appressed, entire, lanceolate-ovate, with areolation of the *Bryum* type; costa strong, excurrent to a short, erect apiculus; limbidium of one row of shortly rectangular cells; distal and median cells short-rhomboidal, the walls incrassate; basal marginal cells rectangular; principal rhizoids low-papillose; rhizoidal gemmae small, round, pale brown, the outer cells protuberant. The most characteristic field characters are the tightly packed slender stems, about 1 cm tall, that tend to break apart in solid bricks. BL: Flagstaff Mt. saddle, 6600 ft, on seeping granite outcrops, with *Imbribryum alpinum, Weber, Wittmann, & Lehr B-113917*; Ouzel Falls, Rocky Mt. National Park, *Miller & Wittmann B-115138, B-115612*; ME: Kodel Canyon, Colorado Nat. Monument, *Weber, Rector, & Siplivinsky B-81042*; ST: Blue Lake, in crevices of calcareous schist, 11400 ft, *W&W B-115165*.

Gemmabryum

Gemmabryum, as its name implies, is a genus characterized by a wide variety of gemmae of various types: Leaf axil gemmae, rhizoidal tubers, stem tubers, flagelliform branchlets and uniseriate gemmae on the rhizoids. The native species are most common as ephemeral spring annuals that are especially diverse in California. All are short-lived perennials that survive above ground hardly more than a year; in dry areas they survive primarily as rhizoids and tubers. Our *Gemmabryum* species have loosely aggregated stems that have scarlet rhizoidal tubers. The leaves are ovate, broadest at midleaf, distal margin serrulate, limbidium of 2–3 rows of very narrow cells; costa excurrent; cells thin-walled, basal cells short and broadly rectangular, median cells narrowly rhomboid, distal cells not appreciably shorter than the median cells.

1a. Short-stemmed, often cabbage-like, tiny plants, usually with large axillary bulbils that fall away easily when dissected; not a weedy species. **G. bicolor**
1b. Not as above . . . . (2)

2a. Weeds of cultivated or disturbed sites such as lawns and under cultivated trees; leaves bright green, not convex; slender plants with red or brown rhizoidal tubers . . . . (3)
2a. Native plants of calcareous alluvial soils or travertine deposits, lacking axillary bulbils; leaves pale, convex; rhizoids with or without tubers . . . . (4)

3a. Rhizoidal tubers 50–100 μm diam. **G. klinggraeffii**
3b. Rhizoidal tubers over 120 μm long. **G. subapiculatum**

4a. Costa weak, not reaching the leaf apex. See **Imbribryum gemmiparum**
4b. Costa strong, red, percurrent or slightly excurrent, the tip slightly recurved. **G. valparaisense**

G. bicolor. A small species with leaves no more than 1 mm long, broadly ovate, with the costa not reaching the leaf apex or only percurrent, somewhat resembling *B. argenteum* but lacking the hyaline leaf apex, or *G. calobryoides* because of its somewhat julaceous stems. However, the leaves are dull, the stems are plump because a leaf axil encloses a single large green gemma with red base; the green leaf primordia are well-developed and arise from near the base and upward, curving over the apex. Collections: LR: Spring Canyon, *Hermann 24576 (B-45270)*; SJ: Deadwood Gulch, near Silverton, *Hermann 24397 (B-45279)*.

G. klinggraeffii [for H. von Klinggräff]. BL: Boulder, a weed in a flower bed, *Wittmann B-115223*; LO: South Platte River, 2 mi S of Crook, 3700 ft, *Rolston B-81977*; WL: West Pawnee Butte, *Weber, Hermann & Arp B-37484b*.

G. subapiculatum. A weed moss growing in urban settings, known from Boulder, but undoubtedly present elsewhere. BL: U. of Colorado campus, alongside Denison Building, surrounding a manhole cover.

Rhizoids brown, beautifully papillose; gemmae red, spherical, 150–160 μm diam. The species also occurs in other shaded garden grounds in the city.

G. valparaisense [of Valparaiso, Chile]. We have a single collection of this recently discovered species. It is very similar to *"Imbribryum" gemmiparum*, and is probably overlooked. GF: Rifle Falls State Park, NE of Rifle; at base of limestone cliffs at the Ice Cave, on saturated travertine mounds under dripping cliffs, *Weber B-63573*.

Excluded Taxa

G. (Bryum) barnesii (Wood) J. R. Spence was attributed to Colorado by Vanderpoorten & Zartman (2002), but no specimen was cited. This has several gemmae per leaf axil; however, the dot on the map is at the same spot as that for *G. bicolor*, which may have more than one gemma. Spence (*in litt.*) agrees that *G. barnesii* is not likely to be found in Colorado.

Bryum gemmilucens Wilczek & Demaret. There is a dot on

the distribution map in Vanderpoorten & Zartman (2002). We do not know the source of the report.

Haplodontium [single row of peristome teeth]

H. macrocarpum. Shaw, et al. (1992) reports this from the San Juans (Red Mountain Mine area). Jamieson has it from Cascade Creek Trail at Engine Creek Falls, on the damp, deeply shaded face of an eroded limestone grotto behind the plunge basin of the foot of the falls (*Jamieson 13703*). Grout (1928–1939, 2:261) claimed that *Bryum nelsonii* Kindberg, type from Gunnison, collected by N.L.T. Nelson, is *Mielichhoferia macrocarpa*, but the Gunnison County populations all belong to *M. mielichhoferiana* and *M. elongata*. Utah specimens have come from limestone cliffs on Mount Timpanogos. This is an exceedingly rare and disjunct species, described first from Disko Island, known also from Greenland and a few localities in western Canada.

Imbribryum [imbricate leaves]

1a. Leaves strongly imbricate, lanceolate; distal laminal cells long, incrassate and somewhat vermicular, >6:1; always sterile **I. alpinum**

1b. Leaves somewhat loosely imbricate, the distal laminal cells shorter, not vermicular, mostly 3–6:1 .... (2)

2a. Leaves straw-colored, flat or weakly cochleariform, narrowly lanceolate, the apex acuminate, drawn out into a slender point; costa inconspicuous, not reddish; older stems almost naked, devoid of rhizoids. **I. mildeanum**

2b. Leaves distinctly cochleariform, the apex broadly acute to obtuse; costa strong and reddish, not reaching the leaf apex or rarely percurrent .... (3)

3a. Stems elongate, in very dense clumps; leaves bright green or red-green, rigid and imbricate, crowded, concave and cucullate, spirally twisted when dry; costa strong, red; always sterile. Plants of seeping granitic outcrops. **I. muehlenbeckii**

3b. Stems very short, scattered; leaves green to yellow-green, lacking red tints, loosely arranged, somewhat distant below; costa brown or red only at base; usually richly fruiting, with bag-shaped urns, broadest at the apex. Plants of western canyon bottoms. **I. gemmiparum**

I. alpinum. A beautiful species, forming large soft deep turfs trending downslope on smooth, sloping surfaces of irrigated granite outcrops mostly in the Front Range, from the base of the mountains to the alpine. Only one collection was made on sandstone: BL: On sandy thin soil, Laramie Formation, White Rocks, 8 mi NE of Boulder, 5500 ft, *Weber B-27930*. The plants are hardly anchored to the rock but are held together by eroded gravel and sand. They are bright green when fresh, but when dry they become golden-brown, with a metallic sheen. According to Crundwell & Nyholm (1964) the rhizoids always have small round gemmae attached, but this is not always easy to demonstrate, since the papillose rhizoids are so mixed with detritus. The anatomy of the leaf does not suggest a *Bryum* because the cells are long and narrow, as in *Pohlia*! Although this appears to be a plant of the foothills, we have one alpine specimen, unusual in having bright red foliage, and smaller leaves (0.5–0.7 mm long). ST: Blue Lake, on seeping rock terraces, 2550 m, *Spribille 9912 (B-111475)* (!Spence).

I. gemmiparum. We believe that this species does not fit morphologically into *Imbribryum* and probably is better placed in *Gemmabryum*. The stem is less than a centimeter tall, the habitat is on alluvial soil, the leaves are minute, the rhizoids are pale and very minutely papillose, and it is readily recognizable in fruit by the baggy pear-shaped, almost neckless capsules. It is distinguished by having oblong leaves that are not at all acuminate or attenuate. The costa is weak and does not quite reach the leaf apex, brown or red at the base only. The distal cell are very short rhomboid, the proximal leaf cells elongate-hexagonal and with somewhat thickened walls. The outer peristome is yellow, not divided along the mid-line, smooth except for the papillose apex; the endostome basal membrane is high, and the cilia swollen at the nodes but not appendiculate. This is a species of calcareous soil along flood plains, particularly common in the canyon country of western Colorado. Collections: ME, MF, SH.

I. mildeanum [for Julius Milde]. Somewhat similar to

G. alpinum, but the leaves loosely imbricate (2–2.3 mm long), pale to golden green, lanceolate, with little curvature to the margins, while in *G. alpinum* the leaves are 1.5 mm long, more ovate, the margins are curved. Rhizoids are difficult to find, and are usually broken. The stems are very slender in looser tufts. The species ranges from montane to alpine, while the more common *G. alpinum* is a plant of the outer foothills. See Spence (2007b). CC: Summit Lake, Mt. *Evans, Jones 19957 (B-16524)*; LR: Greyrock Mountain, *Hermann 23649 (B-57240)*; ME: Grand Mesa, *Hermann 16521*; MF: Douglas Mountain, *Weber B-101111* (!Spence).

I. muehlenbeckii [for H. G. Mühlenbeck]. A beautiful species with massed, erect, julaceous stems; leaves strongly convex and apically cucullate; costa red, percurrent; rhizoids coarsely papillose. The basal cells are short and widely rectangular grading into more elongate-rectangular median cells and shorter, hexagonal apical cells. A rare plant of the Front Range outer foothills, the habitat being similar to that of *I. alpinum* on seeping granite rocks. At this locality *I. muehlenbeckii* and *I. alpinum* grow adjacent, providing a fine opportunity to learn to contrast them in the field. Known in Colorado from a single collection: JF: Coal Creek Canyon, *W&W B-114862* (!Spence). Lesquereux (1874) cited a collection, without data, from the Sullivant herbarium, communicated by General Palmer.

Mielichhoferia [for Mathias Mielichhofer]

Mielichhoferia is a genus that is characteristically found on mine tailings containing copper. The plants can be mistaken for a small *Pohlia*, since the leaves are serrulate at the apex. It has been considered a rare plant in Colorado, but it is likely that examination of the numerous mine tailings, especially those that are permanently moist, in the mineral areas will yield many more populations. Capsules are rare.

1a.   Leaves brilliant turquoise blue when growing, less than 1 mm long, ovate; the cells wide and transparent; costa slender; stems more or less julaceous, the leaves appearing blunt; rhizoids smooth. **M. elongata**
1b.   Leaves yellowish-brown, more than 1 mm long, narrowly triangular-lanceolate, tapered to a point, the cells narrow and densely packed; costa strong; leaves not tightly appressed, and commonly somewhat secund; rhizoids densely papillose. **M. mielichhoferiana**

M. elongata. Although the individual stems of this moss are tiny, their mass effect is overwhelming. When moist, the great turquoise carpets of *Mielichhoferia* on the mine tailings can be seen from the main highway! The leaves are ovate, green, minute (<1mm), appressed to the slender (up to 0.3 mm wide) stems; stems without rhizoidal mat; rarely fruiting, the capsule elongate-pyriform, the operculum conical; inner peristome well-developed, papillose. This is extremely abundant around the mines of Red Mountain Pass, in San Juan County. A second site is in Gunnison County, where Hermann collected both species, of which *Hermann 26235 (B-57071)* contains parts of two polsters, the largest one being *M. elongata*. A third site is at the outlet of Blue Lake, Summit Co., where the moss occurs at the mouth of a prospect hole. The most surprising site is GL: Central City, in horizontal cracks of a metamorphic outcrop at the parking lot of the Red Dolly Casino, *Evenson B-115431, Weber B-115526*.

M. mielichhoferiana [for Mathias Mielichhofer]. Our major locality is in the vicinity of Red Mountain in the San Juan Range of southwestern Colorado where both species of *Mielichhoferia* grow together. We have other collections from limestone cliffs: GN: 3200 m, just below Schofield Pass, *Lehr B-111886, Gradstein B-56659, Hermann B-9494*. One collection, *Hermann 27235*, from the same locality, consists of two pieces, one of which is *M. elongata*. General accounts of the copper mosses are given by Persson (1956) and Shacklette (1967); the ecology of this species in Colorado was discussed by Hartman (1969). The papillosity of the rhizoids was first noted by Nyholm (1993).

Plagiobryum [obliquely inclined capsules]

Plagiobryum is an exclusively alpine genus, not too well separated from *Bryum*, sensu lato. However, sterile plants of *P. zieri* are easily recognized by the pale pinkish leaves; when present, the sporophytes are elongate and horizontally oriented, with a long neck. Long-capsuled species of *Pohlia* have narrow elongate cells in the distal portion of the leaf and the leaves are deep green. *P. demissum* is unmistakable if it has capsules, for the seta is stout and curved, and the urn is widest at the top (bloated), and the operculum is not continuous with the axis, looking somewhat like a species of *Funaria* or *Meesia*. Sterile plants tend to have a plump rosette of broad leaves that usually are quite reddish or brown.

la.   Seta up to 2 cm long; capsule 5–7 mm long, curved and more or less horizontal, with the shriveled neck about as long as the oblong urn; sterile branches elongate, julaceous, with broad, cochleariform, pink to silvery leaves. **P. zieri**

1b.   Seta less than 5 mm long; capsule 3 mm long, horizontal from the bend in the short seta, narrowly cylindric, only slightly curved, the neck not much differentiated; sterile branches short, with ovate-lanceolate, more spreading leaves; new leaves green, the older ones brown. **P. demissum**

P. demissum [drooping]. Unlike *P. zieri*, this species forms dense sods, often on rather hard turf. Rare or infrequent, in crevices of tundra rocks, on hummocks of solifluction lobes, in soil pockets of screes, etc., in the alpine, 12000–13500 ft. CC: Summit Lake, Mt. Evans, sterile, *W&W B-114282*; McClellan Ridge above Santiago Mine, 12500 ft, *Weber, Koponen, & Nelson B-41407*.

P. zieri [for John Zier]. Extremely rare species of subalpine seeping cliffs and moist tundra. The stems are julaceous, usually sparse, with broadly ovate, pale green, or pinkish, delicate leaves with slightly recurved points. The cells are easily seen with a strong lens, and are swollen, reminiscent of a bowl of cut glass! BL, CC, SJ, ST.

Pohlia [for J. E. Pohl]

How is *Pohlia* distinguished from *Bryum*? Lanceolate, often serrulate leaves, usually without borders; costa ending before or in the apex; long, generally narrow and uniform leaf cells; the peristome is double and the endostome is well-developed although cilia, when present, generally lack appendages. Because most of our species are usually without capsules, this is a rather weak combination of distinctions! Spence points out that the genus has a rather definite 'Gestalt'. The plants and leaves have a stiff, rigid 'hygrophobic' (not easily wettable) aspect they share with *Mnium*. The gemmiparous species are usually delicate and pale yellow-green.

The genus has recently been placed in the Mniaceae (Cox & Hedderson, 1999), but unfortunately one needs to forget about morphology and rely on nucleotide sequences from nuclear, plastid, and mitochondrial genomes! Spence (*in litt.*) suggests that the Pohlioid genera might better be segregated as a new family.

Key to sterile plants

1a.   Plants bearing axillary gemmae . . . . (2)
1b.   Plants without axillary gemmae . . . . (8)

2a.   Gemmae arising singly in the leaf axils, cylindrical to oblong, reddish when moist; leaf primordia green and conspicuous, scattered on the body and at the apex. **P. drummondii**
2b.   Gemmae arising in clusters . . . . (3)

3a.   Gemmae spherical to short-oblong, hardly longer than wide . . . . (4)
3b.   Gemmae oblong, obconic or more or less elongate or vermicular . . . . (5)

4a.   Leaf primordia of gemmae triangular-laminate (leaf-like), forming a dome over the apex. **P. bulbifera**
4b.   Leaf primordia of some or all gemmae peg-like (just 1–2 cells broad at the base), erect or incurved over the apex. **P. camptotrachela**

5a.   Gemmae small (70–120 μm) and orange-red, with the primordia peg-like and incurved. **P. camptotrachela**
5b.   Gemmae larger, oblong, obconic, or linear-vermicular . . . . (6)

6a.   Gemmae large (200–500 μm long), oblong to linear; leaf primordia laminate, at least 3–4 cells broad at the base . . . . (7)

6b.   Gemmae small (up to 100 μm long), linear-vermicular; leaf primordia peg-like, 1–2 cells broad at the base. **P. proligera**

7a.   Gemmae oblong or obconic. **P. andalusica**
7b.   Gemmae linear-cylindrical. **P. tundrae**

8a.   Leaf cells broad (13 μm or more wide) .... (9)
8b.   Cells narrow, 6–10 (–12) μm wide .... (10)

9a.   Plants dull whitish, in calcareous seepy or wet areas. **P. wahlenbergii**
9b.   Plants yellow-green, reddish or blackish at least below, in calcareous or acidic, usually wet, sites. **P. obtusifolia**

10a.   Cells with thin to firm walls .... (11)
10b.   Cells thick-walled; leaves ± serrate at the apex .... (14)

11a.   Plants pale, with the leaves often erect and seriate .... (12)
11b.   Leaves shiny or dull green, the leaves sometimes erect but not whitish .... (13)

12a.   Leaves erect and often seriate (the shoot appears pentangular). **P. bolanderi** var. **seriata**
12b.   Leaves more or less spreading, not seriate, broadly lanceolate to elliptic; leaves with an iridescent sheen. **P. cruda**

13a.   Leaves long-lanceolate, erect, coarsely serrate near the apex. **P. longicolla**
13b.   Leaves broadly lanceolate, spreading to erect-spreading, serrate or serrulate near the apex; leaves with an iridescent sheen. **P. cruda**

14a.   Rare plants of wet roadside ditches and swampy ground, subalpine; rarely fertile; new growth bright yellow-green, the older growth black; cushions often deep. Leaf cells broad for a *Pohlia*. **P. obtusifolia**
14b.   Common plants of various habitats, usually in forests, usually fertile; leaves dark green. **P. nutans**

Key to fertile plants

1a.   Endostome segments narrowly or not at all split along the keel; cilia rudimentary or absent, rarely long (2)
1b.   Endostome segments broadly split along the keel; cilia long .... (5)

2a.   Plants dull; leaf cells thick-walled .... (3)
2b.   Plants ± glossy; cells thin-walled .... (4)

3a.   Capsule 2–4 mm long; endostome segments keeled; cilia present. Subalpine fen plants. **P. greenii**
3b.   Capsule 4–7 mm long; endostome segments scarcely keeled, often irregular; cilia lacking; forest plants. *P. elongata* (not yet found here but to be expected in the lower forests)

4a.   Plants dioicous; leaves serrulate to serrate at the apex; capsules rare. **P. bolanderi**
4b.   Plants paroicous; leaves coarsely serrate at the apex; capsules common. **P. longicollis**

5a.   Basal membrane low; segments narrow, scarcely keeled though broadly perforate; cells lax and thin-walled, 10–20 μm wide. **P. obtusifolia**
5b.   Basal membrane ½ the exostome length; segments broad and keeled; cells linear, 6–10 μm wide ..... (6)

6a.   Leaves dull; cells with thickened walls. **P. nutans**
6b.   Leaves slightly to strongly glossy, yellowish, iridescent; cells with thin walls. **P. cruda**

P. andalusica [from Andalusía, Spain]. A gemmiparous species occurring on disturbed sites in the subalpine. It has a slender, glossy appearance and large green gemmae (200 μm or more) arising in clusters from its upper leaf axils, tapering to the base, with 3–6 leaf-like primordia. CC: 7 mi SW of Georgetown, 13000 ft, dry, rocky tundra below Argentine Pass, *Hermann 29065 (B-80209)* (distributed as *P. annotina*); GL: Yankee Doodle Lake, *W&W B-110942.*

P. bolanderi var. **seriata** [for H. N. Bolander]. An alpine species abundant in snow-melt areas. The type locality is on the road to Corona (Rollins) Pass. It is a dull, pale green species in dense tufts. The leaves are ± distinctly arranged in 4–5 vertical rows. *P. bolanderi* has recently been discovered relictual in Europe for the first time, in southern Spain at 2000–3000 m, on Mulhacén, the highest peak in the Iberian Peninsula (Shaw et al. 2004). The variety *seriata*, which is distinguished by its vertically seriate leaves, probably is a trivial variety.

P. bulbifera. A gemmiparous species of disturbed places where the soil is compacted and hard. The gemmae in the leaf axils are spherical, with the laminate leaf primordia closing over the top as a dome. It is not necessarily an alpine species but found along slumping forested slopes. ST: McCullough Gulch, 12450 ft, below a saddle, on a rocky, steep slope near a snow bank, *Redner B-112718.*

P. camptotrachela [bent neck]. A gemmiparous species characterized by its dull rather than glossy appearance, and gemmae that are roughly spheroidal and less than 150 μm diameter with 1–4 peg-like leaf primordia at the apex that are incurved over the apex. It occurs in the forested areas along unstable soils.

P. cruda. Not as common as *P. nutans* and found in more moist sites, especially on little cut-banks along trails in the forests. The pale green leaves with an iridescent sheen are distinctive. It has long-cylindric capsules and well-developed endostome with broadly perforate segments and long cilia.

P. drummondii [for Thomas Drummond]. A frequent species on tundra, but rather difficult to recognize unless the large red gemmae with 3–6 green leaf-like primordia scattered mostly at the stem apex, are present.

P. greenii [evidently for Thomas Green]. A small species of peat hummocks at the edges of subalpine fens. The capsules are small and not strongly inclined.

Probably this is common in subalpine and alpine fens. We have it abundantly at Guanella Pass (*W&W B-112510*). Shaw considers this to be a variety of *P. elongata*. However, from his discussion (below), we feel that it is probably justifiable to recognize var. *greenii* (Bridel) Shaw at the species level until more is known about the variability of both. Shaw (1982) wrote: "Rock crevices in tundra areas, less frequently on exposed soil, Alaska, Washington, California, Colorado, Utah, Newfoundland, Greenland. Reported from northern Europe . . . Although no single character is diagnostic, several morphological attributes distinguish the var. *greenii* from so-called typical expressions: Smaller overall size, shorter, broader leaves, more strongly inclined, broader, shorter capsules, low basal membrane, irregular segments, and complete lack of cilia . . . Sporophytic and gametophytic differences between the varieties, coupled with geographic differences, warrant separation, at least on the basis of North American plants. . . The plants are minute, with capsules hardly longer than wide." Shaw's figure 5, p. 136, shows these features clearly.

P. longicollis [long-necked]. A rare, robust, shiny green plant occurring in subalpine peatlands and streamsides at high altitudes. It fruits frequently. The stems are comose, with a terminal leaf cluster; the stem is almost naked below the comal tuft. The leaves are yellow-green, but lacking the opalescent sheen of *P. cruda*. The capsules are elongate-cylindrical with a constriction below the mouth and a well-developed neck. BL: Trail to Diamond Lake, *J.L. Smith 407 (B-114802).*

P. nutans [nodding]. A common moss in shaded areas throughout the mountains, except for alpine tundra. It usually fruits abundantly, with hanging, oblong capsules. The thick-walled cells give the plant a dull appearance. When sterile, it is difficult to distinguish from *P. elongata*, which, fortunately, is rare or infrequent and confined to the foothills. The color and stiffness of the seta suggests an apt name for this, the Copper-wire Moss.

P. obtusifolia. This is often a large plant up to 4 cm tall. The leaves are somewhat cucullate, with broad cells (12–13 μm or more wide, suggesting a *Bryum*). In these tall forms the plants are black except for the short apex of fresh, pale green leaves. Common along ditches along roads in the subalpine where it receives snow-melt water in the early spring. In an iron fen in upper Geneva Park (CC) it forms extremely tight, dense, black mats on the edges of limonite sheets.

P. proligera [proliferous gemmae]. A species of roadsides and paths in the forested regions. The gemmae occur in dense clusters in the upper leaf axils. They are linear-vermicular to narrowly oblong, with 1–2 peg-like leaf primordia often bent at an angle.

P. tundrae. A gemmiparous species of the alpine tundra that has narrowly cylindrical gemmae with leaf primordia. BL, CF, CC. GA. GL, PT, SJ. Specimens were originally identified as *P. bulbifera, P. proligera,* and *P. annotina* var. *loeskei*. A Gilpin County collection *(Weber B-3471)* contains *P. andalusica, P. cruda,* and *P. tundrae* mixed in the same tuft!

P. wahlenbergii [for G. Wahlenberg]. Characterized by whitish or pale green, erect-spreading leaves on long, slender stems. It is a beautiful moss resembling *Bryum weigelii* but never reddish or with decurrent leaves, and grows in dense, tall tufts in seeps at high altitudes.

Excluded Taxa

Pohlia elongata Hedwig. Our specimens have been re-examined and we find that they are all *Ptychostomum cernuum*. The species differs from *P. nutans* in having a narrow, elongate, not nodding capsule; the endostome segments are narrowly, rather than widely, perforate.

Ptychostomum [puckered capsule mouth]

1a. Stems mostly evenly foliate, not comose; leaf base same color as rest of leaf, usually green; limbidium indistinct, bistratose; distal laminal cells lax, wide and short, 2–4:1, rhomboidal; proximal laminal cells longer and narrower, rectangular . . . . (2)
1b. Stems evenly foliate to comose with upper leaves distinctly enlarged; leaf base mostly reddish; limbidium distinct to indistinct, unistratose; distal laminal cells not lax, 3–5:1, rhomboid to hexagonal; proximal laminal cells similar in width and length but rectangular, occasionally quadrate . . . . (7)

2a. Plants small to medium, monoicous; capsule elongate, with a long neck, curved, peristome reduced; cilia short to absent; spores generally large, >25 μm). **P. cernuum**
2b. Plants medium to large, dioicous; peristome mostly perfect; spores small (<22 μm) . . . . (3)

3a. Leaves strongly and broadly decurrent; costa percurrent; leaves generally pink in living plants, but this color does not persist in herbarium specimens; never with capsules. **P. weigelii**
3b. Leaves not or weakly decurrent, obtuse, acute, or acuminate; costa not reaching apex to excurrent . . . (4)

4a. Leaves broadly ovate to sub-orbicular; leaf apex rounded to obtuse; costa not reaching apex to percurrent . . . . (5)
4b. Leaves ovate to ovate-lanceolate; leaf apex acute; costa typically percurrent to short-excurrent . . . . (6)

5a. Plants often red; leaves strongly cochleariform, not strongly contorted when dry; costa often percurrent; rare alpine. **P. cryophilum**
5b. Plant dull green, lacking red tints; leaves weakly cochleariform or flat, strongly contorted when dry; costa not reaching apex; rare, mostly in subalpine peat fens. **P. cyclophyllum**

6a. Capsules present, short-pyriform, constricted under the mouth; plants dull green-brown, lacking red tints; alpine. **P. turbinatum**
6b. Capsules not present in Colorado material, longer; subalpine; young leaves red-tinged; leaf margins recurved. Propagular branchlets with minute leaves occur in the leaf axils. **P. pallens**

7a. Plants small, densely caespitose; leaves ovate-lanceolate, limbidium indistinct or absent; proximal laminal cells shorter than median and distal cells; dioicous. . . . . (8)
7b. Plants medium to large, caespitose; leaves broadly ovate, ovate-lanceolate to lanceolate, flat or cochleariform; proximal laminal cells same length as distal ones or somewhat longer; sexuality various . . . . (9)

8a. Plants small, leaves <1.0 mm; distal lamina in older leaves becoming hyaline, the proximal laminal cells short-rectangular to quadrate; costa variable, from not reaching apex to excurrent in a slender hyaline hair-point. **P. kunzei**

8b. Plant medium, most leaves 1.0–2.5 mm long; distal lamina green; proximal laminal cells long rectangular; costa strong, excurrent in brown or yellow hair-point. **P. imbricatulum**

9a. Plants comose to caespitose; monoicous; peristome reduced, the cilia short or absent; basal membrane low, mostly 1/2 the height of the exostome; spores large, 18–50 μm .... (10)

9b. Plants evenly foliate to loosely comose; sexuality various; peristome perfect; basal membrane less than half the height of the exostome; spores small, 10–22 μm .... (12)

10a. Endostome strongly adherent to exostome, giving exostome teeth a chambered look; leaves acuminate; costa excurrent in a short arista; capsule short-pyriform only 1 mm long, not constricted at the mouth; operculum sharply conic; spores 22–35 μm; alpine. **P. pendulum**

10b. Endostome not fused to the exostome; leaves acute to acuminate; costa not reaching apex or long excurrent; capsule various; spores 25–50 μm .... (11)

11a. Leaves lanceolate, not cochleariform or keeled; arista long excurrent, denticulate; mouth of mature capsule red; endostome membrane about half the height of the exostome; spores small, 18–28 μm; cilia rudimentary. **P. inclinatum**

11b. Leaves ovate, strongly cochleariform, keeled, acute; costa percurrent to short excurrent; mouth of mature capsule yellow or pale orange; endostome membrane low, less than half the height of the exostome; spores 25–50 μm; cilia absent. **P. knowltonii**

12a. Stem elongate; leaves distinctly decurrent; limbidium strong; costa prominent but lacking a long arista
.... (13)

12b. Stem short, or if elongated, the stems tightly packed; leaves not decurrent; limbidium weak; arista well-developed .... (14)

13a. Leaves long-decurrent and usually remotely spaced on the stem; dioicous; long-stemmed plants of fens. **P. pseudotriquetrum**

13b. Leaves inconspicuously decurrent and closely spaced, overlapping on the stem; synoicous; relatively short-stemmed plants of wet streamsides, often on rocks. **P. bimum**

14a. Stems elongate, branched (the branches ascending with the main stem) forming deep sods bound by rhizoids; new growth often contrasting in color with the older growth; leaves of the season turquoise green, the arista weak and not darkened; autoicous; spores about 18–22 μm. **P. pallescens**

14b. Stems short, unbranched, not forming sods; leaves of the year darker green, often brownish, the arista strong, brown; polyoicous, synoicous, or dioicous; spores 12–16 μm .... (15)

15a. Limbidium narrow, yellowish, 2–3 cells wide; plant synoicous. **P. creberrimum**

15b. Limbidium wide, 3 or more cells wide, the same color as the lamina; plants polyoicous. **P. lonchocaulon**

P. bimum [biennial]. This has usually been considered a variety of *P. pseudotriquetrum*, but the leaves are shortly and faintly decurrent, and not widely spaced along the stem. The habitat is quite different from *P. pseudotriquetrum*. Not a fen-plant, it forms clumps on stones along running streams and is never tall. Ranges from alpine tundra down into the lowest foothills.

P. cernuum. Easily recognized by its narrow elongate capsule. The leaves are well-separated on the stems,

erect-spreading when wet and shriveled and incurved when dry. The cells are quite short and broad, and the margin is strongly bordered and narrowly revolute. There is a superficial resemblance of the capsule to that of *Pohlia elongata* and specimens have been misidentified as such. Frequent in places subject to light flooding or on dripping rocks. CC, GN, LR, MZ.

P. creberrimum [tightly packed]. Evidently a common species, going under various names in the herbarium. It is most closely related to *P. pallescens*. Well-developed plants have short, unbranched stems, and the leaves are less appressed, with stouter excurrent costae, and the leaves are more brown, whereas in *P. pallescens* the leaves of the season are almost turquoise.

P. cryophilum [cold-loving]. An alpine species, disjunct from the Arctic. The plants are delicate, often strongly reddish, the leaves rounded-oval, cochleariform, and not strongly contorted in drying. BL, CC, LR.

P. cyclophyllum [circular-leaved]. A rare species of subalpine peatlands that is characterized by slender stems; dull, bluish-green, oval leaves, rounded-obtuse at the apex, remotely scattered on the stem, and greatly contorted when dry; a costa that does not quite reach the leaf apex. The occasional presence of cylindrical, filamentous-branched gemmae in the upper leaf axils is diagnostic. One verified collection: BL: Peat bog source of Left Hand Creek, 10600 ft, 3 mi W of Ward, *Weber & Dahl B-6983* (!Spence). The site was destroyed by peat mining and vandalism.

P. imbricatulum. A weedy moss, usually sterile, common in sandy soil on the eastern plains and outer foothills. It probably does not occur in the high mountains. It is a plant with low stems, only a few mm tall, forming dense turfs. The yellowish leaves are broadly ovate, with intensely red broadly rectangular basal cells (often with conspicuous oil droplets).The costa is long excurrent into a hair-point. The species is dioicous, and the male plants occur mixed in with the female sporophyte-bearing ones. The (smaller) male plants terminate with a bulbous cluster of leaves with flaring points. It has long been called, incorrectly, *Bryum caespiticium* (which name is a synonym of *P. pendulum*).

P. inclinatum. Common in subalpine wetlands and alpine tundra, where it may replace *B. pallescens* on drier sites. Although characterized technically by having no cilia and a low endostome membrane, we believe it can be recognized on sight. The leaves are long and narrow, loosely spreading and twisted, with a long pale excurrent costa, which on close examination is denticulate. The innovations are weakly julaceous and their leaves imbricate. The limbidium is prominent and revolute from leaf base to apex. The capsule is straight and elongate.

P. knowltonii [for F. H. Knowlton]. A distinctive moss because of the extremely low basal membrane, the extremely slender endostome segments and totally lacking cilia, but it is easily recognized when sterile, as is almost always the case. The plants form dense cushions and are foliose to the base. The stems are red. The leaves are broadly oval, rounded or obtuse at the apex, distinctly keeled, not decurrent, with a stout, percurrent costa, indistinctly bordered and strongly recurved in the lower half. The median cells are shortly rhomboid, 3–5:1. The rhizoids are densely papillose. An infrequent species of moist tundra. CC: Saddle between Mt. Evans and Mt. Epaulet, 13500 ft, *Weber, Miller, & Wittmann B-11510*; Loveland Pass summit, 12500 ft, *Smith B-114807*; ST: Blue Lake, 3000 m, *W&W B-111211*.

P. kunzei [for G. Kunze]. Rare or under-collected. Stems only 2-3 mm tall, julaceous; leaves less than 0.5 mm, bright green, broadly ovate and cochleariform, abruptly rounded at apex, with a short, few-celled apiculus; median cells elongate-rectangular. The minute size and general appearance suggests a small species of *Gemmabryum*, but the leaf areolation is typically that of *Ptychostomum*, and there are no rhizoidal gemmae. MZ: San Juan River north of Four Corners, 100 ft W of Hwy 160 bridge, on exposed dry soil on sedimentary rock, (sparse, with dominant *Syntrichia caninervis*) *Spence 5629 (B-113914)*.

P. lonchocaulon [lance-shaped stem]. This is so similar to *P. creberrimum* that it can easily be overlooked. It should be sought at low elevations in the western counties. We have one collection: MZ: Mesa Verde National Park, Chapin Mesa, seep draining water tank, *Welsh & Moore 2113 (B-55123)* (!Spence) (previously determined by Flowers as *Bryum angustirete* Kindberg).

P. pallens. Infrequent in subalpine wet cliff crevices, foothills to subalpine. Evidently also on ground in subalpine willow carrs and pond shores. The leaves are broad up to the apex, shortly decurrent, and either obtuse or with a slight one-celled apiculus. Immature sterile plants may have small pink, cochleariform leaves, 0.5–1.0 mm, appressed to the stem, unchanged in drying, while fertile plants have longer leaves (up to 2 mm). The areolation is transparent and the leaf margin is narrow. The capsule is about 2 mm long, plump pyriform. The peristome is pale yellow, the endostome segments narrowly perforate, cilia are short, not appendiculate. Sterile plants are not hard to recognize. The distantly spaced leaves that are not much changed on drying. The abruptly short-acuminate ovate leaves that are not at all decurrent; the short and broad, clear leaf cells; and the red stem are good characters. OR: Ouray; Imogene Creek, *Hermann 23250 (B-37739)*; SJ:

Cumberland Basin, La Plata Mts., 11600 ft, *Jamieson 13474 (B-114396).*

P. pallescens. One of the most common species of wet ground, especially at lower altitudes. The tufts are often deep, and the stems tightly packed so as to obscure the branching, which is erect. The current year's shoots have pale green leaves that have long excurrent tips but they are not stout or reddish. The colonies often form distinct tiers of annual growth, and the lower stems are densely clothed with brown rhizoids. Commonly the leaves of previous years are bleached, and the seta of previous years' sporophytes persist. The median leaf cells are short, about 3:1, and the limbidium is prominent. The costa is strong and excurrent, and red at the base (but the leaf base itself is not red). Sporophytes are usually abundant. The capsule is cylindric, with a distinct narrow and shrunken neck about as long as the urn. The exostome is yellow, and the endostome segments have large and broad gaps in the middle; cilia are well developed and appendiculate. This species is similar to *P. creberrimum* but has larger spores (18–22 μm).

P. pendulum. Abundant on moist tundra. It is usually abundantly fertile and, except for this, might be overlooked because of its small size. The leaves are only 1 mm long, in a densely crowded rosette, broadly ovate, abruptly aristate, red-based, widest at the middle, the median cells about 40 x 15 μm. The capsules are short, fusiform-ovoid, (if opercula are present the capsules are pointed at each end). The seta is curved just below the urn, and the peristome teeth are unique. The endostome is adherent to the exostome. The lamellae (divisions separated by horizontal walls) have several cross-walls that are irregularly widened and not transverse (showing diagonal lines); this is most easily seen from the ventral side. In age the capsules are bleached. This can hardly be confused with any other

alpine species, being recognizable on sight without having to examine the peristome.

The habitus depends a good deal on the substrate and growing conditions. On gravelly polygonal ground the plants are extremely short with the stems crowded at the summit of the extremely short stem (*B-16824* from Summit Lake, Mount Evans), On well-developed moss tundra the plants form a relatively deep mat with several growth periods as in *P. pallescens.*

P. pseudotriquetrum [falsely three-ranked]. A common species of wet places from the foothills to the alpine, not usually occurring in large pure stands but consisting of scattered stems in amongst other wetland mosses. It is handsome, with elongate stems and large leaves distinctly decurrent on the stem. It does not seem to fruit regularly here. See discussion under *P. bimum.*

P. turbinatum [top-shaped]. An uncommon species thus far found only in the high alpine area. It is distinguished by the short-pyriform capsule with a small diameter mouth, the urn constricted under the mouth, somewhat resembling that of *P. pendulum*, but the peristome is not unusual for the genus. BL: Green Lakes Valley, 10500–11500ft, on rocks in snow melt drainage, *Weber B-41730, B-42005.*

P. weigelii [for J. Weigel]. A species with distant leaves, growing in springs, wet roadside depressions, or half-submerged in willow fens of the subalpine. It is always sterile, and may be mistaken for *Pohlia wahlenbergii*, but the leaves of *B. weigelii* are greatly shrunken and contorted when dry, and strongly decurrent, and usually pink when fresh, rather than pale glaucous green. The tufts are loose, and the stems not at all crowded.

Roellia [for J. Röll]

R. roellii. This handsome moss occurs on saturated ground among streamlets on the forest floor under spruces in the subalpine of Rocky Mountain Park and Middle Park. It can be mistaken for nothing else but the large species of *Rhodobryum*, which do not occur in Colorado. The old report of *Rhodobryum roseum* (Porter

& Coulter, 1874) check reference, which is now referred to *R. ontariense* (Kindberg) Kindberg, probably represents a misidentification of *Roellia*. *Rhodobryum* has erect stems arising from horizontal underground stems, while *Roellia* lacks rhizomes.

Rosulabryum [with rosulate leaf arrangement]

1a.   Innovations rosulate, most leaves clustered at the stem apex; leaves obovate, flat when moist . . . . . . (2)
1b.   Innovations evenly foliate; leaves ovate, cochleariform when moist. **R. flaccidum**

2a.   Leaves in a tight spiral, appearing narrow and dark green when dry, with a strong excurrent costa; leaf axil
        gemmae lacking; stems lacking brown micronemata; fertile, capsule with a distinct neck **R. capillare**
2b.   Leaves in a loose, irregular spiral, not appearing narrow when dry; filamentous red-brown gemmae in the leaf
        axils, sometimes copious; stems with numerous micronemata. **R. laevifilum**

R. capillare. Uncommon in the foothills of the Front Range. The tightly massed, strongly curved, and spirally arranged (as if brushed) rosulate leaves with stout aristae are clearly different from those of *R. laevifilum*. Our specimens are often fruiting, while *R. laevifilum* is usually sterile. The rhizoids are papillose, and the small rhizoidal gemmae are concolorous. BL: 8 mi NW of Nederland, 10300 ft, *Flowers 9833 (B-55251)*; DA: Castlewood Canyon State Park, 6400 ft, *Lederer B-102658*; LR: Big South Trail, under *Alnus*, bank of Cache La Poudre River, 8700 ft, *Hermann 28159*.

R. flaccidum [limp]. This is a rather tiny relative of the next, characterized by more slender leaves which clothe the stem, not gathered into terminal rosettes. It is abundant on granite outcrops in the outer foothills. BL, JF, LR.

R. laevifilum [smooth threads, i.e. gemmae]. A common moss in the outer foothills, characterized by the leaves with excurrent costae, distinctly twisted spirally around the stem when dry, a capsule that leans but rarely is really pendent, and the characteristic green or brown filamentous and papillose gemmae that are usually found in clusters in the axils of the upper leaves. This species seems to prefer growing on thin soil over rock rather than on deep soils. This species has gone incorrectly under the name *Bryum flaccidum* in North American literature.

Genus and Species Incertus

Brachymenium vinosulum Cardot. Spence (2007b) has identified the following collection as this species: GA: Hot Sulphur Springs; on limestone rock, in a small cavern at the hot springs, just where light becomes obvious, *V. Evenson (B-115430)*. Our diagnosis is as follows: Tuft very densely caespitose, upper stem leaves 0.5–0.8 mm long, ovate; lower stem leaves 1.0–1.2 mm long, oblanceolate. narrowed to the base; basal and median cells elongate-rectangular, distal cells rhomboid, leaf margin with a somewhat distinct limbidium; costa strong, red, leaf apex broadly acute, not apiculate. Sterile.

BRYOXIPHIACEAE (BRX)

A monotypic family consisting of the single genus, *Bryoxiphium*, with one or possibly two, species, depending on the taxonomic interpretation. Probably one of the rarest and most sought-after mosses in the world; each new discovery seems to justify a publication.

The moss is unique in its morphology. Arising from a small bulb-like base, the single stem has two rows of leaves. "Usually they are closely imbricated, erect, and rather folded, but at the same time they are so small that in a superficial observation they might be taken for appressed hairs. The lower leaves . . . are usually characterized by a blunt apex, and the longest leaves are clearly apiculate and tapering" (Löve & Löve 1953). The sporophyte is rare, undoubtedly because male and female plants do not occur anywhere near each other. The shiny, light green or golden little stems with their unusual leaves are impossible to mistake for any other moss.

Bryoxiphium is undoubtedly a Tertiary relict, occurring as it does in highly disjunct areas of the world, including Iceland, Greenland, Japan, Kamtchatka, the American midwest, Mount Rainier in Washington, Mexico, and now, most recently, Colorado.

Bryoxiphium [sword-shaped moss]

B. norvegicum. Until recently, only one collection was known: MZ: on overhang of cliff, ca. 6 mi NNW of Dolores along Beaver Creek, 0.5 mi NW of junction of Dolores River and Beaver Creek, 7500 ft, 12 June 1958, *Pursell 3246a (B-65336)*. This colony was destroyed by high water level in the McPhee Reservoir. Jamieson has found a second colony in CN: Upper tributary of Navajo River, 11500 ft, *Jamieson s.n. (B-113241)*. Botanists desiring to find this species might well be guided by the Löves' observation: "*Bryoxiphium* seems to prefer a substratum porous enough to hold water and at the same time avoid calcareous soils. Sandstone or volcanic material might be preferred in most areas." The leaves are so closely appressed that the stem looks like an undivided flat ribbon!

The type specimen came from Iceland; this taxon is not known to occur in Norway.

BUXBAUMIACEAE (BXB)

Buxbaumia [for J. C. Buxbaum]

la. Capsule dark chestnut-brown, glossy, broadly ovoid, the face broad and flat, nearly perpendicular to the seta. **B. aphylla**
lb. Capsule green or becoming pale brown, not highly glossy, narrow, sub-terete and elongate, erect or merely oblique to the seta. **B. piperi**

B. aphylla [leafless]. Both species are rare or at least well-camouflaged, and occur as scattered individuals in virgin spruce forests with a moist to wet and boggy floor, but also in lodgepole pine forests. *B. aphylla* often occurs on mounds of clay soil probably raised by the falling of a tree and stabilized by algae and moss protonema, a precise habitat and locally developed. *Buxbaumia* species are usually associated with small liverworts.

B. piperi [for C. V. Piper]. This species occurs on soggy, rotting wood. GA: Middle Park, West St. Louis Creek, 6 mi W of Fraser, *Weber & Vaarama B-11086*; GN: Gothic Natural Area, 10000 ft, on rotting logs, *Weber B-17070, Khanna 15328*; SJ: San Juan Mts., Weminuche Wilderness, confluence of Vallecito and Trinity Creeks, *Jamieson 11441, 11490*.

CAMPYLIACEAE (CMP)

Campylium and its relatives

1a. Costa lacking or short and double . . . . (2)
1b. Costa single, ending at or somewhat above the leaf middle . . . . (5)

2a. Leaves squarrose-recurved; yellow-green; leaf bases closely overlapping; margins revolute below. **Campylophyllum halleri**
2b. Leaves wide-spreading to more or less squarrose; margins not revolute . . . . (3)

3a. Plants slender, green, with creeping stems; leaves smooth, minute (0.5 mm), finely serrulate all around especially at the base; alar cells small and quadrate. **Campylophyllum sommerfeltii**
3b. Plants robust, with erect-ascending stems; leaves larger (2 mm), striolate when dry, entire; alar cells somewhat inflated, sub-rectangular . . . . (4)

4a.   Plants usually erect; stem leaves 1.7–2.8 mm long, the acumen constituting 40–65% of leaf length. **Campylium stellatum**

4b.   Plants usually creeping; stem leaves 1.0–2.3 mm long, the acumen constituting 55–77% of leaf length. **Campylium protensum**

5a.   Leaves erect-spreading, gradually acuminate; alar cells enlarged and inflated, becoming thick-walled in age. See *Drepanocladus polygamus*

5b.   Leaves 1.0–1.2 mm long, wide-spreading to squarrose, abruptly acuminate; alar cells not or only slightly inflated. **Campyliadelphus chrysophyllus**

Campyliadelphus

C. chrysophyllus. Forming dense golden-green patches on rocks along streams in the foothills canyons, but also in the upper subalpine, nowhere common. AA: T36N R1E S5, *Lyon B-114201*.BL: Boulder Canyon, 7000 ft, *Weber B-10559*; Eldorado Springs, *Weber, Wittmann & Mazurek B-111741*; Gregory Canyon, *W&W B-112134*.

Campylium [curved, capsules?]

C. protensum [stretched out]. Similar to *C. stellatum*, but ecologically quite distinct from it. It grows on moist ground on the forest edge where water percolates from snow patches under the trees. The leaves are small (less than 2 mm long) compared to the wetland *C. stellatum* and the leaf base more suddenly narrows to the long acumen. BL: 8 mi NW of Nederland, 10260 ft, *Flowers 9777, 9820* (!Hedenäs); GA: Head of Willow Creek, trail to Lost Lake, 8500 ft, *Weber, Wittmann, & Tidball B-112466*.

C. stellatum [star-like]. Common in wet ground of subalpine and alpine willow and peat fens. Easily recognized by the large (2 mm), stiffly-spreading leaves with a rich russet-gold tint. Hedenäs (1997) distinguishes this thus: "When seen from above in the field, shoots of *C. stellatum* often look like small stars. It is closely related to *C. protensum,* from which it differs in its more erect growth, its slightly larger size, and its relatively shorter leaf acumen. Paraphyllia have never been seen in *C. stellatum,* whereas scattered plants of *C. protensum* have a few."

Campylophyllum [squarrose leaves]

C. halleri [for A. von Haller]. A subalpine species growing on rocks. It has minute bronze-colored leaves, is autoicous and usually fertile, and differs from *C. sommerfeltii* by having the leaf bases overlapping and not exposing the stem, and by the short or absent costa. Our only record of this is from SJ: Deadwood Gulch, 2850 m, *W&W B-111922*. However, the species is evidently common in the San Juan Mountains, judging from Jamieson collections.

C. sommerfeltii [for S. C. Sommerfelt]. An infrequent or easily overlooked species on rocks, rotting wood, or duff near stream sides in the Front Range from the outermost base of the foothills to the montane (6000–8900 ft). The plants are deep green to yellow-green, with minute strongly squarrose leaves, inserted with their bases not overlapping, The plants are autoicous and are usually richly fruiting. BL: Bluebell Canyon, trail to Royal Arch, *Roehrick B-6111*; bottom of Skunk Canyon, *Weber, Nelson & Litvak B-35450*; 2 mi S of Ward, 1 mi E of Peak to Peak Hwy, close to Fourmile Creek, 8900 ft, *Shultz B-46217;* GF: Rifle Falls State Park, *Weber B-115581*; MZ: Mesa Verde National Park, *Colyer B-113281*.

CATOSCOPIACEAE (CTS)

A monotypic family consisting of a single genus and species.

Catoscopium [looking down, alluding to the capsule]

C. nigritum. A rare moss of wet tundra, this occurs also in at least one calcareous subalpine fen, forming dense turfs on the vertical sides of shallow channels where obvious associates include small *Picea* trees, *Betula glandulosa, Salix candida, S. myrtillifolia,* and *S. serissima, Kobresia myosuroides, K. simpliciuscula,* and *Trichophorum pumilum*. The shoots are undistinguished, with leaves resembling those of *Ceratodon*, but the sporophyte is unique in the mosses. The capsule is dark, minute, orbicular, and bent down or to the side, resembling a miniature golf club. PA: South Park, High Creek Fen, *W&W B-112549*. CC: Mount Evans, Summit Lake, *Weber, Wittmann, Lehr, & Tidball B-114276*. In the West it is also known from Montana. The Colorado sites are the southernmost localities known for this moss in North America.

CLIMACIACEAE (CLM)

Climacium [ladder-like, the form of the inner peristome]

C. dendroides [tree-like]. Easily recognized, being our only moss with a dendroid habit. The primary stems are creeping and embedded in the soil. Erect secondary stems are dark and clothed with appressed, often brownish or pale leaves, and branch at the top, the radiating branches suggesting the fronds of miniature palm trees. *Climacium* is a common moss of wet streamsides from the lower foothills on up to the subalpine. The plants are sterile in our area, since they are dioicous and only female plants occur. The branch leaves are ovate, strongly irregularly plicate, with a prominent costa disappearing just below the tip. The margin is coarsely serrate at the apex, the median cells are narrowly rhomboid to linear, and the basal cells nearest the costa are thick-walled, pitted, and orange.

DICRANACEAE (DCR)

This key includes some genera of the closely related Ditrichaceae, Seligeriaceae, and Orthotrichaceae. Before trying to use the key one should become familiar with illustrations of the various genera. This is a difficult group.

1a.   Plants less than 0.5 mm high, difficult to see without a hand lens! Capsules erect, goblet-shaped. On sandstone cliffs, often on shaded under-hangs. See **Seligeria** (Seligeriaceae)
1b.   Plants larger . . . . (2)

2a.   Leaves distichous, lying in one plane when wet, with a broad, sheathing base suddenly narrowed into a divergent, filiform lamina consisting mostly of the costa. **Distichium** (Ditrichaceae)
2b.   Leaves spirally arranged around the stem, not lying in one plane . . . . (3)

3a.   Cells of the distal half of the leaf papillose or mammillose . . . . (4)
3b.   Cells of the distal half of the leaf smooth or only slightly mammillose . . . . (8)

4a.   Papillae or mammillae low and rounded, never projecting saliently as teeth; plants forming dense mats on shaded cliffs; capsules with short seta, the urn strongly ridged and constricted below the mouth. **Amphidium** (Orthotrichaceae)
4b.   Papillae or mammillae high, sharp-pointed, directed forward and projecting from the leaf margins and back .... (5)

5a.   Leaf margin bistratose .... (6)
5b.   Leaf margin unistratose .... (7)

6a.   Distal leaf cells coarsely and broadly conical-papillose; gametangia in a lateral perigonium, a bud on a short stalk; rare, on granite outcrops, outer foothills. **Cnestrum schisti** (Dicranaceae)
6b.   Distal leaf cells inconspicuously low-papillose; gametangia enclosed in a sessile perigonium, not a stalked bud; on rocks in subalpine forests. **Cynodontium** (Dicranaceae)

7a.   Leaves broadly lanceolate; capsule not striate; on wet rocks on cliffs and along snow-melt streamlets. **Dichodontium** (Dicranaceae)
7b.   Leaves narrowly lanceolate; capsule distinctly striate-ribbed; on vertical faces of granite rocks, subalpine. **Cynodontium** (Dicranaceae)

8a.   Alar and basal cells sharply delimited, thick-walled, orange .... (9)
8b.   Alar and basal cells not sharply delimited or, if so, not thick-walled or orange .... (10)

9a.   Leaves coppery-brown or dark green, leaf tip blunt, rounded (microscopically); costa broad, occupying most of the distal half of the leaf; plants of seeping rock slopes and gravels. **Blindia** (Seligeriaceae)
9b.   Leaves lively green; leaf tip sharp; costa narrow; plants of dry rock surfaces or on peaty soil, at least not directly on flow lines. **Kiaeria** (Dicranaceae)

10a.   Alar cells not or only slightly differentiated; leaf margin recurved, plane, or incurved .... (11)
10b.   Alar cells differentiated (enlarged-quadrate, thin-walled); leaf margins not recurved .... (17)

11a.   Leaves pale bluish-green, always partly with a loose weft of fine white flexuous organic crystals (resembling 'mold' or 'cotton candy'); plants of a wide altitudinal distribution, never common. **Saelania** (Ditrichaceae)
11b.   Leaves never intertwined with white wefts .... (12)

12a.   Cells of the distal part of the leaf mostly longer than broad .... (13)
12b.   Cells of the distal part of the leaf mostly quadrate .... (14)

13a.   Capsule erect, strongly ribbed, not strumose; peristome teeth 16, not forked; seta yellow. **Rhabdoweisia**
13b.   Capsule curved, smooth or ribbed, commonly strumose; peristome teeth forked; seta red. **Dicranella**

14a.   Capsules short, strongly ribbed, on a short, curved seta and usually hidden among the leaves, mostly not fruiting at all; plants forming compact, deep golden-brown sods on alpine slopes. **Oreas** (Dicranaceae)
14b.   Capsule cylindric, on an elongate seta, well exserted above the leaves, commonly fruiting; plants of various habitats .... (15)

15a.   Leaves narrowly lanceolate or linear; capsule erect, cylindrical, not strumose. **Ditrichum** (Ditrichaceae)
15b.   Leaves broadly lanceolate; capsule strumose .... (16)

16a.   Leaves lacking a sheathing base; capsule straight but more or less horizontal, purple-brown, with a few prominent longitudinal grooves; operculum conic; plants of dry, disturbed sites (probably the most abundant weed moss everywhere in our area). **Ceratodon** (Ditrichaceae)

16b. Leaves with a prominent sheathing base, the lamina abruptly spreading above it; capsule curved and inclined, furrowed when dry and empty; operculum prominently beaked; plants of fens, dripping ledges, or wet streamsides. **Oncophorus** (Dicranaceae)

17a. Plants short-stemmed, forming loose tufts on decaying wood and occasionally on rocks in rather dry forested sites; leaves strongly crisped when dry; capsule symmetrical, erect, not strumose. **Dicranoweisia** (Dicranaceae)

17b. Plants often long-stemmed, forming compact sods on the ground . . . . (18)

18a. Costa narrow, thin or thick, in cross-section convex, without thin-walled hyaline cells on the ventral surface; laminal cells homogeneous from the costa to the margin. **Dicranum** (Dicranaceae)

18b. Costa broad and flat, often occupying the greater part of the leaf width; hyaline cells on one or both surfaces of the costa; laminal cells decreasing in size from the costa out to the margins . . . . (19)

19a. Plants small, with leaves less than 3 mm long; hyaline cells present only on the ventral side of the costa; branch tips deciduous, commonly seen lying on the tuft. **Campylopus** (Dicranaceae)

19b. Plants robust, with leaves commonly over 5 mm long; hyaline cells on both sides of the costa, enclosing central chlorophyllose cells. **Paraleucobryum** (Dicranaceae)

Campylopus [bent seta]

Campylopus is a large genus principally limited to the tropics but, like *Sphagnum*, it has a distribution in the Arctic and alpine areas of the world. Ours is the only boreal species.

C. schimperi [for W. P. Schimper]. A common but easily overlooked species of the subalpine and alpine snow melt areas, occurring on saturated soil around lakes and solifluction terraces. It is an inconspicuous and nondescript sod-former without any obvious field characters except that, with its broad costa, it looks like a small, non-falcate *Paraleucobryum enerve*. However, since the tips of the shoots break away from the stem to form vegetative propagula, it is easy to recognize the species in the field because the tufts have detached yellowish shoot-apices scattered over the surface.

Cnestrum [*knestron*, a rake, the allusion unclear]

C. schisti [from the stone, schist]. A rare species, known in Colorado from a single site. It resembles a small *Cynodontium,* but the antheridia are born in a bud on a short stalk, not sessile. The leaves are falcate when dry, and the distal ends are coarsely papillose. The leaves have papillae rather like those of *Dichodontium,* but in that genus the leaf margins are unistratose, and the habitat is wet cliffs. The capsules are strongly 8-ribbed. The upper half of the calyptra is acicular, the lower half envelops the capsule and is narrowed at the base. (We haven't seen a published description or illustration of the calyptra of this species.) BL: Spring Gulch, west slope of Steamboat Mt., foothills, 6000 ft, ca. 2 mi NW of Lyons; in crevices of shaded granite ledges, up the main draw west of the old city dump, *Bowers B-10567, W&W B-114126* (!G. Mogensen).

Some authors retain this in *Cynodontium* but we prefer to follow Ignatov & Ignatova (2003).

Cynodontium [dog's-tooth, for the peristome teeth]

1a. Capsule straight, erect, deeply furrowed, indistinctly or not strumose, the red, erect peristome teeth evident; the seta twisted; leaves strongly contorted. **C. gracilescens**

1b. Capsule curved, furrowed but not as strongly so, distinctly strumose at the base; leaves only slightly contorted and erect. **C. strumiferum**

C. gracilescens [rather slender]. A fairly common species on moist cliffs in the middle altitudes (Rocky Mountain National Park westward). The leaves are crisped when dry, coarsely papillose-mammillose, with sharp points directed forward. The capsule is erect and strongly furrowed, not strumose, on a twisted seta. The red peristome teeth are erect and split part way down the middle. LR: Rocky Mt. Nat. Park, Odessa Gorge, 3000 m, *Weber & Hermann B-50209.*

C. strumiferum [with a goiter]. Here the capsule is curved, asymmetric, furrowed, and strongly strumose at the base. In both species the operculum is beaked and diverges from the capsule urn at an angle. The habitat is similar to the last but the species is also reported from sandy river banks, perhaps on drier sites. GA: Along Fraser River at Winter Park, 8500 ft, *Vitt 15360 (B-58534).*

Excluded taxa

C. polycarpon (Hedwig) Schimper is attributed to Colorado in the Flora of North America (2007), but we have no evidence supporting this.

Dichodontium [bifid teeth]

D. pellucidum [translucent (the marginal leaf cells)]. Infrequent, occurring on wet stones in snow-melt streams and cliffs from the foothills to the upper subalpine or alpine. The strongly papillose or mammillose cells and often toothed margins of the leaves of these genera may lead students astray, since the leaf morphology strongly suggests the Pottiaceae. BL: Boulder Mountain Park, upper Bear Canyon, vertical cliff face, 7000 ft, *Mazurek B-112321.* CC: Saddle between Mount Evans and Mount Epaulet, 13200 ft, *W&W B-115160.*

Dicranella

The species of *Dicranella* are small, forming little tufts with short stems and narrow leaves. They are usually fruiting. The seta is red in our species, and the capsules small, oblong, either straight or curved, and the operculum is long-rostrate. Unlike the fairly similar *Dicranoweisia*, they grow on soil along trails rather than on rock.

1a.   Perichaetial leaves scarcely differentiated from stem leaves; leaves wide-spreading to secund, not squarrose. **D. varia**
1b.   Perichaetial leaves or both stem and perichaetial leaves with obovate, sheathing bases and spreading to squarrose leaves . . . . (2)

2a.   Leaf cells rectangular, 7–15 μm wide, 2–3:1; annulus lacking. **D. schreberiana**
2b.   Leaf cells elongate-rectangular, 4–5 μm wide, 6–8:1; annulus present . . . . (3)

3a.   Capsule erect and symmetric; leaves squarrose distally, spreading-flexuous proximally (sterile plants can be distinguished by the attitude of the leaves). **D. crispa**
3b.   Capsule inclined or nodding, the peristome appearing overly large for the urn; leaves erect or falcate-secund. **D. subulata**

D. crispa [curly]. When fertile, this species is easily identified by its straight capsule that is slightly furrowed. The differentiated perichaetial leaves are not so easily recognized as they are in *D. schreberiana* because they are small, and form a tight 'knot' that needs to be dissected out. At first glance they seem to be no more differentiated than those of *D. varia*. However, upon dissection they are suddenly narrowed from an ovate base. The leaf cells are long and slender; this, in combination with the erect and straight, furrowed capsule, make the identification sure. One collection, GA: Rocky Mountain National Park, Adam's Falls, *Vitt 15318.*

D. schreberiana [for J. Schreber]. The obviously broadened leaf bases and squarrose leaves form a 'flower-like' perichaetium. Uncommon; known from a few collections: LR: Jinks Creek, Chambers Lake, 8400 ft, *Hermann 28064*; Rocky Mountain National Park, on gravelly, peaty creek banks in spruce-lodgepole forests, Hermann 26596; SJ: Deadwood Gulch, 2.2 mi S of Silverton, *W&W B-111930.*

D. subulata. This may be our most common species. On soil in subalpine spruce-fir forests. Most of our specimens are fertile. The red seta, curved capsule, and conspicuous and surprisingly oversized peristome with incurved teeth with gaping spaces between them, are diagnostic. The perichaetial leaves are small and inconspicuous but distinctly broad at the base on dissection, as in *D. crispa*. The leaves are merely somewhat curved or secund, not spreading in all directions, and the cells are long and narrow. LR: Rocky Mountain National Park, trail to Lake of Glass, 10000–11000 ft, *Weber & Pontecorvo B-18193*.

D. varia. A trailside species, not restricted to a specialized habitat. The plants are small but usually heavily fruiting. The leaves are yellow-green, narrow, and merely somewhat wavy rather than straight but not at all curled. The capsule is asymmetrical and not strictly erect, and the operculum is rostrate. It is fairly common mostly in the northern counties (BL, JA, GA, LR), as well as AA and LP, but is not often collected, probably because the plants are so small and inconspicuous: BL: from a *Salix-Betula* thicket beside a peat fen at 10000 ft, *Vaarama B-11082* (!Crundwell); GA: West St. Louis Creek trail, *Weber & Dahnke B-91858*; LP: along the Animas River (*Jamieson 10639*).

Dicranoweisia [composite of two generic names]

D. crispula. Undoubtedly one of the most common mosses in the forested areas, occurring on boulders and decaying logs. The old capsules persist on the tufts for a long time but lose their peristomes and become bleached. It is a relatively nondescript moss and for some reason has been often mistaken for *Weissia controversa*. With careful manipulation of light the longitudinal striations of the laminal tissue are diagnostic.

Dicranum [split peristome teeth]

1a. Plants bearing short, stiff, branchlets with minute leaves in the terminal leaf clusters; rare plants of the outer foothill canyons. **D. flagellare**

1b. Distinctive brood branches not produced . . . . (2)

2a. Leaves strongly curled when dry, minute, dark green; surface papillose from the apex to middle; frequently consisting of ball-like unattached plants. **D. montanum**

2b. Not as above . . . . (3)

3a. Plants with most leaf apices broken off; leaves straight when dry; on wood. **D. tauricum**

3b. Plants with most leaf apices present, rarely with a few broken; leaves straight or curved when dry; not restricted to wood . . . . (4)

4a. Leaves undulate or rugose, especially near the apex; stems clothed with white tomentum; setae often aggregate; a fen species. **D. polysetum**

4b. Leaves not undulate or rugose; setae solitary; plants of montane forests and alpine tundra . . . . (5)

5a. Capsules essentially straight; leaves straight or slightly curved, not changed on wetting. **D. rhabdocarpum**

5b. Capsules curved; leaves usually curved and spreading . . . . (6)

6a. Leaves strongly falcate-secund; costa with denticulate distal abaxial ridges. **D. scoparium**

6b. Leaves variously curved, spreading, or appressed with spreading tips; costa without abaxial ridges . . . (7)

7a. Distal leaf cells usually elongate, sinuose, pitted; subalpine or alpine wetland mosses . . . . (8)

7b. Distal leaf cells usually short (quadrate, rectangular, or irregularly angular), neither sinuose nor pitted (or with few pits); forest mosses . . . . (10)

8a. Leaves keeled distally. **D. acutifolium**

8b. Leaves not keeled, tubulose distally . . . . (9)

9a. Leaves narrow (2.0 mm), not broader at the base, appressed, the stems with dry leaves slender (ca. 1mm wide when dry); proximal leaf cells usually less than 40 μm long, the median cells pitted mainly proximally to mid-leaf; some bistratose lines on the leaf margins. **D. elongatum**

9b.   Leaves wide (3–4 mm), from a broad base, curved with often somewhat spreading tips; proximal leaf cells more than 50–100 μm long; cells pitted well distal to the middle of leaf; bistratose lines lacking. **D. groenlandicum**

10a.   Leaves relatively unchanged when dry, only lightly crisped, appressed, not standing out at right angles, the stem slightly tomentose but not forming a dense covering; common plants of dry forest floors. **D. muehlenbeckii**

10b.   Leaves strongly crisped and narrowly tubulose when dry, the middle stem leaves standing out at right angles to the stem; stem tomentum white in young growth, becoming red-brown below; rare plants of the outer foothills. **D. brevifolium**

D. acutifolium. A rather nondescript species restricted to the alpine peaty tundra. According to Ireland *(in litt.)*, it is recognized by its erect-spreading leaves, slightly curled when dry, keeled distally, lanceolate, narrowly acute to acuminate, by its percurrent to shortly excurrent costa, smooth to lightly papillose above, and by its leaf cross-section that shows only the abaxial row of differentiated cells, a few bistratose marginal cells and the cell walls between lamina cells smooth to slightly bulging. The proximal cells are extremely incrassate and linear-rectangular. Among the few alpine species, this one has short leaves that are slightly falcate-secund but more obviously curled at the tip; the stems are red-tomentose and in tight, short clumps, with only the current growth green. Two fragmentary specimens (evidently the same collection) collected "mostly within 100 miles of Canyon City," *Brandegee* (NY), were determined from time to time as *D. schraderi*, *D. muehlenbeckii,* and *D. undulatum (D. bergeri)*. CC: 11 mi SSW of Georgetown, 12000 ft, peaty tundra in open willow carr east of Silver Dollar Lake, *Hermann 27980* was verified by Trucco and Lawton; ST: Blue Lake, 11700 ft, on bands of calcareous schists separating tundra solifluction lobes, *W&W B-115174.* Not reported for Colorado in the Flora of North America (2007); however, these collections serve to reinstate this species in our flora.

D. brevifolium. This is close to *D. muehlenbeckii;* in fact it was formerly considered to be a variety of that species. In *D. muehlenbeckii* the leaves are much smaller, tightly tubulose all the way to the apex and the costa is not prominent abaxially in the lower half of the leaf. *D. brevifolium* has leaves that are not as tightly tubulose near the apex, tong-shaped in cross-section, and the costa is prominently bulging and shiny on the abaxial surface. Uncommon or rare in the eastern foothills. BL: Little Thompson Canyon, *Weber B-6086;* LR: Spruce-fir woods, 9800 ft, 10 mi SW of Rustic, *Hermann 27808 (B-60367* (!Ireland); Twin Cabin Gulch, 1 September 2002, *Mazurek & Wittmann B-112844;* SJ: Coal Bank Pass, *Young B-35552.*

D. elongatum. Restricted to alpine and subalpine fens or wet moss tundra. Easily recognized by its almost straight leaves and densely packed stems. See discussion of *D. groenlandicum.* For illustrations showing the differences in the leaf areolation, see Hegewald (1972).

D. flagellare. A species unlike any other, characterized by its production of short microphyllose stems in clusters at the stem apex. The plant is small (about 1–2 cm), with densely tomentose stems and narrow, tubulose leaves that spread at right angles to the stem, becoming loosely crisped. The cells of the lamina are quadrate and thick-walled. Those of the broader, somewhat sheathing base are rectangular, and the basal and alar cells are large and thin-walled. We have only one station: LR: Buckhorn Canyon, northwest of Masonville, 7000 ft, *Rolston 81018 (B-93702), W&W B-114328.* This is an eastern species, evidently relictual in the Front Range.

D. groenlandicum. Nyholm (1954–1969 p. 69) writes: "The tufts of subsp. *groenlandicum* are somewhat more robust than [those] of *D. elongatum.* The leaves are broader below [from an ovate-lanceolate base], the nerve narrower. The best character for distinction, however, is afforded by the elongate, strongly incrassate and porose cells in the upper part of the leaf." Hegewald's (1972) plate has beautiful cell detail of this and its close relative *D. elongatum.* In *D. groenlandicum,* not only are the stems coarser but the pitting of all the leaf cells is clearly defined. In *D. elongatum* the pitting is hard to see but in *D. groenlandicum* it is obvious because the cell walls are exceptionally thick. CC: Mount Evans, *Weber, Porsild, & Holmen B-4464;* ST: Blue Lake, *W&W B-115100.* JA: Upper Slide Lake, Rainbow Lake Trail, 10720 ft, *Rolston 82197 (B-93870).*

D. montanum. A tiny species, inconspicuous and easily overlooked, thus far only found in north-facing canyonsides of the eastern Front Range foothills. Only a few collections are known. The plants are small, rarely forming intact, dense tufts (variety *pulvinatum* Pfeffer).

Crum and Anderson (1981) describe this modification as follows: "It . . . produces weak branchlets with minute, strongly crisped, spreading, roughish leaves. Such branches are, in fact, scarcely different from the ordinary plants except in size." BL: Boulder Mountain Park, 7500 ft, *W&W B-111481*; Gregory Gulch, *Mazurek B-113061*.

D. muehlenbeckii [for H. G. Mühlenbeck]. The most abundant *Dicranum* throughout the forested regions of Colorado. It ranges widely in forests of the middle altitudes. Some specimens were determined by Lawton as *D. fuscescens* Turner. The latter species is common in the Pacific Northwest but its presence in Colorado is highly doubtful.

D. polysetum. In this fen species, the leaves are rather stiffly spreading and have a golden sheen. Unfortunately the plants are without sporophytes; however, the vegetative characters are convincing. BL: Depressions in a *Sphagnum fuscum-Carex aquatilis* fen, Silver Lake Valley, 3000–3500 m, *Weber B-44158*; JA: Big Creek Lake, Park Range, *Popovich B-115233*; DT: Grand Mesa, *G. Austin B-115371*.

D. rhabdocarpum [ribbed capsule]. Abundant in the foothill canyons, especially on shaded north slopes. In fruit it is easily known by its erect, straight capsules, and otherwise by its dense tufts with non-falcate leaves held together by a rich development of reddish tomentum. In winter these tufts fill up solidly with ice! This is an endemic of the southwestern U. S. and Mexico.

D. scoparium [broom]. Common in forests of the foothills and montane, easily recognized when typical by the strongly falcate-secund leaves as if combed in one direction. The Rocky Mountain plants usually are smaller and less strongly falcate than those elsewhere in the U.S. Our specimens were verified by Peterson (1979).

D. tauricum [from Tauris, ancient Greek name for the Crimea]. Frequent in the outer foothills (*Pseudotsuga* forests), almost always on decaying wood. The straight, yellow-green leaves, many of them broken at the tip, and its habitat, make this an easy plant to recognize.

Excluded Taxa

D. bonjeanii De Notaris *ex* Lisa. The numerous Colorado reports of this species have been misidentification of various species.

D. fuscescens Turner. Crum & Anderson (1981) and the Flora of North America (2007) accept the species as occurring in Colorado, but we consider the reports unverified and doubtful. Until the western American collections are more carefully studied, we are placing all of the suspect *D. fuscescens* under *D. muehlenbeckii*.

D. spadiceum Zetterstedt. One collection, CC: Summit Lake, Mt. Evans, *Weber, Vaarama, & Khanna B-11025*, was identified by W. Peterson in 1977. This specimen has short distal leaf cells, tubular and acute distal leaf apex, ruling out *D. spadiceum*. The non-keeled leaves also rule out *D. acutifolium*. We conclude that the specimen belongs to *D. elongatum*.

Kiaeria [for F. C. Kiaer]

This is not a difficult genus to recognize in the field, although the technical features that characterize it are primarily microscopic. Its aspect is that of a small *Dicranum*, about a cm tall. The leaves are about 3 mm long, suddenly narrowed from a broader base, tubular-involute, circinate, with an excurrent costa. Instead of having a relatively long shoot with leaves evenly spaced along it, as in *Dicranum,* there are rather discrete clusters of leaves. The basal cells are rectangular-quadrate, and the lowest tiers are orange. In cross-section the narrow costa is little differentiated, with a median row of large guide cells, on either side of which are thick-walled cells not organized in stereid bands (unlike *Dicranum*, which always has stereids). The capsule is curved, weakly furrowed, with a rostrate operculum. The peristome teeth are wide-spreading when dry, and divided about half way to the base. Crum & Anderson (1981) point out that the best way to be sure of this plant is to see that there is a short cluster of leaves (bud) that contains the male antheridia, just below the perichaetium, bearing the capsule.

K. starkei [for J. C. Starke]. Common in peaty soil along subalpine trails and on the borders of small lakes in the northern counties. The regularly circinate-falcate leaves are distinctive. It is commonly found with sporophytes. LR: Rocky Mountain National Park, *Weber & Hermann B-57051, Crypt. Exs. Vindob. 4785*; BL, EA, JA, RT. This is not reported for Colorado in the Flora of North America (2007).

Oncophorus [with a goiter]

1a.   Leaf limb not sharply differentiated from the sheathing base and not diverging sharply away from it when dry; upper portion of leaf not narrowly subtubulose, not strongly crisped; alar cells somewhat enlarged. **O. virens**

1b.   Leaf base clearly sheathing, strongly clasping the stem, the upper portion abruptly narrowed into a subtubulose limb which diverges sharply 90 degrees away from the sheath; leaf lamina strongly circinate-crisped; alar cells not enlarged. **O. wahlenbergii**

O. virens [green]. Common in subalpine springs and fens. Both species have goiters at the base of the capsule, but the leaves of this species are never strongly curled. The leaves are revolute in the basal half, the distal costa and leaf margins are bistratose and coarsely toothed. The alar cells are orange, but they are not morphologically differentiated from the basal ones.

O. wahlenbergii [for G. Wahlenberg]. Common in subalpine and alpine snow-melt basins. Sometimes the two species grow together, but they are easily distinguished. In places that dry out in summer, the growth is limited, the tufts compact, and the leaves small (forma *compactus*). In the field this may be confused with *Tortella arctica*.

Oreas [of mountains]

O. martiana [for C. von Martius]. This must be one of the rarest mosses in North America. It is known from several localities in the Colorado Front Range, including Mount Evans. where it occurs on the wet shore of Summit Lake at 12800 ft, and on the saddle between Mount Evans and Mount Epaulet. It is locally abundant in wet alpine situations, but rarely produces sporophytes. The seta curves back into the leaves, hiding the capsule. It is easily recognized in the field by its tight mats of a rich golden color. Kjeld Holmen, who came to Mount Evans in 1961 after seeing the moss a few days before at Lake Peters, Alaska, and in North Greenland, discovered this in fruiting condition along the roadside at Summit Lake see Steere (1975b) and Warncke (1975). Many specimens were collected by Vera Komarkova in her survey of the Indian Peaks region (Komarkova, 1979).

Paraleucobryum [like *Leucobryum*]

la.   Leaf margin smooth or with a few teeth at the apex; leaves straight or somewhat curved; costa smooth, broad and thick, filling almost the entire leaf. **P. enerve**

la.   Leaf-margin serrulate; leaves circinate; costa rough on the back, not filling or thickening the entire leaf
.... (2)

2a.   Costa 1/2 or more the width of the leaf base; margins distinctly denticulate. **P. longifolium**

2b.   Costa only 1/3 the width of the leaf base; margins with small, scattered teeth or nearly entire. **P. sauteri**

P. enerve [without a costa]. Usually a plant of moist tundra, we have records from the foothills, where the species occurs in compensating environments. It characteristically forms solid tufts, several inches deep, that are tightly compacted. The color varies from glossy pale green to greenish-black. Sporophytes have not been found in any of the species.

P. longifolium. Common, forming large clumps in the foothill canyons in cool, north-facing forested slopes along with such common species as *Rhytidium rugosum*, *Dicranum rhabdocarpum*, and *Timmia*. At first glance it would appear to be a *Dicranum*, possibly *D. scoparium*, but *Paraleucobryum* is more distinctly gray-green and the costa less distinct. The leaves are also more finely attenuate than in *Dicranum*.

P. sauteri [for A. E. Sauter]. This occurs in the same areas as *P. longifolium*. Usually the tufts are dense and the leaves are not curved. When dissected and placed on a slide, leaves of *P. sauteri* almost never lie flat, so that one sees the costa and only one marginal area. The marginal areas are about as wide as the costa. In *P. longifolium* the wide costa appears to force the leaf to lie flat, so that the entire costa and the narrow laminal margins are visible.

Rhabdoweisia [ribbed *Weissia*]

This genus is generally accepted as belonging to the Rhabdoweisiaceae. We support this reclassification and include the genus here for convenience. See also *Amphidium*.

R. crispata. A small Dicranoid moss with crisped leaves, smooth laminal cells, and erect, straight, bowl-shaped, 8-ribbed capsules. Unlike *Dicranella*, the peristome teeth are not quite separate but joined at the base to form a low basal membrane. We have a single collection. SJ: Summit of Coal Bank Pass, on NE-facing base of vertical gneissic cliff, north end of Spud Mt., head of S Fork Coal Creek, T10N R81W S32, 10500 ft, *Jamieson 13952 (B-113398)*. This was reported by Jamieson (1986b).

DITRICHACEAE (DTR)

1a. Leaves in two ranks, broadly sheathing at the base, the rough, subulate lamina consisting mostly of the costa, spreading outward. **Distichium**
1b. Leaves in spiral ranks, not sheathing at the base . . . . (2)

2a. Leaves ovate-lanceolate to narrowly lanceolate, not linear; laminal cells quadrate; mature capsule strongly ribbed and inclined; weedy species. **Ceratodon**
2b, Leaves linear; laminal cells rectangular; mature capsule not ribbed; mostly alpine plants. **Ditrichum**

Ceratodon [horned (forked) peristome teeth]

C. purpureus. One of the half dozen abundant weedy mosses in the world and no less so in Colorado. It occurs on packed earth, recently disturbed forest soils, burned areas, sidewalk cracks and neglected ground at all altitudes. It should be one of the first mosses learned by a student. When it is fruiting there is nothing that can be confused with it. The capsules are purple-brown, prominently ribbed, with a conical operculum, and the urn is strumose at the base, where the capsule leaves the seta at a definite angle. In the sterile condition, *Ceratodon* can be mistaken for almost anything, and usually is. The leaves may be broadly triangular-ovate or quite attenuate depending on conditions of soil, light and moisture. It resembles *Didymodon* particularly, but the rather uniformly quadrate cells in the distal half of the leaf have a characteristic appearance that will be recognized with experience. Nevertheless, sterile *Ceratodon* is always troublesome.

On limestone cliffs just below Schofield Pass in Gunnison County, a strange form was found that produced a deep turf of slender shoots that we thought, without microscopic examination, might be *Didymodon vinealis*. It grew in crevices along with *Mielichhoferia*. Richard Zander has determined this to be a form with short-ovate leaves and filamentous gemmae (Zander & Ireland, 1979). Of our collection, Zander writes: "Note that it is yellow in KOH and no *Didymodon* has such a yellow hue. The serrulate upper margins and square upper laminal cells are distinctive."

Distichium [two-ranked leaves]

The genus is almost unique in its distichous leaf arrangement and need never be confused with any other except *Bryoxiphium*, which is also distichous (but rare, on sandstone cliffs in southwestern Colorado). The species of *Distichium*, however, seem to be impossible to distinguish without fruit, and even with fruit, it takes some experience and intuition to decide whether the capsules are really straight or inclined, short or long. There does not seem to be any distinctive ecology to separate the two. These are abundant in wet or moist situations in the subalpine and alpine, occurring in fens, on solifluction terraces, and swampy streamsides. *D. capillaceum* is the common species, while *D. inclinatum* is infrequent and possibly restricted to calcareous alpine wetlands.

1a.   Capsule erect, oblong-cylindric to ovoid-cylindric; spores 17–20 μm diameter. **D. capillaceum**
1b.   Capsule inclined, ovoid and somewhat asymmetric; spores over 25 μm diameter. **D. inclinatum**

D. capillaceum. Abundant in high altitude wetlands and seeping cliffs throughout the mountains. It often occurs with *Ditrichum gracile* but when one examines a solitary stem, the different leaf arrangement is obvious.

D. inclinatum. Uncommon in calcareous snow-melt basins. The capsule of *D. inclinatum* tends to be shorter and plumper than that of *D. capillaceum* and, in most instances the leaves, which usually diverge widely from the stem in *D. capillaceum*, are more erect. The spore size is evidently a reliable character. Spores in most instances range from 30–45 μm and seldom down to 20 μm.

Ditrichum [two-haired (forked peristome teeth)]

1a.   Stems 1–4 cm long; leaves up to 3 mm, from an ovate-sheathing base abruptly contracted to the subula; costa abaxially strongly convex; laminal cells near costa with weakly nodulose longitudinal walls; commonly fruiting. **D. flexicaule**
1b.   Stems to 7 cm or more; leaves from an elongate-ovate base, tapering gradually to the long, slender subula; costa abaxially weakly convex; basal laminal cells with weakly to strongly nodulose longitudinal walls; rarely fruiting. **D. gracile**

D. flexicaule. Infrequent and restricted to low altitudes where it forms compact mats or sods in seepage areas over rocks. The leaves are short and rather abruptly narrow to the upper half of the lamina. It might be mistaken for *Didymodon*. BL: Boulder Canyon, 1.3 mi W of Four Mile Canyon junction, 6000 ft, *Weber et al. B-35501*; JF: Between Plainview and Eldorado Springs, 7500 ft, *Weber & Anderson B-10380*; MF: Green River, Sandy Canyon, on shaded sandstone ledges, 6200 ft, *Flowers 8861 (B-66780)*. In Western Slope material, the leaves are short, the sheathing base about the same length as the distal lamina, and not long-attenuate. The Jefferson County plants are sparse and minute, on bare sandstone. We find occasional stands that have terete julaceous innovations. Here there is never any difficulty in distinguishing the two species.

D. gracile. Abundant in subalpine and alpine; large and lax, with long, filiform leaves in which the spiral twisting is well marked. It often has a golden color. When mixed with *Distichium*, one has to look closely to see the difference. In *Distichium* the leaves are green, two-ranked, the sheath appressed to the stem, lamina suddenly narrowed and spreading, not spirally twisted.

Saelania [for A. T. Saelán]

S. glaucescens. This is an example of a species which is better recognized by a metabolic byproduct than by its morphology. *Saelania* is a relatively nondescript plant with lanceolate leaves and an oblong-cylindric capsule as in *Ditrichum* or *Distichium*. The leaves are always a pale blue-green, the color being caused by blue-gray threads loosely adhering to the leaves as a tangled weft of cobwebby material. This is not living stuff but a chemical substance. In the leafy liverwort *Anthelia*, at least, Dr. Siegfried Huneck (*in litt.*) has identified the threads as a diterpene compound.

ENCALYPTACEAE (ENC)

Encalypta [alluding to the mitrate calyptra]

One genus with vegetative appearance much like *Tortula* or *Syntrichia* and difficult for the beginner to distinguish in the sterile condition, although it may be possible to distinguish *Encalypta* by its having the end walls of the basal cells thickened or colored. Fortunately, one nearly always encounters it with sporophytes, and in fruit *Encalypta* is the easiest moss to recognize because of its unique cylindrical-mitrate calyptra with an abruptly narrowed tubular apex (resembling a partly inflated sausage-balloon). Because of the resemblance of the calyptra to some old-fashioned candle-snuffers, the plants are called 'extinguisher' mosses.

Encalypta usually occurs in small tufts of a dozen or so fruiting stems in rock crevices or over poorly developed flakes of mineral soil on rock ledges or cliffs. It never forms extensive sods like *Syntrichia* or *Tortula* (*E. procera* may be an exception). The basal cancellinae (hyaline windows) of the leaves are seldom as clearly differentiated as in *Syntrichia*. *E. procera* is the only species likely to be found without fruit most of the time, and may be separated from similar species of *Syntrichia* by its production of multicellular, filamentous reddish-green or brown, papillose gemmae on the stems. The leaves of *Encalypta* are bright green when moist, and never strongly revolute, as they are in many *Syntrichia* species. Mature and over-ripe capsules, as well as fresh calyptrae, are needed for correct identification

la.   Plants relatively robust, usually sterile and commonly bearing abundant reddish-green to brown, filiform gemmae on the stems; capsules spirally furrowed. **E. procera**
1b.   Plants small, usually with capsules, lacking gemmae; capsules smooth or merely straight-ribbed . . . . . (2)

2a.   Leaves gradually narrowed, acute, the apex stoutly apiculate. **E. alpina**
2b.   Leaves oblong, obtuse or broadly acute . . . . (3)

3a.   Calyptra fringed, the teeth clearly differentiated as triangular, often darker units, their bases marked by rows of small quadrate cells. **E. ciliata**
3b.   Calyptra not fringed at the base . . . . (4)

4a.   Capsule distinctly ribbed (see old capsules); peristome present. **E. rhaptocarpa**
4b.   Capsule smooth; peristome absent **E. vulgaris**

E. alpina. This relatively rare species is exclusively alpine, on wet solifluction slopes. PA: Hoosier Pass, 12500 ft, *Weber & Holmen B-4417*; GN: North Italian Peak, 11000–13000 ft, *Weber & Langenheim B-18775*.

E. ciliata. Frequent in the foothill canyons and up to the tundra, nowhere common. The calyptra has a regular fringe of narrowly triangular segments. Before the capsule is mature enough to push the calyptra upward, the fringe of broad triangular flaps is neatly tucked in under the capsule base. The capsule is smooth but fragile. In fresh material the leaves appear to be light green in comparison to those of the other species.

E. procera [elongate]. Fairly common but restricted to the San Juan Mountains. Distinguished from all others in having filamentous gemmae in the leaf axils. It is rarely fruiting, but a few collections from the San Juans have sporophytes. The stems are usually elongated compared to the other species. Although it is supposed to be autoicous, our material is almost always sterile.

The leaf tips are bluntish with a long reddish point. The basal cells of the 'window' have orange thickened end-walls, and the leaf bases have inter-tangled slender rhizoids associated with them. The calyptra tapers gradually rather than suddenly to the apex.

E. rhaptocarpa [with ribbed fruit]. Montane and subalpine, possibly largely replacing *E. vulgaris* on limestone at high altitudes. We find it often difficult to determine whether the capsules are really ribbed or not, but in fresh, not old capsules the ribs are quite obvious. Peristome presence or absence is, in our estimation, an unreliable character. ST: Blue Lake Dam, Monte Cristo Creek valley, on limestone, *W&W B-111125*.

E. vulgaris. Common in rock crevices in the foothills canyons and going over to bare ground in the tundra. Mature capsules may be somewhat wrinkled, but do not have strong, reddish ribs. And, while the calyptra may be somewhat ragged at the base, it is not really ciliate.

ENTODONTACEAE (ENT)

A heterogeneous group of genera sharing the following characters: Leaf cells elongate-linear, costa short, double or lacking; alar cells often well-developed. The technical characteristics, exclusive of the sporophyte, suggest a close relationship with *Pleurozium*!

Entodon [peristome teeth inserted below the capsule mouth]

E. concinnus [pretty]. The similarity of the vegetative parts with those of *Pleurozium* is striking although the species are easily distinguished. In *Entodon* the color of the plant is green or a bronze yellow; the stem is not red, and has scattered stiff tufts of rhizoids; the leaf base is abruptly narrowed, the alar cells are numerous in several rows (about 8 across and ten or more high), bistratose, quadrate and little thickened. The stem leaves are closely imbricate and do not expose the stem. In *Pleurozium* the stem leaves are more distantly attached and spread out, revealing the red stem which lacks rhizoids. The alar cells are few (about 3 across and high), with thick, yellow walls.

Except for one or two occurrences, discussed below, this is an alpine wet tundra species, rare in western North America, usually occurring mixed with other mosses. The plant, always sterile, is recognized by its yellow-green color and julaceous stems with rounded leaf-apices. The green alar cells stand out conspicuously because of their bulging, densely bistratose tissue.

Since we had always encountered it at high altitudes, it was astonishing to find it in the outer foothills: BL: Boulder Mountain Park; Mesa Trail, at about 6500 ft, *W&W B-112178*. The plants were sparse, growing in a loose tuft of *Hypnum revolutum*. We tried in vain to place it in some other eastern species, but the alar cells are clearly bistratose. Another montane record is HN: Rhyolite cliffs, Cebolla Creek Campground, 9400 ft, *Weber B-13067*. Sayre reported her only collection, No. 169, as *Entodon compressus*.

FABRONIACEAE (FBR)

Fabronia [for Fabbroni]

Fabronia is the tiniest and most delicate of our pleurocarpous mosses. The creeping stems, leaves and all, are only about 0.3 mm wide when dry (to about 0.5 mm when wet). The plants can be recognized easily under the microscope, however, because of the large teeth which project outward from the margins of the leaves. These teeth are commonly hyaline, contrasting with the green cells of the leaf lamina. The leaf-tip is also hyaline, abruptly attenuate from the ovate blade, and frequently forked at the apex. This unusual serrate or even fimbriate condition is unique among our mosses. Because of the ciliate margins and the long, slender, often colorless leaf-tips, the branches often appear fuzzy under the hand-lens. *Fabronia* almost always grows in the innermost recesses of rock crevices of cliffs in the foothill canyons. Our collections are all from siliceous rocks; in more humid regions it occurs on tree-trunks. The tufts are loosely attached to the substrate, sometimes almost lying free, and range in color from bright green to pale yellowish-green. The capsule is goblet-shaped on a well-exserted seta.

1a.   Leaf apex acute or acuminate but not long and narrow; marginal teeth one-celled, often blunt. **F. ciliaris**
1b.   Leaves long attenuate, usually 2 or more cells in the apical part and basally with a double row of cells. **F. pusilla**

F. ciliaris. Frequent in crevices of granite cliffs in the foothills of both Eastern and Western slopes. The hair-like leaf apices are conspicuous, but not elongate, and the general color of the plants is pale or yellowish-green.

F. pusilla. Rare in similar sites, evidently restricted to more mesic habitats in foothill canyons of the Eastern

Slope. In our experience the plants are darker green, the attenuate leaf apices are long, often much longer than the leaf itself, and they tend to spread out, creating a spider-web appearance. LR: Twin Cabin Gulch, 7000 ft, with *Porella, Pseudoleskeella sibirica,* and *Orthotrichum hallii, Wittmann et al. B-115215.*

FISSIDENTACEAE (FSS)

Fissidens [split peristome teeth]

The leaves of the Fissidentaceae are in two ranks, lying in a single plane and thus giving the shoot a frond-like appearance; they have a unique structure. The lower half of the leaf has a large, single lamella extending from the costa, forming in effect a second leaf-blade lying over and partly covering the inside half of the lamina. This smaller 'half-leaf' is called the dorsal lamina. The 'sheath' formed by this structure is called the 'vaginant lamina'. The sporophytes are minute, with a well-developed seta and a cylindrical urn that is constricted below the rim; the operculum is rostrate. The species of *Fissidens,* our only genus, are limited to wet ground, dripping cliffs, and shelving streambanks and road-cuts of wet clay. A few species even live submerged in pools and streams.

la.   Completely submerged in pools or pot-holes in sandstone stream-channels; plants slender, with long, narrow, rather distant leaves. *F. fontanus*

1b.   Terrestrial but often in wet places; leaves not as above, usually touching each other and forming a flat frond-like shoot . . . . (2)

2a.   Large plants several cm long; leaves ribbon-shaped, multistratose, the vaginant lamina similarly shaped, running along the leaf and not easily seen except along the margin. A plant of waterfalls. *F. grandifrons*

2b.   Minute plants less than a cm long; leaves not as above . . . . (3)

3a.   Leaves not bordered by narrower elongated cells. **F. osmundoides**

3b.   Leaves wholly or in part bordered by a band of narrow elongated and colorless cells . . . . (4)

4a.   Leaf apex rounded or obtuse, not apiculate, **F. obtusifolius**

4b.   Leaf apex ending in a short apiculus . . . . (5)

5a.   Leaves crispate when dry. **F. crispus**

5b.   Leaves flat or nearly so when dry . . . . (6)

6a.   Leaves lanceolate to ligulate, acute; apiculus poorly developed. **F. bryoides**

6b.   Leaves ovate to lanceolate, rounded to obtuse-apiculate; apiculus a single sharply pointed cell. **F. sublimbatus**

F. bryoides. Stems with closely overlapping leaves, flat when dry. The limbidium is narrow and often disappears near the leaf apex, the laminal cells are up to 16 μm long, and the their walls are distinctly squared (angular). Capsules are usually present. In the other species, capsules are rarely produced. Frequent, but inconspicuous, on moist banks from the foothills through the subalpine. This species is not substrate-restricted, occurring on soil over seeping granite rocks. BL, GL, GA, JA, LR, ME.

F. crispus [curly]. Similar but the leaves are longer, more slender, less closely overlapping than in *F. bryoides,* and curl downward rather than lie flat when dry, the cells are small (6–10 μm long), rounded, and the limbidium is strong and broad, developed on both laminae. BL: Boulder Canyon, *Sayre 212a, Weber B-80111;* JA: Big Creek Falls, 2800 m, *Rolston 81120 (B-93577).*

F. fontanus [of springs]. A distinctive species, forming frondose tufts in pools of intermittent streams. The leaves are long and narrow, up to 3-4 mm, and except for the distichous arrangement, might be thought to resemble a small *Fontinalis* rather than a *Fissidens*. In fact it is claimed by some to be a special genus, *Octodiceras julianus*. We have no Colorado record, but *F. fontanus* occurs in Mora County, New Mexico, and should be expected in the limestones of Mesa de Maya, southeastern Colorado.

F. grandifrons. The fronds are quite long and pendent on seeping canyon walls, and the leaves are multistratose. This extraordinary species is common in the canyon country of Utah. We have one collection that occurs within easy walking distance of the Colorado state line: Uintah Co.: Jones Hole Creek Campground, Yampa River, in the direct spray of a waterfall, *Lehr 2036 (B-113306)*. The species must occur in Colorado, but most likely will be discovered by a boating party.

F. obtusifolius. Easily distinguished by the obtuse or rounded leaf apices. It grows on limestone and calcareous sandstone along streams and waterfalls. ME: Unaweap Canyon, *Weber et al. B-43555*; MZ: Beaver Creek, 10 mi NNW of Dolores, 7500 ft, *Pursell 3247*.

F. osmundoides [like the fern, *Osmunda*]. Easily recognized by the lack of a leaf border and the crenulate margin caused by the protruding corners of the marginal cells. *F. osmundoides* is not uncommon but is found only on peaty banks in the alpine and subalpine.

F. sublimbatus [partly margined]. A plant of arid areas of the desert-steppe, often under overhanging rimrock and boulders. BL: Hall Ranch Open Space, *W&W B-114729*, Steamboat Mt., N of Lyons, *Wittmann & Lehr B-114761*.

FONTINALACEAE (FNT)

The Fontinalaceae are large mosses attached to rocks in mountain streams. The leaves are tristichous (no other mosses in Colorado have three-ranked leaves) and the branching is irregularly pinnate. The sporophytes are borne on short lateral shoots and the capsule is almost hidden by the closely enveloping perichaetial leaves. The peristome of the Fontinalaceae is also unique, consisting of an outer row of 16 teeth and an inner row of 16 cilia variously united to form a cone-like trellis.

la.   Leaves without a costa, dark green to blackish, plane or keeled, not falcate-secund. **Fontinalis**
lb.   Leaves with a costa, always keeled and folded, reddish- or golden-brown, strongly falcate-secund. **Dichelyma**

Dichelyma [cleft calyptra]

Dichelyma is a large, semi-aquatic moss of the size of *Fontinalis*, but the strongly falcate leaves are reminiscent of *Scorpidium;* the leaves, however, are keeled, folded, and three-ranked.

1a.   Leaves strongly falcate, folded lengthwise (conduplicate), the costa merely percurrent; the stems somewhat wiry; in swiftly flowing water. **D. falcatum**
1b.   Leaves straight or slightly falcate, not folded, the costa long-excurrent; the stems flaccid; in quiet water of fens. **D. uncinatum**

D. falcatum. Attached to stones in cold, rocky streamlets in the subalpine and alpine zones. JA: Trail to Big Creek Falls, *W&W B-113487*; CC: Summit Lake, *W&W, et al. B-114268, B-115064*.

D. uncinatum Mitten [hooked]. Evidently restricted to quiet water of fens. SJ: Chattanooga Fen, *Cooper 2368 (B-110449)*; Upper Vallecito Creek, *Jamieson 11637 (B-114404)*.

Fontinalis [of fountains]

la.   Leaves usually plane and distant along the stem. **F. hypnoides**
lb.   Leaves (examine the median cauline leaves) usually keeled or keeled or folded, usually overlapping . . . (2)

2a. Ends of leafy stems and branches conspicuously elongated, triangular-pyramidal in shape, with closely imbricate leaves, the branch clearly triangular in gross form; leaf keel straight or slightly curved, apex acute; perichaetial leaves apiculate. **F. neomexicana**

2b. Ends of leafy stems and branches not as above; keels moderately to strongly curved (from base to apex), leaf apex blunt to broadly obtuse; perichaetial leaves obtuse. **F. antipyretica**

F. antipyretica [preventing fire]. This seems to prefer slow-moving water. Common in the subalpine, but also occurring at lower altitudes. LR: Rocky Mountain National Park, trail to Fern Lake, *Weber & Grove B-36608.*

F. hypnoides. This species prefers slow-moving water of fens, and ditches, and ox-bows. It has not been collected enough to estimate whether or not it is rare. GN: Doyles, *C.F. Baker 631 (B-64272)* (!B. Allen); ME:

Grand Mesa, in a small, shallow wetland with *Isoetes bolanderi*, *G. Austin 158 (B-115657).*

F. neomexicana. Common in swift-moving water in the foothills subalpine. The strongly triquetrous and dense terminal shoots are usually quite evident especially in the fresh condition. This species has an unusual disjunct distribution, mostly Rocky Mountains but occurring on the Upper Peninsula of Michigan!

FUNARIACEAE (FNR)

1a. Capsule elongate, asymmetric and curved, the mouth oblique, not perpendicular to the axis. **Funaria**

1b. Capsule straight, goblet-shaped, erect, smooth, with the mouth perpendicular to the axis . . . . (2)

2a. Capsules urceolate and short-necked. **Physcomitrium**

2b. Capsules globose-pyriform and long-necked. **Entosthodon**

Entosthodon [sunken peristome]

E. tucsonii. A minute, evidently rare, moss on eroding sandstone ledges. The leaf rosette is only a few mm high, the leaf cells large and hyaline, the costa ending well below the leaf apex. When dry, the leaves are contorted. The leaf cells are hyaline and irregularly hexagonal to rectangular; the cells of the margin are distinctly longer and narrower. The distal leaf margin is coarsely serrate by the blunt ends of protruding marginal cells, or serrulate by the bulging short tips of

the cells. The seta is slender and elongate. The capsule is oblong, with a long neck, and the peristome is absent. This species is otherwise known from Arizona and Mexico. The southwestern corner of Colorado is evidently the farthest north it comes in North America. MZ: On soil beneath a large overhanging boulder, mouth of Yellowjacket Canyon at the junction of Yellowjacket and McElmo creeks, ca. 30 mi W of Cortez, E-facing slope, 4800 ft, *Pursell 3224* (!B. Allen).

Funaria [twisted, alluding to the seta]

1a. Annulus lacking . . . . (2)

1b. Annulus large, revoluble (turning inside out when freed) . . . . (3)

2a. Costa ending in a narrowly acuminate leaf apex or excurrent; endostome segments broadly triangular proximally and nearly as long as the exostome teeth. East of the Rockies. **F. americana**

2b. Costa ending before the slender, filiform acumen; endostome segments narrowly lanceolate and about 2/3 the length of the exostome teeth. Western Colorado. **F. muhlenbergii**

3a. Endostome short and truncate; plants minute; rare, on wet tundra. **F. microstoma**

3b. Endostome of lanceolate segments, at least half as long as the exostome; common, large (over 1.5 cm tall), weedy plants especially in burned areas, at any altitude; also abundant in greenhouse pots. **F. hygrometrica**

F. americana. An eastern species reaching as far west as Nebraska and probably present on sedimentary outcrops on the High Plains of eastern Colorado. Our only collection is from BL: Steamboat Mountain, N40° 15′ W105° 17′, on a loose stone on W-facing rimrock slope, *Wittmann & Lehr B-114764*.

F. hygrometrica [water-measuring, for the hygroscopic twisted seta]. A abundant moss of sandy soil along streams and in burned areas in forests, where the burned wood and depressions where trees once grew provide seasonally moist sites. The offset operculum is unmistakable. The species commonly volunteers as a weed in flower pots in greenhouses and nurseries.

F. microstoma [for the small capsule mouth]. A small plant, rarely occurring on the tundra. Fife (1990) cited a specimen from the vicinity of Pikes Peak, *Schantz 47* (FH), evidently collected on the first International Phytogeographic Excursion in 1913. We have it also from ST: Blue Lake, *Weber, Wittmann & Spribille B-11112*. Better specimens are needed.

F. muhlenbergii [for G. H. Mühlenberg].We have two certain records: ME: On arid benches above East Creek N of Gibbler Gulch, in deep recess, on calcareous soil at base of overhanging sandstone rock, *Weber, Kunkel & LaFarge B-43493*, and from DT: Unaweap Canyon. An old report of "*Funaria hybernica* Hooker" (Porter & Coulter, 1874) may have referred to this species but the specimen has not been found.

Physcomitrium [inflated mitre (the calyptra)]

This genus is easy to recognize by its small size, broad, translucent leaves with large *Bryum*-like cells, and its small, erect, goblet-shaped capsules on a short or longer seta. The operculum is flat when attached, and has a distinct erect rostrum. The plants are evidently annual, so small and often mixed with common species (weedy *Bryum*, for example) that they are only collected by mistake.

1a.   Annulus present, large, revoluble, of two or three rows of bloated cells. **P. hookeri**
1b.   Annulus remaining attached to the capsule mouth, consisting of a row or two of small, brown cells. **P. pyriforme**

P. hookeri [for W. J. Hooker]. We have no modern collections. *P. coloradense* E. G. Britton (1894) was described from "Colorado", based on a type collection of Brandegee, April 13, 1877, probably from the Arkansas River basin. Crum & Anderson (1955) saw a possible fragmentary type specimen of this on a Grout slide at DUKE. They concluded this taxon is synonymous, and that "seta length is of no great importance in *P. hookeri*, where the seta ranges from 2 to 4 mm in length, so that the capsules may be either emergent or exserted." This is evidently a moss that is most common on the high plains and should hardly be expected to occur in high mountain wetlands. We have no records of this since Brandegee's day. However, the plains of Colorado have not been well-explored bryologically.

P. pyriforme [pear-shaped]. The only relatively common species, *P. pyriforme* has been found only a few times, once in a chain of beaver ponds in Routt County, and once in a sandy flood plain near Boulder. It frequently occurs in greenhouse pots, so is likely to escape into adjacent wet areas almost anywhere. In this species the annulus is small, not revoluble, consisting of a single row of thick-walled, orange cells, remaining attached to the capsule; the capsule mouth is bordered with about 6-8 rows of horizontally elongate cells. BL: Rocky Flats, flood-plain of South Boulder Creek, *Weber B-39084, Kunkel & Shultz B-46180*; LK: Twin Lakes, 1873, *Wolf & Rothrock* (as *P. latifolium* = *hookeri*, but the annulus is not large and revoluble)*;* RT: Diamond Park Road, 28 mi N of Steamboat, *Weber B-23608*; Hinman Park, 8000 ft, *Flock B-96023*.

Excluded Taxa

Physcomitrium immersum Sullivant. T. P. James (1871), reporting on the collections of Wolf and Rothrock at Twin Lakes, listed *Aphanorrhegma serratum* (obviously a misidentification of a *Physcomitrium*, possibly *P. immersum*). This plant of the Great Plains is unlikely to occur in the high mountains. The Twin Lakes lie in the upper valley of the Arkansas River. The locality is now destitute of mosses because of development and seasonal recreation.

Pyramidula tetragona (Bridel) Bridel was reported by Porter (1876) from "moist, sandy soil on the plains, *Hall*." The specimen, in the T. P. James Herbarium at Harvard is from "moist sandy soil on the Platte, *E. Hall*." This probably came from the Scotts Bluff area of Nebraska, but the eastern tier of Colorado counties has never been explored for mosses and some of the ephemeral spring-growing species are likely to be found here. It is a common weedy moss in the eastern United States.

GRIMMIACEAE (GRM)

Grimmia and its related genera form one of the most diversified groups of mosses in our flora, and the species have been notoriously difficult to identify.

With a little practice, the various genera presently segregated out of the old *Grimmia* may be recognized in the field. *Coscinodon* forms neat hoary tufts (with long hair-points) on hard or granitic rocks. It is easily recognized, when it fruits, by its large, mitrate calyptra. *Jaffueliobryum* grows in small tufts in crevices of sandstone cliffs or on bare vertical faces of bad-land 'fingers' in the badlands of the northeastern plains. It has broad, thin, light green leaves and long hair-points. The *Grimmia anodon* group with its two species have immersed capsules and enlarged perichaetial leaves; they form low, short-stemmed dense tufts with variably hair-pointed leaves, and usually grow on sandstone or limestone in the desert-steppe. *Schistidium* grows on various kinds of rocks, sometimes in or near shallowly flowing rills. The plants are nearly always fruiting, and have immersed capsules with a red operculum. In size they range from a few mm high tufts to some several cm long.

1a.   Stem with short lateral branchlets arising along its length and creating the impression of a main stem with horizontal side-branchlets; basal and usually median leaf cells with strongly sinuose-nodulose walls; rarely fruiting in our region; plants of rocks and moist or wet depressions in the subalpine or alpine. See '**Racomitrium**' for a key to **Bucklandiella**, **Codriophorus**, and **Niphotrichum**.

1b.   Stems simple or branched more or less dichotomously, the branches all ascending together; leaves without elongate sinuose-nodulose cells or with this character not strongly developed; most species found fruiting here; plants of various sites . . . . (2)

2a.   Leaves cochleariform, obtuse, oblong-lanceolate, lacking hair-points; lamina unistratose throughout, margins incurved; costa narrow, ending below apex; leaf cells rather uniform, quadrate to short-rectangular with smooth thin walls throughout; rare, forming dull green tufts in icy water from melting snow-banks in the alpine region. **Hydrogrimmia**

2b.   Leaves not as above, if broad then acute with or without hair-points, or with revolute margins, and in different habitats . . . . (3)

3a.   Leaves distinctly channeled distally, in adaxial view the costa appearing channeled distally; hair-points always well developed; calyptra mitrate, plicate . . . . (4)

3b.   Leaves not or indistinctly channeled; hair-points various; calyptra cucullate or if mitrulate, small and smooth . . . . (5)

4a.   Leaves broadly ovate or obovate, bright green, unistratose; leaf apex rather suddenly narrowed to the hair-point; calyptra smooth; plants restricted to arid sandstone or calcareous sandstone on the plains and outer foothills. **Jaffueliobryum**

4b.   Leaves lanceolate, deep green, partially or entirely bistratose; hair-point arising gradually from the leaf apex; widely distributed in montane habitats, less common in the outer foothills. **Coscinodon**

5a.   Capsules usually present, red-orange, sessile and partly hidden by the often larger perichaetial leaves; columella not falling the with operculum; plants forming dense or loose tufts, usually on vertical rock surfaces. **Schistidium**

5b.   Capsules variably present or absent, usually green or yellowish, with a seta, although in two species this is short and bent; columella falling with the operculum; plants brownish or black; hair-point lacking or short; tufts small, often loose; common from the plains to the alpine tundra. **Grimmia**

Bucklandiella [for Mt. Buckland, Tierra del Fuego]

All of our *Bucklandiella* collections were verified by Halina Bednarek-Ochyra.

1a.   Leaf margins regularly 2(–3)-stratose in 2–4 rows; costa 3–4(5)-stratose throughout; hair-point short, less than 200 μm, dense, yellowish. **B. macounii**

1b.   Leaf margins regularly 1–2 -stratose in 1 row; costa 2–3-stratose proximally; hair-point more than 200 μm, hyaline . . . . (2)

2a.   Costa weakly convex abaxially in the upper part, flat abaxially in the middle and lower parts; innermost perichaetial leaves hyaline, strongly modified. **B. affinis**

2b.   Costa strongly convex abaxially in the upper part, keeled throughout on the abaxial side; innermost perichaetial leaves similar to the vegetative ones. **B. sudetica**

B. affinis [close to]. This is a near relative of *B. heterostichum* of the Pacific Northwest. In the field the leaves are obviously hair-pointed. One collection, LR: Rocky Mountain National Park; trail from Odessa Lake to Fern Lake, subalpine, on granite cliffs, locally humid sites, 3000–3400 m; forming huge bulging cushions about 6 m from the ground, *Weber & Hermann B50211.* In our specimen, the stems are slender and loose.

B. macounii [for John Macoun]. Forming relatively loose mats. Leaves larger than *B. sudetica*, 1.0-1.5 mm,

obviously tapered from a broader base, gracefully arcuate when wet. Frequent in the subalpine, but not as common as the next. GA, JA.

B. sudetica [of the Sudetenland, Europe]. The leaves are minute, 1.0-1.2 mm, linear-lanceolate, spreading but rather more stiff than arcuate when wet. Common in upper subalpine and alpine, near streams or in seepage on granite rocks, usually forming rather tight mats. BL, EA, GA, JA, LR, RT, SJ.

Codriophorus [error for *Codonophorus,* a bell carrier]

The genus, according to Bednarek-Ochyra, et al. (2001–2006) is typified by several characters: The calyptra is densely papillose and the costa ceases well below the leaf apex. Large, flat papillae are distributed on both abaxial and adaxial surfaces of the leaf over the longitudinal walls and most parts of the lumina, leaving only a narrow depression in the middle. The cell wall features are difficult to make out except by making a cross-section. However, the two species are easy to tell apart, and both are rarely encountered.

1a.   Leaves broad and elliptic, obtuse or rounded, often dentate at the apex. **C. acicularis**

1b.   Leaves lanceolate, acuminate, narrow and awned or awnless. **C. fascicularis**

C. acicularis [needle-like]. All collections are from JA: Roxy Ann Lake Trail about halfway down, 1 mi S of Roxy Ann Lake, 2 mi NE of Mt. Ethel, 3292 m, *Rolston 82185;* Upper Slide Lake, Rainbow Lake Trail, 0.5 mi NE of Mt. Ethel, 3267 m, *Rolston 82205;* between Rainbow Lake and Slide Lake, 0.5 mi below Slide Lake, 25 mi W of Walden, 3170 m, *Rolston 81099;* Lone Pine Creek trail, Mt. Zirkel Wilderness, 2950 m, 18 mi W of Walden, *Hermann 26734.*

C. fascicularis [in bunches]. A plant with slender, elongate stems with numerous short branchlets. This branching type differs clearly from that of *Bucklandiella sudetica*, in which the branches are about equal to the main stem. Its semi-aquatic habitat differs from that of the superficially similar *B. affinis.* Our two records are from CC: Shallow tundra pools just below south slope of Mount Evans at 13500 ft, *Weber B-24866, Weber & Bujakiewicz B-36515.*

Coscinodon [lattice-tooth]

Sterile tufts of *Coscinodon*, especially *C. cribrosus*, will be mistaken for *Grimmia* species, such as *G. alpestris* and *G. sessitana*; however, they have a more shaggy appearance because the hair-points are often as long as the leaves. The leaves do not spread out arcuately when wet, as they do in the common *G. longirostris*, and the leaf margins are not recurved. The leaf anatomy is critical. In *Coscinodon* the hair-point is commonly long, slender, flexuous, and practically without teeth. The distal leaf lamina is 'crimped' and its leaf cells are small, irregularly rounded-quadrate, and often sinuose. The median cells are quadrate-rectangular, thin-walled, and the basal cells are rectangular, the basal marginal ones more elongate-rectangular. The costa is commonly distally grooved.

1a. Leaves 1.1–1.9 mm long, bistratose distally; margins involute above; capsules immersed. **C. cribrosus**
1b. Stem leaves 1.4–2.4 mm long, unistratose or patchily bistratose above; margins plane or revolute on one side; capsules exserted. **C. calyptratus**

C. calyptratus [with a cap]. Abundant and characteristic of granitic boulders in the outer foothills in the ponderosa pine stands, but less commonly occurring up to the highest elevations. Fruiting abundantly in early autumn and easily recognized in the field by the hoary aspect, exserted capsules, and large mitrate calyptra. The species also occurs in central Asia (Muñoz, 1998c).

One must be aware that there are also *Grimmia* species with hoary tufts and long hair-points. The leaves of *C. calyptratus* are plane, with thickened, bistratose but not recurved margins. The distal lamina is partially bistratose, while the rest of the lamina is unistratose with rounded-quadrate cells, and quite translucent. The basal juxtacostal cells are elongate-rectangular, those more toward the margin short-rectangular, and the few marginal rows elongate-rectangular and hyaline. The leaf cells are not thickened or sinuose-walled.

C. cribrosus [perforated, alluding to the peristome teeth]. Hastings (1999) says: "Its bi-plicate, bistratose leaves with strongly incurved margins and immersed capsules with distinctly perforated peristome teeth separate *C. cribrosus* from other species in the genus." This is a distinctly smaller plant than *G. calyptratus*. The species is evidently minerotrophic, restricted to a particular rock formation associated with mining areas. CF, CN, GN, HN, OR, SJ, and ST. Thus far, this has only been found on the Western Slope.

Grimmia [for J.F.K. Grimm]

1a. Leaves muticous, with gemmae on their distal portions. **G. anomala**
1b. Leaves not muticous, or if so, lacking gemmae .... (2)

2a. Leaves, even when dry, widely spreading and contorted, or spirally twisted around the stem; hair-points short or absent .... (3)
2b. Leaves never wide-spreading or crisped or spirally twisted around the stem when dry; with or without hair-points .... (5)

3a. Leaves, when dry, spirally twisted around the stem, the tips often coiled into a circle, erect-spreading when wet; upper leaves sometimes with a short hair-point; gemmae often present on the back of the costa of the upper leaves; rare, forming continuous sods on cliff faces, almost always associated with *Amphidium*. **G. torquata**
3b. Leaves crisped and contorted when dry, linear-lanceolate. .... (4)

4a. Leaves ca. 2 mm long, elongate-lanceolate, tapering and gracefully cygneous when wet, yellowish, the older ones brown; hair-points well-developed. **G. trichophylla**
4b. Leaves up to 1.5 mm long, abruptly narrowed, generally blackish; hair-point short or lacking. **G. incurva**

5a. Stems slender, less than 0.5 mm wide, about 2 cm long, almost unbranched; leaves 1 mm long or less, the upper ones with a spinose hair-point; always sterile here .... (6)
5b. Stems stouter, usually more or less branched, with longer leaves .... (7)

6a.   Hair-point coarse and finely denticulate, or sometimes without a hair-point. **G. teretinervis**
6b.   Hair-point weak, often flexuose, short and strongly spinulose. See **Schistidium tenerum**

7a.   Leaves plane or cochleariform, dark green or blackish, bistratose, margins not recurved, apex with a long, flat hair-point; costa lying in the plane of the lamina, hardly visible. . . . . (8)
7b.   Leaves not as above in all respects . . . . (9)

8a.   Leaves ovate-triangular; basal marginal cells oblate (transversely elongated), hair-points stout, strongly denticulate and decurrent; plants forming low tufts which separate into individual stems upon handling; the leaves immediately spreading widely upon wetting. **G. laevigata**
8b.   Leaves oblong-ovate, the basal marginal cells quadrate-rectangular with thickened end walls; hair-points only slightly decurrent. **G. ovalis**.

9a.   Robust; stems up to 10 cm long, in loose clumps with the stems usually trending downslope; leaves 3-5 mm long, linear-lanceolate, elongate, revolute; lower leaf cells rectangular, with thickened sinuose walls; capsules uncommon, on short side branches, strongly longitudinally ribbed; abundant on massive granite outcrops, foothill canyons to alpine. **G. elatior**
9b.   Stems shorter (usually up to 2.5 cm); leaves shorter; basal cells usually not sinuose-walled; usually fertile (except *G. pilifera*) . . . . (10)

10a.   Seta shorter than the capsule, which is therefore immersed . . . . (11)
10b.   Seta equaling or longer than the capsule, which is exserted . . . . (14)

11a.   Leaves 2 mm long; hair-point often long (see discussion under *G. elatior*); seta straight, centrally attached to the capsule. **G. pilifera**
11b.   Leaves about 1mm long; hair-points medium or short; plants of level rock surfaces; seta arcuate to sigmoid, attached off center to the capsule . . . . (12)

12a.   Leaves keeled distally, unistratose; peristome well-developed, cribrose; hair-points poorly developed. **G. plagiopodia**
12b.   Leaves cochleariform distally; bistratose on the margins and toward the leaf apex; peristome absent or rudimentary; hair-points strongly developed on perichaetial leaves; peristome absent . . . . (13)

13a.   Tufts dense, the stems not easily separating; leaves broadest at the base; basal marginal cells rectangular; median cells quadrate with sinuose walls. Plants commonly fruiting. **G. anodon**
13b.   Tufts very easily disintegrating; leaves broadest at the middle, narrowing distally and proximally; basal marginal cells short-rectangular with thicker end-walls. Plants not fruiting here. **G. crinitoleucophaea**

14a.   Tufts low, dense, dark green or blackish; leaves usually less than 2 mm long . . . . (15)
14b.   Tufts taller, loose, green or brownish-green; leaves usually over 2 mm long, ... (20)

15a.   Leaves short, broad, broadly keeled, curved toward the stem when dry; seta cygneous or simply curved. **G. sessitana**
15b.   Leaves elongate, narrow, not keeled or curved toward the stem when dry; seta straight . . . . (16)

16a.   Basal half and margin of the leaf with elongate rectangular clear cells, not thickened transversely; costa strongly raised abaxially. **G. donniana**
16b.   Basal and marginal cells otherwise . . . . (17)

17a.   Leaves short and stiff, not at all flexuose when moist. **G. montana**
17b.   Leaves not short and stiff, usually flexuose when moist; costa weak proximally, channeled distally and raised I-beam-like on the abaxial surface . . . . (18)

18a. Basal marginal leaf cells rectangular; one leaf margin sometimes slightly recurved; capsules striolate, yellowish, with thin-walled exothecial cells. **G. sessitana**

18b. Basal marginal leaf cells quadrate; leaf margins always plane; capsules smooth, brown, with thick-walled exothecial cells .... (19)

19a. Leaf apex cucullate; lamina plicate distally; the cells mammillose; hair-point short. **G. caespiticia**

19b. Leaf apex not cucullate; lamina not plicate distally; cells not mammillose; hair-point well-developed. **G. alpestris**

20a. Leaf lamina distally unistratose; seta curved downward; leaves broad, abruptly narrowed to the broad-based hair-point; green apex of leaves distinctly rounded or even emarginate; hair-points long, making the tufts hoary-looking; calyptra mitrate or mitrulate. **G. pulvinata**

20b. Leaf lamina bistratose; seta straight; leaves narrow, the green apex acute and grading into the narrow-based hair-point; hair-points evident but relatively short and inconspicuous; leaf lamina bistratose; calyptra cucullate .... (21)

21a. Leaf-margins plane or incurved; leaves channeled above, thickish and waxy-looking; costa broad, not terete; leaves not keeled; plants dioicous. **G. ovalis**

21b. One or both leaf margins revolute; costa terete; leaf keeled; plants autoicous. **G. longirostris**

G. alpestris [of high mountains]. This somewhat resembles *G. montana*, but is always distinctly glaucous-green, with rather long hair-points. It also forms a softer tuft, and is never the deep black color that one associates with *G. montana*. It closest relative is *G. sessitana*, which differs in its rectangular basal marginal cells. GN, MN, SJ, ST. In MN it occurs in Paradox Valley, between Naturita and Bedrock, on a gypsum-salt dome in center of the valley in an area vegetated chiefly by lichens and small black mosses, 5200 ft (!H. Greven). Otherwise, it usually occurs at high altitudes in the mountains. Some sterile collections have minute leaves, almost too difficult to dissect; these are evidently male plants.

G. anodon [lacking teeth]. *G. anodon* and *G. plagiopodia* are dark green or brownish rather than green mosses forming low, tightly attached tufts on sedimentary and calcareous rocks at low altitudes. The immersed capsule and the enlarged perichaetial leaves serve to mark the group. Plants low, the stems decumbent, 'brushed' to one side, the perichaetial leaves broadly oblong, copper-colored distally, green below, hair-point broad at the base, flexuose and serrulate; distal laminal cells rounded-quadrate with sinuous walls; margins nearly plane, bistratose; basal marginal cells quadrate, the inner rectangular (2:1), not sinuose. Capsule immersed, globose; operculum conical; calyptra mitrate, attenuate. A short seta is present, but it is attached to one side of the capsule base, and is bent sideways and upward.

G. anomala. An uncommon species on siliceous rocks in the forested regions, not forming discrete mounds,

but with more or less loosely spreading, forked stems, resembling less a *Grimmia* than a sterile *Orthotrichum*. The leaves are dull apple-green, lanceolate but blunt-tipped, with an abortive hair-point. When dry they are appressed, but when moistened they are immediately wide spreading and the stem is densely foliated. The leaves are broadly revolute along one side, the cells quadrate, hardly lengthened near the leaf base. The costa is strong and bulging on the dorsal side, with a dorsal epidermis of elongate cells. The leaf cells appear in face view to have one or two small papillae, but in side view they are merely low-mammillose. On most of the leaves there are spherical, multicellular gemmae, at first green and becoming orange-brown, along the costa. The leaf margin is bistratose, and the lamina, under the proper illumination, has longitudinally striate cuticular ridges.

RT: Trail above Seed-house Campground, *Weber & Nelson B-49639*; BL: On the side of a shallow gulch at the base of the Flatirons, on boulders adjacent to stands of *Rhytidiadelphus triquetrus* and *Pleurozium schreberi*, 6000 ft on west edge of Boulder, *W&W B-112124, B-113377*; Middle Boulder Creek NW of Hessie, *Hermann 25935*; SJ: Coal Bank Hill Pass, *Hermann 23326*. The type collection is from the Swiss Alps. The species occurs in much of Europe, and is rare in North America, occurring in the Great Lakes area, the Rocky Mountains and isolated occurrences in California and Colorado. It has also been reported from Trans-Caucasia and Kashmir.

G. caespiticia [forming clumps]. Greven (2003) states that this species is closely related to *G. alpestris* and *G.*

sessitana, and that all may be found in the same habitat. "Characteristic are plicae on each side of the costa, and mammillate distal lamina cells. The hair-points are usually inconspicuous, short, and under the microscope they appear as if implanted on the cucullate apex." Even when sterile, it is quite distinct even in the field. The stems are short, the leaves small and convex, and somewhat catenulate, with short, weak, and sometimes absent hair-points. The foliage is of a grayish color in contrast to the black of *G. montana*, which is somewhat similar.

In our experience, the species occurs on slightly sloping smooth granite faces of huge tors, usually in the protection of overhangs. Large stands have a distinctive grayish-green color, and tend to fill with fine soil accumulated from above. Such stands are dense and packed with soil, and crack to form areolae. When no material washes down to fill the colonies they may be short, flat disks about 1 cm in diameter. All of the plants we have seen are sterile.

BL: Ceran St. Vrain trail, 8300 ft, in deep recess in a granitic alcove, 8300 ft, *Wittmann & Lehr B-114896*, *W&W B-114912*. Hastings has provided us with citations for three additional collections: CR: South Colony Creek, *Kiener 10252* (CANM); ML: 2 mi W of Wolf Creek Pass summit, *Griffin 356* (UBC); SJ: Graysill Mine, *Schofield 83408* (UBC).

G. crinitoleucophaea [with pale gray hairs]. Both this *G. anodon* occur at low altitudes in *Pinus ponderosa* woodlands and desert steppe rimrock. The sessile sporophytes are common. *G. crinitoleucophaea* is rare and forms low, flat, very easily disintegrating tufts. The median leaf cells are rounded, the basal marginal cells are short-rectangular with thickened end walls. *G. anodon* is abundant on both sandstone and granite, and forms convex tufts that hold together when collected. BL: Boulder, Flagstaff Mt., Tenderfoot Trail, 6700 ft, *W&W B-114732*; DA: 1 mi S of Indian Creek Campground, N39°23′ W105°06′, *Hastings C90.2.398* (PMAE); FN: Phantom canyon, between Pueblo and Canyon City, *Weber B-36533*; *W&W B-114732;* LR: 5 mi SW of Livermore, 6000 ft, *Hermann 25934* (MICH)

G. donniana [for George Don, Scottish botanist, 1789–1856]. Infrequent in subalpine and alpine sites. Forming dark cushions with exserted yellow-green capsules on straight setae. The leaves are oblong-lanceolate, with plane, bistratose margins. The costa is strongly raised abaxially. The hair-point is almost or quite smooth. The distal leaf cells are small, isodiametric and incrassate. The proximal and basal cells are elongate-rectangular and clear, the marginal

cells. especially, are elongate and thin-walled. When associated with *G. sessitana*, the capsules are larger, the leaves are longer and straight, and the basal cells are clearly different.

CC: Summit Lake, Mt. Evans, 12500 ft, Weber *B-19692*, *W&W B-114289* (mixed with *G. sessitana*). Hastings (*in litt.*) reports collections from GA: S of Winter Park, spruce-fir forest, N39°50′ W105°45′, 2900 m, *Hastings C90.2.238* (PMAE); CN: Between Platoro and Summitville, 5.7 km S of crossing of Alamosa River, on granite boulders in Tertiary intrusive outcrop, *Populus-Picea* forest, *Hastings C91.6.254* (PMAE). She writes that these specimens are indeed clearly *G. donniana* with large pellucid basal regions. From the common *G. longirostris, G. donniana* is distinguished by its plane, non-revolute leaf margins and a hyaline basal lamina with long, thin-walled cells.

G. elatior [taller]. One of the dominant and characteristic species of sloping, irrigated rock outcrops from the foothill canyons to the alpine. It is our largest *Grimmia*, but varies a good deal in its development, and is rarely fertile, hence can be troublesome until one knows it. Characteristically, it forms loose tufts with long naked lower stems that trail downwards over the cliffs. When fertile, the clearly exserted seta distinguishes the species from *G. pilifera*.

The long leaves, about 3 mm long excluding the hair-point, are gracefully curved and spread at a wide angle from the only slightly broader sheath-like, hyaline base and are not greatly changed by wetting. One leaf margin is broadly recurved near the base. The coarseness of the plant, and the large, spreading leaves with strongly sinuose basal leaf cells are almost suggestive of a species of *Racomitrium*. The hair-points vary greatly in length, sometimes being as long as the lamina. It rarely fruits. The capsules are produced on shorter branches, not clearly terminal, the seta is short and twisted and often curved, and the capsules are strongly ribbed in age. See comparison with *G. trichophylla*. Widespread on vertical cliffs throughout the mountains.

G. incurva. A rare species, restricted to fell fields in the alpine, where it grows mostly on the overhanging undersides of large boulders. Once recognized, it can hardly be mistaken for any other *Grimmia*. The leaves are contorted and incurved, not at all appressed even at the base, distally linear, suddenly narrowed from a broader base, up to 4 mm long; hair-point is lacking or only a few cells long; the distal cells are bistratose; the basal cells are elongate-rectangular, the juxtacostal cell thick-walled and nodulose, the basal marginal cells elongate-rectangular and thin-walled. The capsules are minute, <1 mm long, ovoid, on a short, somewhat curved seta. The tufts are most commonly black and sterile. The species rarely fruits, but when it does, the plants are green, and the capsules are abundant. BL, CC, EP, GA, GN, TL. We have one fruiting collection: CC: Summit Lake, 12800 ft, on boulders, *Weber & Bujakiewiz B-43344.*

G. laevigata [smooth]. Characteristically a plant of sandstone rimrock of the warmer steppe-desert areas, on sedimentary rocks of the pinyon-juniper belt. The leaves are densely crowded on the stem, especially clustered near the stem apex; when dry, the overlapping, broadly triangular laminae resemble the scales on the back of a lizard. They are extremely hygroscopic; a drop of water causes them to spread instantly to almost right angles to the stem. The foliage is generally brown or blackish. The hair-point is broad, flattened, spinulose even on the flat face, and decurrent. The lamina is bistratose and the basal marginal cells are oblate (wider then long). It is recognizable without capsules, which are rare. EP: Garden of the Gods, *Ries* (in 1893); BL: Hall Ranch Open Space, north access, abundant on low, rounded arkosic conglomerate outcrops in open grassland, 5600 ft, *W&W B-114041.*

This and a few other species produce, within normal tufts, a few scattered and unusual slender, light green stems with appressed, narrow leaves and short hair-points. These were mentioned by Greven (2003) for *G. laevigata* and were the basis of a new species *Grimmia glauca* Cardot. We have observed this curious feature and, upon careful dissection, have found that these are not independent stems, but that they arise from the old, presumably dead, parts of normal stems. It is curious that no one has noted or reported this fact. Whether the slender stems have any function, such as propagula, is not known. It is, nevertheless, interesting to find that the old stems may still retain some meristematic function.

G. longirostris [long-beaked]. One of the most common and consistently misidentified species in the flora. It is distinguished from *G. ovalis* by leaves that are recurved along one margin. The leaf anatomy is distinctive. The upper and median leaf cells are small, the median cells are larger, all irregularly thick-walled (accordion-pleated), and the proximal cells larger and elongate-rectangular, the marginal ones with rounded margins and with clear contents. This variable species has been known under at least 55 different names (see Muñoz 1998a).

G. montana. Probably the most ubiquitous mat-forming moss of boulders and outcrops from the foothills to the alpine. It is a black moss, low and tightly attached to the substrate, dark even when moist under the microscope. The leaves are not long-attenuate but only acute, straight and not arcuate when wet. The hair-point is rather short, fragile. For distinctions between this and *G. alpestris*, see that species.

G. ovalis. Infrequent on granitic boulders in the foothills. It somewhat resembles *G. longirostris*, but its stiffly straight stems with appressed leaves of a dull dark green color (Greven, 2003) says 'shiny'), leaves with no keel (in face view they are like slender sausages), and involute margins, are diagnostic. The juxtacostal basal cells are elongate, pellucid, with thickened, sinuose longitudinal walls; the basal marginal cells are variable, short-rectangular to almost quadrate. In *G. ovalis*, the leaves are sigmoid when moist, and abruptly narrowed to the base. The leaves of *G. longirostris* show an abrupt color change from the dark old leaves to the new bright green ones, whereas in *G. ovalis* there is no such change. *G. bernoullii*, reported by Weber et al. (2003), is considered, correctly we believe, by Muños (*in litt.*) a synonym of *G. ovalis.*

G. pilifera [with hair-points]. This species is somewhat like *G. longirostris*, but has several unique features. It is larger, and its long hair-points gives the tuft a shaggy appearance. The capsule is immersed, the calyptra is mitrate (with a long tubular distal end and a short, spreading, scalloped base). Instead of the stems being rather appressed to the substrate and tightly massed, the tuft has the stems erect and loosely grouped. The hair-point is variable, sometimes almost equaling the leaf length and much longer on the youngest leaves. Whereas the capsule of *G. longirostris* is narrowly oval-ellipsoid and narrowed toward the apex, the capsule of *G. pilifera* is barrel-shaped and widest at the top. The leaf margins are bistratose distally and without the thin, recurved margin of *G. longirostris*. Muñoz has studied the species across North America and treats *G. arizonae* as a synonym. From *G. elatior*, *G. pilifera* differs distinctly in its costa, which is simply rounded in cross-section in *G. pilifera*, and on *G. elatior* is like a partial I-beam in section with the abaxial side forming a distinct horizontal flange. This is also visible without sectioning if one focuses up and down at the abaxial leaf surface.

G. pilifera is a locally abundant species on cliffs of the outer foothills canyons of the Front Range. BL: 12 mi NW of Boulder at Castle Rock, on north-facing granite cliffs and outcrops, 2250 m, *Weber & Schlüter B-67606* (!Greven); JF: Parmelee Gulch near Indian Hills,

7200 ft, *Sayre & McNally B-19563* (!Hastings); LR: Rocky Mountain National Park, Big Thompson River on Bear Lake road, 8500 ft, *Hermann 17665.5* (!Hastings).

G. plagiopodia [oblique-based]. A common species on sandstones on the edge of the Great Plains and in the southwestern corner of Colorado, locally occurring in the east at White Rocks near Boulder and probably on similar outcrops elsewhere. The moss is rather reddish brown, not dark green as in the more common *G. anodon*, and the capsule has a peristome. The distal areolation is unistratose in contrast to *G. anodon*, in which it is bistratose at the margins and toward the apex.

G. pulvinata [cushioned]. A fairly common species on granite rocks, particularly in the foothills. When fertile, this small plant is easily recognized as the only one with a recurved seta. However, it is only the immature capsules in which the seta (bright green) is curved, with the capsule pointing down. At maturity the capsules are erect. The capsule is short and plump, obviously striate or ribbed, and the calyptra is narrowly mitrate, split all around at the base. The tufts are low, and not more than 1–2 cm in diameter. It is often mixed with *G. longirostris*, which has oblong, smooth capsules well-exserted on straight setae.

G. sessitana [from Val Sessia, Italy]. Common on granite in the alpine tundra. The clumps are short, dense, and black. Many of our collections are fertile, and the short-ovoid capsules are borne on a crooked or somewhat curved seta. The leaves are broadly keeled and when dry they curve inward toward the stem. The hair-point is weak and crinkly, short and smooth or weakly denticulate; the distal lamina is bistratose with isodiametric cells with sinuose walls, the proximal lamina is unistratose, and the basal marginal cells are quadrate to short-rectangular with thickened transverse walls. "The best distinguishing characters are found in the basal marginal cells: Quadrate in *G. alpestris* and rectangular in *G. sessitana*. Also, the leaves in *G. sessitana* are narrower, sharper, and more canaliculate, especially in the anterior (distal) part. In *G. sessitana*, sometimes one leaf margin is slightly recurved. The leaf margins in *G. alpestris* are always plane." (Greven, *in litt.*)

The most common alpine rock Grimmias are the 'black' Grimmias, *G. donniana*, *G. montana*, and *G. sessitana*. They all belong to Subgenus Guembelia and are not easy to distinguish without some experience. *G. alpestris* is closely related, too, but is more a moss of desert-steppe and not alpine (despite the fact that Greven calls it a common alpine species). [There is a difference between the terms alpicola, alpina, and alpestre]. *G. alpestris* is not a black moss, but waxy green, with a somewhat glaucous appearance and long, rather stiff hair-points.

G. donniana is recognized by leaves with plane margins and a hyaline, rather uniform basal lamina with long-rectangular, thin-walled cells. *G. montana* has leaves that, while often incurved distally, may also have plane margins. It, however, has quadrate to short-rectangular basal marginal cells that have thick end-walls and are rarely hyaline. *G. sessitana* can also have leaves with plane margins, but often one margin is recurved. Its basal marginal laminal cells are rectangular, like those of *G. donniana*, but they have thick rather than thin walls and are typically not hyaline. The leaf cells of *G. sessitana* are most always bulging, mammillose; those of *G. donniana* are not.

G. teretinervis. This is a species with slender, tightly packed stems less than 1 mm in diameter. It inhabits cracks or soil accumulations in calcareous sandstone in the Front Range. The leaves are unusual in being rather broadly ovate at the base and abruptly narrowed in the upper two-thirds. The costa is prominently raised on both surfaces, yet the leaves are convex rather than keeled. The distal and median cells are small and irregularly rounded, but the lower and basal cells are elongate-rectangular. FN: Phantom Canyon, 6400 ft, 9.6 mi W of Florence, on the face of a sandstone, *Hermann 23616 (B-38353)* (!H. Greven); LR: Stove Prairie Road, *Hermann 27013 (B-94027), 27021 (B-58537)*.

Blom (*in litt.*) writes: "Several specimens of *Grimmia teretinervis* have been named *Schistidium tenerum*. The two species are similar in habit and sometimes grow together in Arctic North America. The leaves of *G. teretinervis* are oblong, ovate, or shortly ovate-lanceolate, and possess a biconvex costa (transverse section), whereas they are mostly ovate-triangular with a planoconvex costa in *S. tenerum*. The hair-point of *G. teretinervis* is coarse and finely denticulate whereas it is weak (often flexuose) and strongly spinulose in *S. tenerum*. The difference in hair-point structure is striking and the best differential character in the field."

Phytogeographically, the species is of extraordinary interest. Hastings (2002) writes: "It appears that in North America the distribution of *G. teretinervis* can be tied to calcareous bedrock exposures created by ancient oceans that periodically inundated the continent from the Ordovician to the Cretaceous periods. In western North America *G. teretinervis* occupies calcareous limestones and sandstones that formed near the oceanic boundary and were subsequently uplifted by tectonic and thrusting events that created the Rocky Mountain Front Ranges."

G. torquata. This species can almost be recognized by its constant association, on cliffs in the upper subalpine or alpine, with *Amphidium lapponicum*. There are many places where the latter grows alone, but I know of none where *G. torquata* occurs without *Amphidium*. *G. torquata* is isolated in the genus. The leaves have a characteristic spiral torsion and curled tips, and the first young growth is green, but immediately back from the tip the stems and leaves become glossy black. Instead of forming hemispherical tufts, *G. torquata* covers the rock in a uniform sod. According to Greven (1999), there are brown gemmae on the adaxial side of the upper leaves.

G. trichophylla. This is a common species in the Pacific Northwest, but in Colorado it is rather rare. Its habitat is in vertical crevices of large granite boulders in the eastern foothills. The leaves are slender, strongly keeled, not appressed but curved out basally, the lamina curved, with a well-developed slender hair-point, or muticous, but the stems are not tightly clumped and they are obviously branched; the lower leaves are brown and dark colored; the upper are yellowish and have a waxy appearance. When moistened they are gracefully cygneous but not recurved.

The leaves are unistratose and translucent except for occasional lines of bistratose cells especially at the margin. The distal cells are quadrate, incrassate, and constricted in the middle, grading into the median cells which are more elongate, sinuose and thick-walled. The juxtacostal and basal cells are elongate and thick-nodulose walled, while the marginal cells are rectangular with distinctly thickened transverse walls.

The capsules are plump, the operculum rostrate, and the seta distinct but short and arcuate, not allowing the capsule to protrude much from the perichaetial leaves. Old capsules are striate. In general aspect, the convex, hoary cushions suggest *Coscinodon calyptratus*. However, the calyptra is slender, with only a short basal portion mitrate, and the leaf anatomy different. BL: Ceran St. Vrain trail, fertile and with muticous leaves, *Weber, Wittmann, & Miller B-115140*; BL: Diamond Lake Trail, 11200 ft, *J. L. Smith 305 (B-114799)*; LR: Mt. McConnell, 8000 ft, Cache La Poudre River Canyon, sterile and with hair-points, *Hermann 27759 (B-60463)*; Fall River Campground, Rocky Mountain National Park, *Hermann 23511 (B-37361)*.

Excluded Taxa

Grimmia elongata Kaulfuss: A specimen from GA: Indian Peaks area, *Komarkova B-42053* was verified by Greven, but the material does not at all compare favorably with European specimens, and appears to be identical to our specimens of *G. incurva*.

Grimmia ungeri Juratzka was cited for Colorado by Muñoz (1998b), based on a collection from GL: Tolland, *Roberts s.n. (FH)*. The species is mapped to include the Pacific states, and Colorado. Greven claims that *G. ungeri* is a Cyprian endemic, and the name is ignored by Hastings & Greven (2007). There is nothing special about the habitats in the Tolland area that might suggest the presence of this species.

Hydrogrimmia [aquatic *Grimmia*]

H. mollis. This is a good example of a moss that is always considered to be a great rarity. For many years the Mount Evans locality was our only station. However, in her exhaustive field work in the Indian Peaks area, Komarkova (1979) discovered that while the micro-habitat may be scattered, the plants can almost always be found in small rivulets of snow-bed slopes. The same can be said of *Oreas martiana*, which had been considered equally rare, but filled several drawers in the herbarium when its micro-habitat was recognized.

We have found that, once the habitat is understood, one can find it above 12000 ft throughout the alpine area of at least the Front Range. In late snow beds the plants are submerged or at least wet throughout the summer, but they are also found quite dry in snow melt channels that exhaust the flow earlier.

Hydrogrimmia can be confused with no other grimmioid moss. We resist the current trend to submerge this unique species within the genus *Grimmia*, and prefer to follow the most recent treatment by Ignatov & Ignatova (2003). This species has yet to be assigned a sub-generic position in *Grimmia*. No other *Grimmia* is semiaquatic, flaccid, or restricted to seasonally running alpine streamlets. Keys eliminate this in the first couplet. Some credence ought to be given to the obvious and complete barriers that have been erected against hybridization (Dobzhansky, 1951, pp. 254–288). The greatly disjunct distribution pattern suggests a long period of isolation.

Jaffueliobryum [for Rev. P. F. Jaffuel, 1857–1931]

1a.   Leaves abruptly contracted to the apex, the margins serrulate; hair-point longer than the lamina. **J. wrightii**
1b.   Leaves more long-attenuate, the margins entire; hair-point only equaling the lamina. **J. rauii**.

J. rauii [for E. A. Rau]. Common on sandstones of the Great Plains to the base of the outermost foothills. In the Weld County badlands the plants are minute, 1–3 mm high, the largest leaves 1.2 mm long and 1 mm wide, mostly embedded in the sand grains and visible only as a greenish 'cast' on the rock surface, not in crevices as in *J. wrightii,* and less obviously hair-pointed. Leaves obtusely narrowed to the broad-based, blunt-toothed, flexuous hair-point less than half the leaf length. Median cells rounded-quadrate, 10–12 μm diam, smooth, a few rows of marginal cells distinctly smaller and more hyaline; basal and marginal leaf cells rectangular, the marginal ones quite narrow and hyaline; costa broader near the apex, canaliculate abaxially and thickened adaxially, the flattened margins sometimes more or less undulate but not lamellate. WL: Pawnee Buttes National Grasslands, *W&W B-113212.*

J. wrightii [for Charles Wright]. In crevices of horizontally-bedded sandstones of the foothills cuestas on the eastern slope of the Front Range. The plants are light green and somewhat resemble *Coscinodon,* or even *Crossidium,* but *Jaffueliobryum* of course lacks the lamellae of the latter. BL: Hall Ranch, *W&W B-114767.*

Niphotrichum [Greek, snowy hair]

This genus is easily recognized by its broadly-triangular hair-point and centrally-located conical papillae on the laminal and costal cells. In aspect it resembles a small *Hedwigia ciliata* but *Hedwigia* lacks a costa. *Niphotrichum* is always found in higher altitudes and on open level ground.

N. canescens [hoary]. This is the most frequently encountered species of '*Racomitrium*', often covering large areas of merely moist ground in the alpine tundra and in open subalpine sites. It does not occur in close proximity to running water.

'Racomitrium' [ragged cap, alluding to the calyptra]

The genus *Racomitrium* has recently been divided into four genera (Ochyra, et al., 2003). *Racomitrium, sensu stricto,* does not occur in Colorado. Our species belong to *Bucklandiella, Codriophorus,* and *Niphotrichum.* We are indebted to the Ochyras for their help with these genera.

1a.   Leaves strongly papillose with high papillae; hair-point usually present, papillose and conspicuous. **Niphotrichum**
1b.   Leaves smooth or with low, inconspicuous-papillae; hair-point variable, not papillose . . . . (2)

2a.   Leaves broad and elliptic, obtuse and rounded, often dentate at the apex. **Codriophorus acicularis**
2b.   Leaves lanceolate, acuminate, narrow, hair-points usually present, at least on the uppermost leaves . . . (3)

3a.   Leaf margin bistratose distally; distal leaf-cells short-rectangular or isodiametric; peristome teeth short, irregularly divided to the middle into 2-3 irregular prongs. **Bucklandiella**
3b.   Leaf margin unistratose throughout; distal leaf cells three times as long as wide or longer; peristome teeth long, split into 2 or sometimes 3 more or less regular filaments. **Codriophorus fascicularis**

Schistidium [perforate, teeth]

No other genus except *Grimmia* deserves such careful attention to detail. Until you have some experience, do not attempt to identify species without capsules; fortunately, they are usually present. Nevertheless, with experience, most species may be recognized in the field. One must be attentive to the following characters: The tuft or mat (dense and rounded, or loose and covering unconfined areas); color (dull or shiny, with or without reddish tones); leaves, (elongate lanceolate or short and broad); leaf margins (flat, or variously recurved from top to bottom or only partially); hair-points (present or not, slender and cylindrical, or short and flat); leaf cell shape (basal marginal, juxtacostal, and median) and cell wall thickness and sinuosity; perichaetial leaves (broad or narrow, hiding the urn or not); the length of the seta; shape of the urn(cupulate or oblong), shape of the exothecial cells; peristome (reflexed or erect when dry, hygroscopic or not); and peristome teeth, (cribrose or solid).

    This is a formidable list of characters to observe and tackling them makes the study of *Schistidium* a challenge. We believe that, with experience, most of the Colorado species can be recognized in the field with only the hand lens. In northern Europe, identifications are difficult because a number of species may occur in mixtures, but here the discreteness of our mostly arid habitats prevents such mixing. The following key is designed to be especially adaptable to field observations.

1a.   Plants loosely organized, freely branched, the branches not dissimilar to the main stems . . . . (2)
1b.   Plants densely tufted, forming mounds or carpets of erect stems; branches short and obviously lateral (7)

2a.   Leaves lacking hair-points, blunt; spores 18–24 µm; empty capsule goblet-shaped or funnel-form; on moist rocks and near running water, frequently in and along streamlets in the higher mountains . . . . (3)
2b.   At least the upper leaves usually (not always) with a short hair-point, acute; spores up to 15 µm; empty capsule oblong to short-cylindric; mostly on dry rocks . . . . (5)

3a.   Leaves not keeled, the margins plane to slightly recurved, not distinctly revolute, lanceolate and somewhat falcate; leaf apex suddenly narrowed to a fleshy blunt point; lamina unistratose, or bistratose only at the apex; capsule distinctly longer than broad, not tapering to the base. **S. occidentale**
3b.   Leaves keeled, often distinctly revolute, variously shaped, unistratose or bistratose; leaf apex not unusually narrowed; capsule not or little longer than broad . . . . (4)

4a.   Leaves strongly and narrowly revolute; lamina and margin bistratose in the upper part; capsule goblet-shaped, with parallel sides. **S. rivulare**
4b.   Leaves plane, the lamina and margin unistratose; capsule funnel-form or conical, especially in age. **S. agassizii**

5a.   Main stems very slender, with appressed leaves, strikingly differing from the larger perichaetial leaves at the stem apices (the apex forming a thick 'flower); capsules widely gaping, funnel-shaped. **S. trichodon**
5b.   Stem with uniform leaves, the branches tending to be curved and the leaves spirally arranged; capsules barrel-shaped, deeply hidden in the appressed perichaetial leaves . . . . (6)

6a.   Hair-point short or represented by only a few hyaline cells. **S. apocarpum**
6b.   Hair-point well-developed on most leaves, stout and conspicuous. **S. ambiguum**

7a.   Shoots slender, julaceous; leaves 0.75–1.3 mm; hair-point narrow, not flattened; plants reddish or olivaceous. **S. tenerum** (never found here with capsules)
7b.   Shoots and leaves otherwise . . . . (8)

8a.   Plants, generally black or dark-reddish, the leaves densely crowded on the stems . . . . (9)
8b.   Plants green, the leaves well spaced along the stem and gracefully curved, particularly so when moist
    . . . . (12)

9a.   Plants with virtually no stems, jet black except for the immature leaves, not closely organized into tufts; very rare, on thin soil of calcareous rocks, desert-steppe. **S. atrofuscum**
9b.   Plants with elongate stems, alpine or subalpine . . . . (10)

10a.   Leaf cell walls and often cell contents pale reddish or deep red. Strictly alpine. **S. boreale**
10b.   Leaf cell walls not reddish; not alpine . . . . (11)

11a.   Leaves minute (0.8 mm long), straight, broad at base and abruptly narrowed distally; margins plane, laminal cells very incrassate, irregular distally and rectangular proximally; hair-points lacking on most leaves; capsules when present <1mm long. On granite gravels of alpine sites, and rarely in canyons near waterfalls. **S. strictum**

11b.   Leaves 1.5 mm long, lanceolate appressed when dry, curved when moist, margins revolute, uniformly hair-pointed with short, broad, and appressed hair-points; generally sterile; plants of medium altitudes. **S. pulchrum**

12a.   Plants tending to form erect clumps of elongate unbranched stems; leaves muticous, somewhat cucullate at the apex; when moist, the leaves are seen to be in vertical lines ("spirally arranged" according to Blom); peristome wide spreading or reflexed; capsule with vertical sides, the urn 1.5–2.0:1. *Species incertae sedis*

12b.   Plants forming low rounded cushions . . . . (13)

13a.   Minute plants only about 2–3 mm high (very rarely much longer); leaves broadly ovate, plane, rounded at apex, lacking hair-point. **S. atrichum**

13b.   Plants larger; leaves narrowed to the apex, the margins recurved, usually with at least a short hair-point . . . . (14)

14a.   Peristome short, inconspicuous, irregularly developed, not nearly large enough to cover the mouth of the capsule; capsule broadly cup-shaped, pale brown with a bright red rim; perichaetial leaves much longer than the broad, rounded stem leaves, sheathing the base of the capsule like corn-husks. **S. flaccidum**

14b.   Peristome larger and completely formed, or lacking . . . . (15)

15a.   Perichaetial leaves narrow (urn much exposed in lateral view); hair-point distinctly narrower than the chlorophyllose part of the apex, up to 0.15 mm; urn finely striated (after dehiscence); plants dull. **S. dupretii**

15b.   Not as above . . . . (16)

16a.   Exothecial cells predominantly isodiametric , but transversely elongated in the central and lower parts of the urn; many leaves with a basal marginal leaf border of 8–30 hyaline, narrowly oblong cells; peristome teeth narrowly cribrose in the upper half. **S. frigidum**

16b.   Exothecial cells predominantly oblong in central and lower parts of urn; basal marginal leaf cells short, not forming a distinct border but often a differentiated square to rectangular alar group of cells with thickened cross walls; peristome teeth broadly cribrose. **S. confertum**

S. agassizii [for Louis Agassiz]. Rare and thus far only found in swift running streamlets in the foothills canyons west of Fort Collins, and in wet tundra on Mount Evans. A black species with slender, acuminate leaves and a characteristically broad funnel-shaped capsule.

S. ambiguum. This may be described as a *"S. apocarpum"* with very well-developed hair-points, whereas in *S. apocarpum* the leaves are muticous or merely tipped by a few hyaline cells. This species was described by Sullivant from material collected by Fendler in northern New Mexico. Although Sullivant said it resembled *S. confertum*, the beautiful plate clearly shows a better resemblance to *S. apocarpum* except for the well-developed hair-point. This is the plant that Blom annotated *"Schistidium taxon W"*. Our collections come from the Front Range foothills, especially in the south. McIntosh (2007) implies that this species may not be in America; however, the type actually is from New Mexico, leg. Fendler. BL, CF, CR.

S. apocarpum [with sessile capsules]. The species is distinguished by having the leaf margins at least slightly denticulate. It is a fairly large species with leaves that are distinctly recurved for most of their length. The hair-point is short or absent. The capsules usually are abundant and broadly oblong (barrel-shaped). The peristome has broad teeth that are rather characteristically perforated, especially in the upper half, with numerous narrow vertical slits. It is common on granite cliffs and ledges throughout. Blom (*in litt.*) considered the species rare in the United States, but evidently he did not see much material.

S. atrichum [lacking a hair-point]. Easily recognized but sparsely represented in herbaria. The plants forms a neat, low tuft on granitic rocks in the montane. The leaves are minute, appressed and only slightly spreading when moist; broadly ovate-lanceolate, with a rounded apex, and with no hair-point. The capsules are minute, and the perichaetial leaves are not enlarged as they are in *S. confertum*. The teeth are not at all cribrose. The peristome was said by the author to be smooth and not papillose, but in the specimen verified by Blom the teeth are definitely papillose. RT: Park Range, 1–2 mi above Slavonia, E of Clark, 2700–3000 m. alt., *Weber & Nelson B-49347*, c. fr. (!Blom); SJ: San Juan Mts., Deadwood Gulch, 2 mi SSW of Silverton, 9500 ft, *Hermann 23287*, sterile.

S. atrofuscum (black-brown). This has hardly any visible stem, and miniature jet-black, broadly ovate-lanceolate, contorted leaves only 1.5–2.0 mm long, muticous or with a short, stiff white hair-point. The capsules are minute, shortly oblong-cylindrical, and slightly exceeded by the perichaetial leaves. The capsule is minute, 1 mm high, cupuliform, brown, the rim very dark; seta straight, not excentric; exothecial cells irregularly quadrate or very short-rectangular, thin-walled; peristome orange, rudimentary, reduced to the first four or five basal cells, the horizontal walls incrassate; perichaetium slightly exceeding the capsule. The species is a calciphile and is disjunct from mostly southeastern Europe and the Middle East, known in America from southern Texas and New Mexico, with an outlier in Alberta. We have a single collection: FN: Phantom Canyon, between Florence and Victor, 5,900 ft. On accumulated calcareous soil over rimrock, *W&W B-112166*. This is not reported for Colorado in the Flora of North America (2007).

S. boreale. A black mostly calciphile species with elongate branches, growing on wet rocks in the alpine tundra. The leaves are gracefully recurved when wet, and strongly keeled, with narrowly revolute margins. The cell contents and especially the thick cell walls are reddish. CC: gravelly tundra, Mount Evans, 12,500–13,500 ft, *W&W B-104085, B-114296*; ST: Blue Lake, 11000 ft, *Weber, Wittmann, & Miller B-115108*. In North America, *S. boreale* is known otherwise from arctic and northern Canada and Alaska.

S. confertum [clumped]. Evidently the most common species on outcrops in the outer foothills and montane zones. Blom (1996) writes: "The combination of a flattened, strongly denticulate hair-point, shortly recurved leaf margins, and a partly bistratose lamina, narrow and slightly sinuose laminal cells, a ridge-like costa (particularly in perichaetial leaves) and strongly perforated peristome teeth distinguishes *S. confertum* from most specimens [of *frigidum*]."

S. dupretii [for Fr. H. P. Dupret]. Plant small and delicate, in low mounds. The hair-point is short or lacking. The capsule is striate and somewhat exposed.

The leaves are appressed, short, ovate, and sub-sheathing. Vegetatively it is somewhat similar to *S. confertum*, but that has a broad and flattened hair-point. The laminal cells are strongly sinuose, constricted in their middles. The peristome teeth spread out horizontally and are not cribrose as they are in *S. confertum*. Our collections are from the subalpine. It appears that *S. dupretii* always occurs on limestone. PT: Castle Creek, *Clebsch 13423 (B-39311)* (!Blom).

S. flaccidum. This species is easy to recognize. The plant forms dense tufts with tightly packed slender stems, growing on vertical rock faces and having a 'combed-down' appearance. The leaves are broad and rounded at the tips. The perichaetial leaves are much larger than the normal ones, are somewhat plicate, and a lighter green. The hair-point may be absent or weakly developed. The capsule is broadly bowl-shaped (cyathiform), deeply sunken in the cluster of perichaetial leaves, and the peristome teeth are short, often withered, not hygroscopic, and in various stages of degeneration. Unlike the bright red teeth of other species, in *S. flaccidum* they are more yellow, and are distinctly and densely papillose. The capsule rim is red, contrasting with the paler teeth. The operculum is unique in being mammillose. We have found this most abundant on volcanic rocks at the Red Mountain mine area of the San Juan Mts. (*W&W B-111942*), where *Mielichhoferia* is so abundant. At first blush one might be reminded of *Grimmia anodon*, but the seta, while short, is straight and not excentrically attached, the capsule is exceedingly minute, and the leaves are revolute distally.

S. frigidum. This belongs to the Confertum group which also includes *S. confertum* and *S. flaccidum*. *S. flaccidum* is easily distinguished by its long, plicate perichaetial leaves and rudimentary peristome. *S. frigidum* is similar to *S. confertum* and grows in similar habitats although it may possibly largely replace *S. confertum* at higher altitudes in the foothills. Blom (1996) writes: "In the Rocky Mountains there is a confusingly large range of variation in leaf size, leaf cell sinuosity, border development, and even leaf shape. On the other hand there is little variation in sporophyte characters. . .However, the majority of specimens from the southern Rocky Mountains . . . are made up of small plants with short leaves (1.25–1.6 mm), a narrow leaf apex with a correspondingly narrow, not or very shortly decurrent hair-point, moderately sinuous leaf cells and mostly distinct hyaline basal marginal leaf border of 2–3 (–5) cell rows."

S. occidentale [western]. An alpine or subalpine species of wet rocks and rills in the northern part of Colorado, thus far known from Rocky Mountain National Park, but probably more widely distributed. The leaves have a fairly prominent costa but they are definitely not keeled, rather, merely rounded-convex. The apex is prolonged into a well-developed fleshy tip. The capsule is about 1 mm long, brown, 2:1, with an obliquely rostrate operculum. The leaf shape is probably the best way of distinguishing this from *S. rivulare*. LR: Rocky Mt. Nat. Park, Longs Peak, 10400 ft, Grout *N. A. Musci Perfecti. 387 (B-50301);* 1/5 mi W of Longs Peak Ranger Station, *Hermann 25990.*

S. pulchrum [beautiful]. A common species in the higher foothills. The stem is rather elongate and the tufts loose. The leaves are dark green, sometimes reddish, and the margins recurved almost along the entire leaf margin. The distal cells are small and strongly sinuose. The basal cells are slightly longer and thinner-walled but not conspicuously rectangular or elongate. The hair-point, which is borne on lost leaves, is short and broad, with coarse teeth and narrowly decurrent along the upper end of the lamia. The capsule is deeply immersed although it is stalked. The perichaetial leaves are broad and their tissue is quite clear compared to the normal leaves. The capsule is short-cylindrical and the seta is about half as long as the urn. The peristome is red, entire or variably perforate, and under high power its cells are finely and densely papillose. The exothecial cells of the capsule are more or less isodiametric and not elongated. Spores 10–12 µm. Evidently this species is not restricted to limestone. OR: *W&W B-111011;* LR: *Hermann 23512;* PT: cliffs above Roaring Fork River canyon 1 mile above The Grottoes, 9 mi E of Aspen, *Weber B-9671.*

S. rivulare. Common on irrigated rocks from the foothills to the alpine. It is similar to *S. occidentale* but the leaves are more ovate-lanceolate and not attenuate. The perichaetial leaves are broad and envelop the capsule in both species. The keeled nature of the leaf is most useful if examined near the leaf apex.

S. strictum. A rare species of wet granitic gravels on alpine saddles in the Gray's Peak area, where it is quite common, forming loose, spreading stem clusters along the edges of tundra frost scars. The stems are very slender and elongate, little-branched, dark green to dull blackish. The leaves are strongly appressed, short and broad, muticous or with a short hair-point. The margins are strongly revolute, the distal lamina partly

or wholly bistratose. The capsules are minute, 0.8 mm long, 0.5 mm side, goblet-shaped with straight sides, and the seta is about as long as the capsule. This was distributed in the *Krypt. Exsicc. Vindobonense, No. 4494,* as *S. apocarpum* var. *gracile.* CC: Mount Evans, *Weber; Vaarama,* & *K. Khanna B-26199;* Argentine Pass, 13000 ft, *Hermann 29077 (B-80128).*

S. tenerum [slender]. Not uncommon on dry or irrigated vertical cliffs in the middle altitudes. In Saguache County it is frequent on vertical faces and cracks in rhyolite boulders. In addition to the slender stems, the tufts have a peculiar tan-brown color which is difficult to describe but once recognized is a good diagnostic feature of an otherwise rather nondescript moss. This moss is even more slender than *G. teretinervis,* and ranges more widely across Colorado than the latter, which is confined to the Front Range. Blom (1996) writes: "The julaceous shoots and the small, predominantly ovate-triangular leaves with sparsely spinulose hair-points and partly bistratose lamina, make *S. tenerum* an easily known species. It should not be mistaken for any other species in the genus." *Grimmia tenuicaulis* R. S. Williams is a synonym, according to Blom (1996). BL, CC, GN, SH, ST.

S. trichodon [slender teeth]. This has elongate and branched but slender stems up to 1 cm long, the leaves lacking hair-points and distinctly secund. Rather than forming neat round patches, this species forms extensive mats. The main stem is slender, with appressed and often secund leaves and is terminated by a conspicuously broader apex bearing a broadly funnel-shaped capsule surrounded by larger long-pointed perichaetial leaves.

Species incertae sedis

Blom (*in litt.*) writes: "This [unpublished] species is easily known in the field by its muticous leaves with a rather broad apex, and the distinctly spiral leaf arrangement." It seems to be quite common in sandstone cliffs in western Colorado. The stems are unbranched, with short, broad, blunt leaves that are incurved and appressed, with revolute margins from base to apex, and dull dark green in color. If one looks at the dry stem, the 'spiral' arrangement is not easy to see but the vertical disposition of the leaves (which indicates a spiral arrangement) is more easily seen on the older bleached lower leaves. When wet, the arrangement is easily seen on the current season's growth. The capsules are short (1.5–2.0:1 with vertical sides) and deeply immersed but the perichaetium does not greatly overtop the urn. OR: 5.5 mi SW of Ouray, *Hermann 23228*. ST: Blue Lake Dam, on limestone terraces, *W&W B-111163* (!Blom).

HEDWIGIACEAE (HDW)

Hedwigia [for Johannes Hedwig]

1a.   Leaves very opaque, with a broad, short-triangular, white, plane tips; leaf margins revolute or plane from base to apex; apical green cells with 2 or more strongly forked papillae; cells of the hyaline tip similar. **H. ciliata**

1b.   Leaves less opaque, with elongate, narrowly triangular, white, narrowly involute tips; basal leaf margins revolute; apical green cells with a single forked papilla; cells of the hyaline tip sparsely papillose with simple papillae, 3–5 or more in a line. **H. nivalis**

H. ciliata [for the ciliate perichaetial leaves]. One of the most easily recognized of all mosses, it should be found by every beginning bryologist. It occurs exclusively on siliceous rocks. The stems are in sprawling tufts, the leaves are dull gray-green to blackish, with narrowly recurved margins, closely imbricate when dry, thick and opaque, ecostate, and tipped with a broad hyaline point. The white tip varies considerably from one population to another. Microscopically the leaf cells have papillae ranging from short, squat ones to tall, branched ones. The capsules are almost sessile, short, with a flat operculum, and are terminal on the branches. The perichaetial leaves are elongate, with ciliate margins. Antheridia occur in axillary bud-like branches. There is no peristome. The species is abundant in the mountains from the foothills to the subalpine.

H. ciliata is plastic. The plants may be very small or very large. The leaves may have short and slender or very broad, appressed or spreading white tips or be muticous, straight or secund. Several strikingly different forms have been given names: f. *secunda*, f. *leucophaea*, f. *subnuda*. Beginners may not recognize the plant when wet if they have seen it only in the dry state, for when moistened it expands immediately and dramatically to produce a bright yellow-green tuft with widely spreading leaves. Curiously, the species is unknown in Utah, possibly because of fewer exposures of granite.

H. nivalis [of the snows, a misnomer]. This is a plant of the dry tropics that has only recently been noted in the United States (Buck & Norris, 1996). We have seen this in the Chiricahua Mountains of southern Arizona and find it to be a slender, much smaller, often minute plant only 1–2 cm high with slender stems. It has an eerie resemblance to *Niphotrichum canescens*.

Typically the plant is very slender, but *H. ciliata* may also be so. In the field, the longer, more appressed, narrower white (isosceles) triangular leaf apex is indicative of *H. nivalis*, but microscopic examination is critical. The leaf apex is narrowly involute, and the basal margins recurved. Because of the fewer laminal papillae, the leaves are less opaque than those of *H. ciliata*.

Buck & Norris (1996) in their paper on two species of *Hedwigia* in far western North America, refer, in passing, to the status of *H. nivalis* in Colorado. We have found one specimen that compares very well to *H. nivalis*, but the plasticity of *H. ciliata* is so great that consistently qualitative characteristics are not easily found, and may be of doubtful taxonomic significance.

We are including the species here tentatively in the hope of stimulating more detailed studies. The only specimens that seem to represent *H. nivalis* have been found in the lowest, most arid areas of the northern Front Range. LR: Big Thompson Canyon, 6200 ft, *W&W B-110694*.

HELODIACEAE (HLD)

Helodium [Greek, *helos*, marsh]

H. blandowii [for O. C. Blandow]. A large and beautifully pinnately branched moss common in fens and willow carrs in the subalpine zone. The stems are usually erect and rather tightly packed, of a pale yellow-green color, and the side branches gracefully curve out and back from the main stem. The main stem triangular-ovate leaves are larger than the branch leaves, and have branched cilia on their basal margins. Although the genus has long been included in the Thuidiaceae, we follow Ochyra (1989) in assigning this genus to the family Helodiaceae.

HYLOCOMIACEAE (HYL)

This family embraces several large moss species, some of which seem to be only distantly related to the others; however, for the amateur at least, they form a nice, easily recognized group. These are 'northern' mosses, which, in Colorado, are relictual and scattered in relatively few mesic forest sites. *Pleurozium* probably is better placed in or near the Hypnaceae.

1a.   Stems regularly 2–3-pinnate, forming broad, flat yellow-green to brownish fronds with minute leaves, with annual innovations often in stair-step manner. **Hylocomium**
1b.   Stems not regularly pinnate, remotely branched; leaves not minute . . . . (2)

2a.   Leaves plicate and rugose, falcate-secund, yellow-brown. **Rhytidium** (in Rhytidiaceae)
2b.   Leaves neither falcate-secund nor rugose, or if slightly so, bright green . . . . (3)

3a.   Stem densely covered by paraphyllia . . . . (4)
3b.   Stem lacking paraphyllia . . . . (5)

4a.   Stems wider than 4 mm including the leaf spread; leaves spreading, bright green. **Rhytidiopsis**
4b.   Stems narrow; leaves cochleariform, incurved toward the stem. **Hylocomiastrum**

5a.   Stems lax; leaves appressed, the apices rounded to obtuse, often appearing bluntly apiculate because of broadly rounded, incurved margins; leaf margins entire except at apex; stems usually red. **Pleurozium**
5b.   Stems stiff; leaves widely-spreading, the apices acute to acuminate, margins serrulate throughout; stems green. **Rhytidiadelphus**

Hylocomiastrum [-*astrum*, resembling]

H. pyrenaicum [of the Pyrenees]. A rare species occurring on the spray-wet borders of cascades in the subalpine forests of the Front Range. Our only collections are from Rocky Mountain National Park and in the San Juan Mountains. LR: Rocky Mountain National Park, Ouzel Falls, 9500 ft, *Weber & G. N. Jones B-20498, Rolston 80021 (B-91805)*; Fern Falls, *Hermann 26878 (B-94806)*.

Hylocomium [*hylocomos*, woodland]

H. splendens. One of the most abundant mosses of the boreal forest floor across northern North America. The species proper is immediately recognized by its beautifully bipinnate stems, with new feathery branches arising from the back of the old, and arching upwards in a stair-step fashion. In Colorado, however, conditions tend to be dry, and the genus is restricted to the most optimum moist subalpine forest areas and, occasionally, to tundra. Sometimes it occurs in deep mats and forms dense cover along with *Pleurozium schreberi.*

In Colorado we rarely find the typical luxuriant form. The common plant is relatively dwarf and is sometimes called var. *alpinum* or var. *alaskanum.* Steere (1978) discussed this as follows: "This name *(Hylocomium alaskanum)* has been long and widely (and loosely) used by many bryologists, including myself.

However, after many years of observation in the field of every possible stage in the intergradation between *H. splendens* and *H. alaskanum,* combined with the convincing reports of his experimental observations by Tamm (1950, 1953, 1964), who observed the rapid reduction in the physiological vigor of plants of *H. splendens* growing at some distance from the tree canopy, I have come to the final conclusion . . . that *H. alaskanum* can only be a stunted physiological-ecological tundra form or ecotype of *Hylocomium splendens* which does not merit nomenclatural recognition at any taxonomic level."

Pleurozium [*pleuron,* rib, alluding to the capsule]

P. schreberi [for J. von Schreber]. A large and coarse, irregularly pinnate moss of relatively dry and duff-covered forest floors. It is one of the most abundant mosses in the subarctic regions of the world, but relatively uncommon here. The plant is easily recognized by its sprawling habit, red stems, and rounded cucullate leaves. The leaf has margins rolled under in the upper half, and the alar cells form a conspicuous group of quadrate, thick-walled, orange-brown cells. The median leaf cells are long and narrow (70–100 x 5 μm) and porose. The leaves are 1.5-2.0 mm long. Our records range in altitude from 6000–9500 ft. BL, GA, LR, SJ.

Rhytidiadelphus [resembling *Rhytidium*]

R. triquetrus [three-ranked]. A moss of humid and oceanic climates, abundant across northern North America, especially in the Pacific Northwest. Its nearest stations are in Arkansas and Montana. Our first and only record is from BL: Outer foothills of the Front Range, Green Mountain, 6500 ft, *Mazurek & Wittmann B-112025.* The total population is estimated to cover no more than 100 square ft. The moss is associated here with *Pleurozium schreberi* and *Grimmia anomala.*

Rhytidiopsis [resembling *Rhytidium*]

R. robusta. From its coarse habit and large size this moss is easily recognized as belonging in the family. It is unique, however, in its pronounced monopodial growth form, in which the main stem is exceptionally elongate and the lateral branches remotely scattered. This is a species of the Pacific Northwest west of the Cascades, with isolated areas in the humid parts of northwestern Montana and northern Idaho. Our single collection is from BL: Ceran St. Vrain trail, 8300 ft, *Wittmann & Lehr B-114904.* The population is estimated to cover about four square ft.

HYPNACEAE (HPN)

In the early years of bryology, most pleurocarpous mosses were called *Hypnum*. In time, many genera were removed from the genus, and even new families were made to accommodate some of them. The process of refinement is still continuing, with the result that some small genera are still shuttled back and forth between the Hypnaceae, the Plagiotheciaceae, and the Amblystegiaceae.

1a. Plants rather robust, strikingly handsome, pinnately branched to form oblong-triangular fronds, tending to ascend, never in tight mats; leaves strongly plicate. **Ptilium**

1b. Plants small to medium-sized, not particularly striking in appearance, variously branched but not frondose; leaves plane or weakly plicate . . . . (2)

2a. Plant small and slender; leaves 0.15–0.5 mm long; rhizoids densely papillose . . . . (3)

2b. Plants larger (small to fairly robust); leaves more than 0.5 mm long; rhizoids smooth . . . . (4)

3a. Leaf cells narrowly rhombic, not linear. **Platydictya**

3b. Leaf cells linear. **Isopterygiopsis**

4a   Leaves erect when dry, narrowly lanceolate, subulate; alar cells scarcely differentiated. **Orthothecium**

4b. Leaves homomallous or falcate-secund, ovate or oblong-lanceolate . . . . (5)

5a. Leaves falcate (sometimes weakly so in *Hypnum cupressiforme*) . . . . (6)

5b. Leaves not falcate, but usually homomallous . . . . (8)

6a. Stem with an outer epidermis of small thick-walled cells; hyalodermis lacking. **Hypnum**

6b. Stem with several outer rows of small, thick-walled cells, plus an outer row of thin-walled, fragile epidermal cells (hyalodermis; section the younger part of the stem) . . . . (7)

7a. Stem flattened, prostrate; leaves complanate, weakly falcate and overlapping; color brownish-green; alar cells weakly developed. **Breidleria**

7b. Stems terete, erect; leaves strongly falcate (like *Hypnum*), not complanate; color lively green; alar cells in a clearly inflated group. **Calliergonella**

8a. Leaves dark, brown, with a sheen; leaf bases recurved at base; clusters of small-leaved brood-branchlets in some leaf axils. **Platygyrium**

8b. Leaves green, often pale; leaf margins plane; brood branches lacking . . . . (9)

9a. Capsules inclined (at least at the apex) to horizontal; leaf cells, especially the distal ones, short-rhomboidal; quadrate alar cells in many rows . . . . **Homomallium**

9b. Capsules erect and symmetric; leaf cells linear-rhomboid; quadrate alar cells few (up to about 12 along the leaf margin). **Pylaisiella**

Breidleria [for Johann Breidler]

B. pratensis [of meadows]. This has been considered to be a *Hypnum*, but Hedenäs (1990b) writes, "The species of *Breidleria* are close to some *Hypnum* species, but due to a combination of characters, including the large size, the flattened foliage of the stem, the hyalodermis of the stem and the tendency to grow in moist or wet habitats, *Breidleria* is considered distinct." The plants are a rich reddish or golden-brown, the stems dorsiventrally compressed, the leaves complanate, in one plane and closely overlapping, not strongly falcate, slightly denticulate apically, the margins plane, alar cells not sharply delimited. It is inconspicuous and rarely collected, but expected to be frequent in willow carrs and wet tundra. BL: In *Carex scopulorum* swale, Niwot Ridge, 11500 ft, *Weber & Nelson B-11376*; GL, RT, SM.

Calliergonella [little *Calliergon*]

The genus *Calliergonella* is based on *Hypnum cuspidatum* and has been known under several names in other parts of the world: *Stereodon cuspidatus*, and *Acrocladium cuspidatum* (Hedenäs 1990b). *Calliergonella* differs from *Hypnum* in having the peristome equipped with appendiculate cilia and in having leaves with large groups of strongly inflated alar cells. Hedenäs believes that this genus belongs in the Plagiotheciaceae.

C. lindbergii [for S. O. Lindberg]. Common in willow carrs and along snow-melt streamlets in the subalpine. The plants tend to be erect rather than creeping, as most *Hypnum* species would be, and always grow in wet places. Its earmarks are that it is a yellow-green falcate leaved '*Hypnum*' that has a row of thin-walled, outer-cortical cells in the stem.

Homomallium [turned in one direction]

Homomallium resembles some small Leskeaceae, and the leaf areolation is quite like Leskeaceae. The leaves have prominent alar groups of quadrate cells extending a third the way up the leaf. The leaves are abruptly acuminate, and the leaf cells are rhomboidal. The costa is short and double or absent. The foliage is dark greasy green in *H. adnatum* and reddish green in *H. mexicanum*. It does not resemble other Hypnaceae!

la. Branches with leaves neatly directed forward, not or slightly homomallous; leaf apex shortly and broadly acuminate. **H. adnatum**

lb. Branches with strongly homomallous leaves, the branches appearing to be shaggy; leaf apex slenderly acuminate. **H. mexicanum**

H. adnatum [tightly appressed]. Infrequent but probably not rare where it occurs on sandstone ledges in the Mesa de Maya region of southeastern Colorado. In addition to the characteristics mentioned in the key, the plants usually have a brown tinge whereas *Pylaisiella* is usually clear green, with larger leaves. In most parts of its range in eastern United States, *H. adnatum* occurs on trees, but here and in other parts of the Middle West it occurs mostly on sandstone and limestone. This could be confused with *Platygyrium repens*. The latter has erect capsules, clustered brood bodies at the stem apices, and leaves with margins that are reflexed along the lower third. In *Homomallium* the immature sporophytes, before the capsules form, are strictly erect, but the mature capsules are distinctly curved at the apex, not bent to one side as in *Brachythecium*.
LA: 15 mi S and 12 mi W of Pritchett, *Maslin*

B-3831. A specimen collected by Grout, labeled *Amblystegiella adnata*, GL: Road to Rollinsville, near Tolland, is a mixture of *Orthotrichum rupestre* and *Hypnum cupressiforme* var. *subjulaceum*.

H. mexicanum. Common in southern and western Colorado, especially on the bases of scrub oaks. The main stems are distinctly homomallous, with tufts of red rhizoids at intervals. Sometimes the short-leaved branches are so numerous that they obscure the main stems and may be mistaken for *Pseudoleskeella*. The leaves have an areolation similar to *Pseudoleskeella*, but there are a large number of cells across the widest part of the leaf between the costa and the quadrate alar group which extends far up the margin. The leaves are never catenulate or so uniform in *Pseudoleskeella*, which is neither homomallous nor with conspicuous tufts of red rhizoids.

Hypnum [*hypnon*, ancient name for a moss]

la. Stem with a rind consisting of several layers of small, dark, thick-walled cells; alar cells usually numerous, quadrate and green, or some of the lower ones larger; plants of relatively dry sites, on rocks or on the ground, or on moist edges of streams, mostly in the foothills, never in fens .... (2)

lb. Stem with an outer layer of large, thin-walled cells (hyalodermis) outside the rind of incrassate cells, these sometimes collapsed and sloughed away, in which case the incrassate layer, with remnants of the walls of the hyalodermis, appears ragged or dentate; alar cells few or, if more numerous, then mostly large and inflated; plants of subalpine fens, never in the foothills. See *Calliergonella* and *Breidleria*.

2a.   Leaf margin recurved or revolute almost its entire length; leaves strongly circinate-falcate; ubiquitous moss, especially on rocks throughout the forested zone. **H. revolutum**

2b.   Leaf margin plane, or slightly recurved only in the lower part . . . . (3)

3a.   Leaves cochleariform, smooth, weakly to strongly falcate-secund, with a glossy shell-like texture . . . . . (4)

3b.   Leaves falcate-secund, appearing somewhat plicate . . . . (5)

4a.   Median leaf cells short and broad (25–40 μm long x 5–6 μm wide); the alar cells numerous, quadrate, green, homogeneous, the lower ones not conspicuously larger; leaves short and cochleariform, broad and relatively short, forming neat cylindric shoots, usually with a good deal of brown color. **H. vaucheri**

4b.   Median leaf cells longer (50–100 μm long); the alar cells numerous, the upper ones quadrate and green, often compressed, the lower ones larger, pale, inflated; leaves long, cochleariform or almost flat and ribbon-like, with a long, tapered apex, usually of a clear, pale green color. **H. cupressiforme**

5a.   Plants freely fruiting (autoicous); leaves weakly serrulate distally. **H. pallescens**

5b.   Plants never fruiting (dioicous); leaf margins smooth. **H. revolutum** var. **ravaudii**

H. cupressiforme [cypress-like]. A common and variable species throughout the mountains, often growing near to *H. revolutum* but in slightly more mesic sites, along bases of boulders where there is often seepage, and along stream-banks. It is easily recognized when large and with its characteristic pellucid shell-like leaves, but develops curious slender forms in which the leaves are hardly falcate.

Another common ecological modification is pale and with long, not clumped stems, leaves not at all falcate or cochleariform, and the branches more or less homomallous. A characteristic that is not mentioned in the treatments is that the median leaf cells are distinctly prorulate (which Grout mentions only for *Hypnum* (now *Ctenidium) molluscum*. We first took this for *Callicladium haldanianum*, but it has been confirmed by Schofield as a little-known variant, *H. cupressiforme* var. *subjulaceum* Molendo. It grows on vertical granite cliffs and never forms dense clumps but often single or few isolated stems. Is any of this variation genetic?

H. pallescens. Evidently rare in Colorado. The capsules are elongate-cylindric, only slightly curved, and pale. See discussion under the next. BL: 8 mi NW of Nederland, *Flowers 9843 (B-66070)*; EA: Ute Trail, 2 mi SW of Big Spring, near Burns, north exposure, 9500 ft, *Morris 66 (B-91859)*; EP: Foothills of Eagle Mountain, 10000 ft? (the summit of Eagle Mountain is about 9000 ft), *Jewett, B-27762* (Grout, North American Musci Pleurocarpae Exsicc. 367); ML: Deep Creek, Creede, 10000–11000 ft, *Stewart B-26264*.

H. revolutum [recurved, the leaf margins]. Probably the most abundant and conspicuous pleurocarpous moss throughout the forested area, from the lower foothills up to the alpine. Typically it covers boulders with a smooth, shiny, green carpet of beautifully 'braided' stems, covering several square ft and visible from a great distance. The size and appearance of the species is extraordinarily varied. It tolerates considerable drought, probably getting only a little melted snow-water and intermittent wetting from rain. In luxuriant masses on granite outcrops, it is one of the most visible of Colorado mosses; yet on forest floors and tree bases, in stressed situations, it is often a minute plant, the variety *ravaudii*.

Flowers was puzzled by the small plant and listed it as *H. hamulosum* of northern Canada and Alaska. The plant is minute; the slender main stems dominate, their branches suppressed; the leaves are minute, weakly or not at all revolute. Vegetatively it is similar to *H. pallescens*, which is autoicous and almost always has sporophytes. The distal leaf margins are serrulate in *H. pallescens*. In fact, Nyholm (1954–1969) logically placed *H. ravaudii* and *H. pallescens* together in the Section Drepanium, and *H. revolutum* in Euhypnum. It appears to be impossible to find reliable characters to separate the intermediate forms, so we are following Schofield's treatment in the BFNA in treating *H. revolutum* as variable.

H. vaucheri [for J. P. Vaucher]. A close relative of *H. cupressiforme*, but recognizable in the field by the short and cochleariform leaves, forming neat cylindric shoots, usually with a good deal of brown color. Under the microscope, the leaf cells are much shorter than those of *H. cupressiforme*. Sometimes the leaves are not at all falcate, and the combination of such leaves, short and forked costa, and large numbers of quadrate cells filling the alar region and going up the sides of the leaf, suggests an *Entodon*. In aspect, such a form suggests a small *Brachythecium* rather than a *Hypnum*. Frequent in the outer foothills and mesas, particularly on sedimentary rocks with some calcium carbonate.

Isopterygiopsis [resembling the genus *Isopterygium*]

Ireland in Sharp, et al., (2:1035, 1994) says: "*Isopterygiopsis*, recently segregated from *Isopterygium* (Iwatsuki, 1970), is distinguished by its stems with small to large epidermal cells with thin outer walls and thick inner ones, papillose, axillary rhizoids, no pseudoparaphyllia, 2–6-celled smooth, cylindric or fusiform propagula clustered in the leaf axils, and annulate capsules."

1a.   Leaves cochleariform, mostly denticulate, cells in mid-leaf 30–40 μm long; rare alpine plant. **I. alpicola**
1b.   Leaves plane or slightly cochleariform, almost entire; cells in mid-leaf 60–100 μm long; frequent in the forested areas up to timberline. **I. pulchella**

I. alpicola [of high mountains]. Ireland (*in litt.*) wrote: "The specimens are typical of *I. alpicola* as you had them named, except that the leaf margins were not denticulate at the apices . . . This is the first record that I have seen of *I. alpicola* from Colorado. It was previously known in North America only from the Northwest Territory and Alaska." ST: Blue Lake Dam, 3000 m, *W&W B-111144.*

I. pulchella [dim. of *pulchra*, beautiful]. This suggests *Amblystegium serpens* superficially, but the median leaf cells are more elongate and there is no costa. Capsules are usually present, and are erect and curved near the top. *I. pulchella* seems to be frequent on rotting logs in spruce-fir forests, but we have found our best specimens in crevices of talus.

Orthothecium [straight capsule]

1a.   Leaves 2 mm long, strongly plicate, narrowly lanceolate, tapering in a straight line from base to apex. **O. chryseum**
1b.   Leaves 0.6–1.0 mm long, not plicate but merely slightly revolute or explanate, rounded to an acuminate apex **O. strictum**

O. chryseum [golden]. A moss with elongate, sparingly branched stems and straight, lanceolate, reddish-golden leaves. It might be mistaken for *Tomentypnum nitens*, but the stems do not have the matted tomentum of that species. *Tomentypnum* has a costa while in *Orthothecium* the costa is lacking or short and double. Our only collection comes from ST: Blue Lake, on limestone terraces, 3000 m, *W&W B-111105, B-111201.*

O. strictum [straight]. Rare, subalpine, in moist rock crevices. One record from SJ: Deadwood Gulch, near Silverton, *Hermann 1976.*

Excluded Taxa

O. diminutivum (Grout) Crum, Steere, & Anderson. The type has been examined, and turns out to be *Isopterygiopsis pulchella.* The type locality is Jenny Lind Creek, 10000 ft, near Tolland, Gilpin County, in a mixed tuft of *Pohlia* and *Drepanocladus aduncus*, July, 1914, *Grout* (DUKE and herb. Dixon), as *Holmgrenia diminutiva* (Grout, 3:172, 1932).

Platydictya [broad net, alluding to the areolation]

P. jungermannioides [from a fancied resemblance to a slender liverwort]. Not rare but so inconspicuous that it is simply overlooked. The plant forms thin mats in the shaded overhangs of cliffs or large rock outcrops, usually in deep spruce-fir forests, not far from streams. The stems are minute, much finer even than those of *Pseudoleskeella tectorum*, and pale green. Under the microscope the few-celled gemmae are usually quite easily located in the leaf axils. One can be forgiven for believing this plant should be in the Amblystegiaceae!

Platygyrium [broad annulus]

P. repens [creeping]. A species of the eastern woodlands, rare or infrequent and restricted to the eastern base of the Front Range. Occurring at the base of *Pseudotsuga* in protected ravines. One collection known: LR: Twin Cabin Gulch, tributary of Buckhorn Creek, 7000 ft, *W&W B-114370, B-114467*. The dark, brownish-golden color of the plants, their plane leaves with elongate-rhomboidal leaf cells, lack of costa, and presence of microphyllous, bud-like branchlets are diagnostic. The main branches tend to be rather tightly affixed to the bark by scattered clumps of rhizoids. We have not found it with sporophytes. Probably the species has been overlooked in many similar sites in the Front Range. It occurs with *Pylaisiella,* which has bright green leaves and branches that tend to be erect, and it usually occurs with erect cylindric sporophytes.

Ptilium [feather]

P. crista-castrensis [soldier's plume]. A beautiful moss, characterized by its flat, pinnately-branched, stiff, ascending organized branches that are not matted like *Hypnum* and of a yellow-green color. Compared to *H. revolutum*, it is more clearly pinnate, and the stem leaves are larger than the branch leaves and broadly cordate-ovate. Rare in Colorado. BL: Ceran St. Vrain trail, *W&W B-114906*; LR: Big South Trail along Cache La Poudre River, 8600 ft, *Hermann 28158*; *W&W B-113492*.

Pylaisiella [for Bachelot de la Pylaie]

P. polyantha [many-flowered, alluding to the capsules]. Common in the foothill canyons, usually at the bases of tree-trunks, occasionally in the subalpine. The plants are yellowish-green with strongly homomallous leaves and branchlets that tend to curve upward. The straight, erect capsules are usually numerous.

LEPTODONTACEAE (LPT)

Leptodon has always been placed in the Neckeraceae, but the leaves are not transversely undulate, the stems are not complanate, and when well-developed they are strongly circinately curled when dry. the branching is pinnate, and the plants are somewhat dendroid, with the main stem becoming naked. The foliage is a dull, dark green. The leaves are rounded apically, with a strong costa

Leptodon [slender teeth]

L. smithii (for J. G. Smith). The single occurrence of *Leptodon smithii* in Colorado is the only record known for North America. This is a calciphile species with a widely disjunct distribution including the Mediterranean region, where it is common on the dusty bark of roadside trees in limestone areas, South and Central Africa, southern South America, eastern Australia, and New Zealand. It belongs to the Tertiary Relict element of the Southern Rocky Mountain flora. JF: Clear Creek Canyon, close to the high water mark at the west end of Tunnel No. 2, with *Fabronia*, in crevices of a metamorphic limestone cliff, *P. Nelson B-40861, Weber & Nelson B-40726*. See Nelson (1973). The specimen consists only of male plants. Additional stations have been sought but never found. The story of this unique discovery is worth repeating here in the hopes that someone will be able to find the 'mother lode'.

Patricia Nelson, one of my graduate students, spent a summer making an inventory of the mosses of lower Clear Creek Canyon, west of Denver, taking advantage of a new paved road. This canyon is noted for its extensive outcrops of calcareous metamorphic rock and, until then, we had few bryophyte collections from the area.

During the following winter, Pat constructed, mostly from the literature, a random access key to the moss genera of Colorado that used punched IBM cards (Weber & Nelson, 1974). Logically, she decided to test her key on the collections she had made the previous summer.

The key was quite a success; however, one day Pat came to me with a moss that did not fit, and that she felt had to be a member of the Neckeraceae. I referred her to Grout's Flora, but there was nothing that matched. She brought me the specimen, and I took one look at it and recognized it to be *Leptodon smithii*, which I knew well from my trips to Greece and eastern Australia. But this was impossible! I suggested that perhaps she had accidentally picked up from the work desk a stray piece that might have dropped from a recently-processed specimen from elsewhere, but she did not feel that was likely.

I asked her, "Do you remember just where you found it?" "Yes." "Can you take me there tomorrow?" "Of course!"

The rest is history. We stopped at the west end of Tunnel No. 2, clambered down the talus to the outcrop that was slightly above flood-line, and there it was. The plants are not typical, being smaller and hardly circinate, but there is no doubt of their identity. (It has been suggested that these are likely male plants.) Try as we might, we have never found another site. Unfortunately, the most promising places are across the often turbulent waters of Clear Creek. My friend Pat Murphy and I have crossed in low water times, but to do repeated crossings is difficult and time-consuming.

We provide here the excellent illustration of Patricia Eckel to help anyone interested in rediscovering this, one of the few species in the North American Flora to be known from a single collection on the continent.

LESKEACEAE (LSK)

This family and its various genera are not easy to characterize. A number of other families have been or still are included within its boundaries by various authors who include in Leskeaceae the Theliaceae, Thuidiaceae, Pterigynandraceae, and Anomodontaceae. Crum and Anderson (1981) place two of the basic genera, *Pseudoleskea* and *Pseudoleskeella* in the Thuidiaceae, basing their contention on the differences in the type of peristome, which in our genera are rarely produced. How then, can we characterize what we call the Leskeaceae?

Our genera are small, creeping pleurocarps with slender, rope-like stems, the leaves usually appressed when dry, sometimes with the apices slightly spreading. The color is dark green or blackish, matte; a few species with lighter green colors usually indicate that the leaves are papillose.

The leaves are ovate and never long attenuate, usually closely overlapping. In *Pseudoleskeella tectorum*, most authors describe the leaves as 'catenulate' (like a chain). In fact, a European species is named *P. catenulata*. The resemblance to a chain is in the aspect of the leaves in which the base is spreading but the body and apex of the leaf are directed up and in toward the stem, thus allowing spaces to show between them. We feel it more appropriate to refer to the leaves as arcuate (bent like a bow).

Plicae of leaves are of two sorts. Many Leskeaceae have bi-plicate leaves, that is, the leaf is cochleariform and the margin is broadly recurved (more of a flange). Ordinary plicae are longitudinal folds between the leaf margins and the costa.

The leaves are often loosely described as papillose, but true papillae are superficial protuberances on the cell walls. In many Leskeaceae what are called papillae are really prorulae, the protruding distal apices of the cells. The leaf cells characteristically are short, seldom more than 5:1, and the alar cells are usually quadrate or rounded and numerous. The costa may be absent, short or sometimes forked, or extending half way up the leaf rarely far up into the apex. Sporophytes have rarely been found here. The substrate is often rock, but sometimes the species occur on dry wood.

1a. Leaf cells with a single large papilla in the center; costa strong, ending near the leaf apex . . . . (2)
1b. Leaf cells lacking papillae or with protruding ends (prorulate); costa various . . . . (3)

2a. Branches with numerous paraphyllia. **Lescuraea patens**
2b. Paraphyllia absent, but branches usually with small, short bud-like gemmae branchlets. **Lindbergia**

3a. Costa strong, never forked, ending in or just below the leaf apex . . . . (4)
3b. Costa short, almost lacking, or extending to about mid-leaf, sometimes forked . . . . (5)

4a.   Leaf cells inconspicuously papillose abaxially, but the wall often merely bulging; costa not reaching into the acumen; stem stout and rigid, with short pinnately disposed branches; stem leaves obviously larger than the branch leaves; microphyllose branches lacking. **Leskea**

4a.   Leaf cells smooth; costa reaching well into the acumen; stems not thick or rigid; stem and branch leaves not differentiated; clusters of small propagating branchlets usually at the branch tips. **Leskeella**

5a.   Stems usually with paraphyllia; leaves irregularly plicate; median leaf cells more elongate than the marginal and basal cells, distally prorulate. **Lescuraea**

5b.   Stems lacking paraphyllia; leaves cochleariform, not plicate in the lamina but with the margins often broadly bi-plicate; leaf cells uniform or not, not prorulate . . . . (6)

6a.   Plants dark green or black; leaves merely acute or acuminate; costa lacking or short and double, or strong only to mid-leaf. **Pseudoleskeella**

6b.   Plants apple-green, usually occurring as slender single strands mixed with other mosses, especially *Neckera*, in overhangs of cliffs; costa forked; leaf cells oblong; the leaf is not strongly bi-plicate and the leaf apex slightly spreading. **Leptopterigynandrum** (See Pterigynandraceae)

Lescuraea [for Leo Lesquereux]

1a.   Leaf cells with a single high papilla over the lumen, on both sides; leaf cells small (about 10 μm), irregularly rounded or almost isodiametric, appearing collapsed. **L. patens**

1b.   Leaf cells with distal prorulae, often not easily distinguished, formed by the projecting ends of the cells.(2)

2a.   Side branches relatively short, the terminal branches not distinctly longer; leaves minute (–0.5 mm), cochleariform, broadly ovate with abruptly acuminate apex; stems plump but not clearly catenulate, appearing julaceous; alar cells very numerous; costa extending nearly to the acumen; median leaf cells variable, but some definitely oblong. See **Pseudoleskeella arizonae**

2b.   Main branches typically elongate; leaves larger, narrower, and appressed or arcuate, the apices more gradually acuminate; alar cells more distinctly confined; median cells mostly oblong . . . . (3)

3a.   Bases of leaves spreading, the median and upper parts incurved (catenulate); median leaf cells not conspicuously elongate; leaves distinctly cochleariform, visibly separate, widely spreading when moist. **Lescuraea incurvata**

3b.   Leaves appressed from the base upwards, the individual leaves not easily distinguished; median leaf cells distinctly oblong, sometimes linear; leaves not distinctly cochleariform, narrow, not wide-spreading when moist, the narrow apices somewhat falcate . . . . (4)

4a.   Stems stout, densely foliate with appressed, merely acute apices and entire leaf margins; inner basal cells almost quadrate, shorter than the median oblong ones; forming massive colonies on granitic rock faces generally in the higher forested areas. **L. radicosa**

4b.   Stems slender, the leaf apices obviously attenuate, the margins serrate; inner basal cells oblong, not different from the median ones. **L. saxicola**

L. incurvata. The distinction between *L. incurvata* and *L. radicosa* is easy because of the gross aspect of the leaves, whether catenulate or appressed. In this species the leaves are distinctly incurved and the leaf apices are not at all falcate-secund. The costa is stout. On boulders, especially in or near streamlets, in the forests, from foothills to subalpine.

L. patens [spreading]. A slender species, and the leaf cells, instead of being oblong, with prorulae, are isodiametric and often appear to be collapsed. The coarse papillae are best seen in side view. In our specimens the slender leaves appear to be characteristically arcuate. It occurs on moist rocks along subalpine rills near timberline. We have relatively few collections. BL: Buchanan Pass trail, 3100 m, *Rolston 82317 (B-84835)* (with sporophytes!); RT: 1–2 mi above Slavonia, 2700-3000 m, *Weber & Nelson B-39367*.

L. radicosa [with roots]. Common on boulders in the

subalpine forests. It may be recognized in the field by its appressed, densely imbricate and slightly falcate-secund leaves. *Lescuraea* may be separated from *Pseudoleskeella* in the field by its light green younger leaves and the older coppery-reddish ones. Also, the leaf cells are much more numerous than in *Pseudoleskeella*. In some specimens the median cells are linear!

L. radicosa var. *pallida* is an extraordinarily slender form. We have a collection from LR: Rocky Mountain National Park, Timber Lake Trail, ca. 9400 ft, in a deep overhang in fractured granite outcrop, *Weber B-112865*. When well developed, the branches are distinctly complanate, with shiny, ovate spreading leaves that end in a slender, twisted tip. The leaves are not changed in

drying. The leaves, merely acute and serrulate at the apex, the elongate narrow median cells, numerous quadrate marginal and alar cells are distinctive. It is unlike the species proper and its habitus reminds one of *Leptopterigynandrum!*

L. saxicola [of rocks]. More slender than *L. radicosa*, and never forming great masses of crowded stems on boulders. The microscopic distinctions, however, are in the narrower leaves, strongly serrate acumen, strongly prorulate distal part of the costa, and longer basal cells. GL: spruce forest at East Portal, Moffat Tunnel, 2700 m, *W&W B-110936*; EP: Williams Canyon, on vertical conglomerate cliff, 7000 ft, *W&W B-113148*.

Leskea [for N. G. Leske]

L. polycarpa. A species of the eastern United States, relictual and rare in the Front Range foothills. The leaves of the stems are homomallous, facing upward and forward; the undersides of the stems have scattered rhizoids. The leaves of the branches are short and broad, and densely arranged, spreading stiffly when moist. In the east it is a plant of streamsides, bases of trees, and only rarely on rocks. JF: Coal Creek Canyon, Twin Spruce Road, *W&W B-114969*; DA: Castlewood Canyon State Park, sandstone cliff, *Lederer 4373*; HF: Cuchara Camps, *McAllister 231* (UBC); PE: Limestone cliffs along Greenhorn Creek, *W&W B-112201*).

We have been impressed with how similar this species is to *Leskeella nervosa*. To our knowledge, they

are not compared in the available floras. They are indeed hard to distinguish if the tips of the branches lack the characteristic bunches of microphyllous propagula. We have found that *Leskea polycarpa* never has propaguliferous branches, the stem leaves are distinctly larger than the branch leaves, more ovate and long-attenuate, the costa does not reach up into the leaf apex but ends below the acumen, and if carefully examined under the microscope the distal leaf cells on the adaxial leaf surface have low, broad papillae that one sees only by focusing up and down; the papillae are broadly dome-shaped and are definitely not cellular inclusions.

Leskeella [little *Leskea*]

L. nervosa. Common on boulders in the outer foothills. In this species the leaf cells are all much alike and short diamond-shaped. The leaf is twice as large as that of the *Pseudoleskeella* species and there are as many as 20 cells from margin to costa. The costa reaches into

the acumen and may even be slightly excurrent. The stems are almost always provided with short terminal microphyllous branchlet clusters (functional propagula), and are not as unbranched as in *Pseudoleskeella* forms in which they may be slenderly filiform.

Lindbergia [for S. O. Lindberg]

L. brachyptera [short wing]. Uncommon on rock in the outer foothills. Inconspicuous, not forming extensive clumps. At the present time we have only two collections: LR: Granite bluff, 1/5 mi E of Glen Haven,

9400 ft, *Hermann 24981*; FN: Phantom Canyon, granite cliff 5 miles up canyon, *Weber, Hermann, & Feddema B-37554*. See the key for distinctions between this and *Lescuraea patens*.

Pseudoleskeella [false *Leskeella*]

One must be sure about the differences between *Pseudoleskeella* and *Homomallium,* which is currently placed in the Hypnaceae. *Homomallium* has a leaf areolation much like *Pseudoleskeella*. See discussion under that genus. Although several authors state that the stem and branch leaves are not differentiated, we find otherwise. They may be hard to find, but once discovered the stem leaves are broadly triangular-ovate with an abruptly narrowed attenuate apex, in sharp contrast to the bluntly ovate branch leaves which are much smaller.

1a.   Stems stout and stiff, the leaves difficult to strip away; stem leaves distinctly larger than branch leaves, broadly ovate-cordate, abruptly attenuate, strongly bi-plicate and flaring at the base; the costa variable, to half the lamina length and often forked; short branchlets produced in great quantities, short and plump, often erect, leaves densely foliate, not catenulate, the apices imbricate, resembling the scaly covering of a lizard. **P. arizonae**

1b.   Stems slender and filamentose, leaves easily stripped away; stem and branch leaves similar, narrowly ovate and blunt tipped; leaves not densely foliate, not strongly imbricate, usually catenulate . . . . (2)

2a.   Costa lacking or short and double; all leaf cells short and rhomboid, not prorulate. **P. tectorum**

2b.   Costa strong, extending to mid-leaf; median leaf cells oblong, prorulate, longer than the marginal and basal. **P. sibirica**

P. arizonae. A fairly common moss throughout the southwest. Until now this has been passing as *Pseudoleskea, Pseudoleskeella,* or *Lescuraea arizonae.*

P. *arizonae* bears some resemblance in its leaf shape and size to *Leskeella nervosa,* and in habit and habitat to *Pseudoleskeella tectorum.* However, *P. arizonae* is recognizable in the field by having short, plump, julaceous branches in which the leaves are not at all catenulate but are densely imbricate and appressed, the surface of the branch suggesting the scales of a lizard's tail. The stems are noticeably thicker than in *P. tectorum* and there are many more laminal cells, from 20 to 30 across from costa to margin. The strongly flared bi-plicate leaf bases are distinctive as well.

This is an arid-land, low altitude species of sunny horizontal rock faces, forming dark green or black flattened patches on horizontal sheets of sandstone rocks. A common characteristic is that at the margins of the patch the new growth extends outward and is distinctly pinnately-branched, with scattered rhizoidal tufts, producing short, plump, and easily detached bud-like branches. In the center of the patch the branches are plump but distinctly elongate and julaceous and commonly erect.

Many collections have been incorrectly identified as *P. tectorum.* The geographic range extends to both sides of the Continental Divide, most commonly on the western sandstone rimrocks, but also in similar desert-steppe sites at the eastern base of the Front Range. Wilson and Norris (1989) treat this as a *Lescuraea,* but the evidence is not convincing.

P. sibirica. Similar to *P. tectorum* but typically forming rather deep polsters of slender, weak, tightly packed,

erect stems that around the edges spread out over the substrate much like in *P. tectorum.* The unexposed parts of the stem (all but the terminal leaves) are pale or orange. Typically this species grows on vertical cliffs, whereas the abundant *P. tectorum* usually grows on flat rock surfaces. The stout, truncate costa and the distinctly acuminate leaf tips, together with the oblong median leaf cells are additional distinctive features. Wilson & Norris (1989) state that "*P. tectorum* . . . is typically very different but grades almost continuously into *P. sibirica.*" We believe that it is wise to recognize this at least tentatively, although some specimens are difficult to be sure about.

In Cement Creek Canyon, Gunnison County, we were puzzled by a minute moss with crowded, erect stems that superficially resembled a dwarfed *Myurella* or even a liverwort, packed between basal squamules of *Cladonia.* Careful dissection revealed that these were erect microphyllous shoots that easily separated from the main creeping stems buried deep beneath. This is an uncommon occurrence although it is mentioned by Crum & Anderson (1981). Thus, *Leskeella nervosa* is not the only species that produces microphyllous propagula.

P. tectorum [of roofs]. This is the smallest and probably the most common species in the genus, occurring on fairly dry sites in forests, often covering large areas of boulder faces. Its small size and short, broad leaves usually with the tips strongly incurved toward the stem (catenulate) are good field characters. The leaf cells, under the microscope, are short-rhombic, chlorophyllose, and peculiarly clear and smooth in aspect. The leaves have relatively few cells (about 10) from costa to margin in the broadest part of the lamina.

MEESIACEAE (MEE)

The Meesiaceae are rather rare fen-mosses with tall, slender setae bearing an erect but curved capsule upon a narrowed elongate neck. The plants are usually turf-forming, with brown tomentum on the stems. The distinguishing features are chiefly in the leaf characters. The leaves are prominently costate, with the costa ending just below the apex.

1a. Stems elongate, with strongly squarrose-recurved leaves; submerged aquatic in subalpine fens. **Paludella**
1b. Stems short, the leaves not recurved; terrestrial . . . . (2)

2a. Leaves narrowly linear-setaceous; capsule pendent, pear-shaped with a narrow neck; common on burned ground in middle and higher altitudes. **Leptobryum**
2b. Leaves lanceolate or ribbon-like; capsule curved, not swollen at base; plants of subalpine fens . . . . . . (3)

3a. Leaf margins plane; the cells large, thin-walled, visible under the hand lens; capsule pale olive-brown. **Amblyodon**
3b. Leaf-margins revolute; leaf-cells small, thick-walled, not distinguishable with the lens; capsule pale when young, reddish-brown, and quite dark when old and empty. **Meesia**

Amblyodon [blunt tooth]

A. dealbatus [whitened]. A calciphile subalpine wetland moss, known in Colorado from only a few collections. It often occurs on the sides of peaty hummocks at the bases of willows, usually not close to rapidly-flowing water. Because of the relatively wide leaves with loose areolation, and the strongly curved capsule, *Amblyodon* will at first glance be mistaken for some sort of *Funaria*, but on examination, the mouth of the capsule is not really oblique, and is perpendicular at least to the distal portion of the capsule. CF: 1 mi W of Fox Lake, Buena Vista, 10500 ft, *Conard 41-109 (B-14751)*; PA: High Creek Fen, *W&W B-112713*.

Leptobryum [with very narrow leaves]

L. pyriforme [pear-shaped]. Frequent in disturbed soils, often on burned ground and peaty banks which have been quarried, definitely a moss of weedy tendencies, mostly found in the middle and higher altitudes. The combination of filiform leaves and nodding, pyriform capsules, is definitive. The leaves are slender, but not secund, just somewhat spreading in all directions. The costa is quite wide and the leaf cells are slender but blunt, not long tapering at the ends. The margin may be slightly denticulate. Farther back from the apex of the shoot, the leaves are rather suddenly small and broad, and the stem is black. The capsule may be only divergent, and has a well-developed slender neck which suddenly widens into a bowl. This plant certainly does not suggest *Bryaceae*, although that was its classification until recently. In the greenhouse, where this species, along with *Funaria*, is a common weed, it displays a unique habit. The slender stems are nearly naked except for the terminal rosette, but a new stem arises from this to produce another shoot and sometimes another, terminated by the seta and wide-mouthed capsule.

Meesia [for David Meese]

1a. Leaves narrowly oblong, with a disproportionately broad, dorsally convex costa; leaf apex blunt; lamina hardly wider than the costa; leaf cells narrow, firm-walled. **M. uliginosa**
1b. Leaves ovate-oblong, narrowed to an acute apex; costa ending below the apex, not disproportionately broad; lamina much wider than the costa; leaf cells large, broad, thin-walled. **M. longiseta**

M. longiseta. This beautiful moss with a long seta is found in calcareous fens in the subalpine. The fresh capsules are erect with a conspicuous narrowed neck; when dry the capsules may curve in almost a complete circle! PA: High Creek Fen, South Park, *W&W B-112557*.

M. uliginosa [of marshes]. Frequent along streamsides or in fens in the alpine and subalpine. It is rarely collected, since it may fruit infrequently and the plants are easily overlooked if sterile.

Paludella [little marsh]

P. squarrosa. A monotypic genus growing half-immersed in shallow water of alpine and subalpine fens. It can be mistaken for no other moss. The leaves are decurrent and squarrose-recurved, with the outer part folded in such a way as to appear heart-shaped. The leaf cells are strongly mammillose, and the margins coarsely toothed. The seta and capsule resemble those of *Meesia* but fruiting material has not been found in Colorado. It is a characteristic moss of the Arctic but rare here. CC: Pools at Guanella Pass, *D. Cooper B-96186, W&W B-110852*; fen along Clear Creek, Herman Gulch, *Lederer & Reid B-101959, A. Dibble B-115126*.

MNIACEAE (MNC)

Medium-sized mosses with spirally arranged leaves and unbranched stems. Leaves mostly broadly oblong or circular, narrow in one species. Leaves commonly bordered, entire or single- or double-toothed. Cells isodiametric or hexagonal. Plants of streamsides, fens, or wet forest floors. Capsules usually pale yellow-green, pendent.

It is not difficult to determine the sexuality of Mniaceae. The sex organs are surrounded by the uppermost leaves of the stem (never on the stoloniferous ones). Dissect away the leaves, and the archegonia or antheridia are readily detected. The antheridia tend to be embedded in a dense mass of brownish paraphyses, while the archegonia are usually dark brown, narrow, with a long neck. Most of our species are dioicous; rarely are they found with capsules.

Marginal teeth that are blunt, consisting of only slightly extruded terminal ends of the cells are termed by Koponen as being 'arrested' teeth. These occur in *Plagiomnium* and frequently in *Mnium arizonicum*.

1a.   Leaves entire, leaf margin strong, partly or entirely bi- or multi-stratose; large plants with stout stems clothed densely with macronemata. **Rhizomnium**
1b.   Leaves toothed or almost entire (sometimes with only 'arrested teeth', the distal cell of a marginal cell protruding bluntly); small plants with micronemata and/or macronemata, but these relatively inconspicuous and only on the lower stem parts . . . . (2)

2a.   Marginal teeth, when present, in pairs (one flush with the lamina and the other slightly on the lamina surface), or almost lacking in the tiny *M. blyttii*; creeping or arching stoloniferous stems not present. **Mnium**
2b.   Marginal teeth, when present, not in pairs; creeping or arching stoloniferous shoots with apparently two-ranked distant leaves present. **Plagiomnium**

Mnium [ancient Greek name for a moss]

1a.   Leaves small (1.0–1.5 mm long), elliptic-ovate, the border weak, of only one or two rows of cells, the teeth arrested or lacking; cell contents turning blue with addition of KOH following heating in alcohol. **M. blyttii**
1b.   Leaves larger (up to 4 mm or more long), variously shaped; no blue color reaction . . . . (2)

2a.   Large plant, up to 5 cm tall; leaf border strong throughout; abaxial teeth of costa large and numerous; laminal cells quadrate to elongate, in diagonal lines from costa toward distal border; rare alpine species. **M. spinosum**
2b.   Small plants up to 2 cm tall; otherwise not as above . . . . (3)

3a.   Leaf cells with obviously thickened corners; cells isodiametric to more or less elongate . . . . (4)
3b.   Leaf cells lacking corner thickening . . . . (5)

4a.   Leaf cells 15–17 μm; cells near the costa and margin homogeneous. **M. thomsonii**
4b.   Leaf cells 17–35 μm; cells near the costa more or less elongate, larger than the isodiametric cells near the border. **M. marginatum**

5a.   Leaves serrate only near the apex, sometimes the teeth arrested or almost lacking. **M. arizonicum**
5b.   Leaves serrate almost from base to apex . . . . (6)

6a.   Stems slender, elongate; leaves narrowly lanceolate, not much altered in drying; triangular, long-acuminate, strongly colored leaves numerous on lower stem; costa rarely reaching the apex; decurrent leaf bases narrow and difficult to discern; cells 17–35 μm. Rare foothills plant. **M. hornum**
6b.   Stem short; leaves dark green, contorted and spirally arranged when dry, commonly with tints of red in the upper foliage; leaves broadly decurrent; cells 13–17 μm. Alpine species. **M. thomsonii**

M. arizonicum. This is the most common dry woodland species, and it is almost a western North American endemic from Yukon to Baja California (but also occurring in Greenland!). The plants are usually densely crowded and short, the leaves are twisted when dry. Margins are recurved at the base and serrate only near the apex (the teeth may be so poorly developed that it is impossible to see if they are in pairs); cells are almost quadrate to elongate-hexagonal, 17–25 x 20–30 μm, with evenly thickened walls (not just in the corners). The border is weakly differentiated, 1–2 cells wide. Dioicous, rarely fruiting.

M. blyttii [for M. N. Blytt]. This is our smallest species and has minute, broadly rounded leaves with poorly differentiated leaf margins. The cell sap takes on a blue postmortem color with the addition of alcohol and KOH. It is quite uncommon and occurs in shaded places from the foothills to rocky alpine tundra.

M. hornum [the current year, significance obscure]. A rare Pleistocene relict species of the eastern American woodlands. The upper leaves are more or less linear, especially on sterile stems, narrowly triangular, long-acuminate, strongly colored, dark green and reddish, numerous on the lower part of the stem. Costa sub-percurrent to shortly excurrent, abaxially toothed; decurrent leaf bases narrow, difficult to discern; distal cell walls firm, only slightly thickened at corners, leaf cells small, 12–21 μm. Plant forming loose cushion of erect stems leafy to the base, reminiscent of a small *Timmia* or *Atrichum*. Dioicous. Our only collection is from BL: North St. Vrain Creek, on grassy terrace, *Weber & Evenson B-112415*. Only one clump found.

M. marginatum. Differing from *M. thomsonii* in having larger rounded-quadrate leaf cells, 25–30 μm, the juxtacostal cells of the proximal part of the leaf somewhat elongate. Both species have red coloration in the stems or foliage. In our specimens the marginal teeth are arrested or absent. Synoicous. Probably this species is overlooked and should be more common at low altitudes. Collection: GN: Cement Creek Canyon, 9700 ft, *W&W B-113685*.

M. spinosum. A large *Mnium* with comparatively broad and coarsely toothed leaves and costa with large abaxial teeth. In addition, the shape of the leaf cells and their arrangement in diagonal rows from costa to border separate *M. spinosum* clearly from the other species of *Mnium*. The distal leaf cells are usually quadrate, 20–30 μm, proximally more elongate and nearly rectangular. Dioicous. Known only from CC: Summit Lake, Mt. Evans, 13000 ft, *Weber B-22449* (!Koponen).

M. thomsonii [for T. Thomson]. This is listed by Crum & Anderson (1981) as an obligate calciphile that grows in crevices of limestone cliffs or on soil rich in calcium. It is an alpine species with dark green leaves and commonly tints of red in the upper foliage. The most reliable diagnostic feature of *M. thomsonii* is the leaf areolation. The leaf cells are smaller than in any other species and their walls are typically straight so that in places the cells are nearly quadrate-rectangular. Dioicous. This differs from *M. marginatum* in having smaller leaf cells (up to 17 μm; *M. marginatum* has cells over 27 μm). BL: Indian Peaks area S of 4th Green Lake, 11700 ft, male, *Komarkova B-43750*; CC: Chicago Creek, 8000 ft, *Sayre 161 (B-22457)* (!Koponen) as *B. marginatum* in 1972; ST: Blue Lake, on limestone terraces, 3000 m, female, *W&W B-111173*.

Plagiomnium [creeping *Mnium*]

Medium-sized mosses with two kinds of stems: Erect with spirally arranged leaves, and arching stoloniferous stems with leaves appearing somewhat two-ranked, larger at the base of the stem and becoming smaller distally. Margins distinct, unistratose, with single teeth. Wetter areas of forests, particularly along small streamlets. Leaf cells isodiametric or early so. In *Plagiomnium*, both macronemata and micronemata develop; the macronematal initials are rounded, in the leaf axils. The micronematal initials are in lines especially along the stoloniferous stems.

1a.   Leaves serrate above the middle, long-decurrent, the apex acute; synoicous, commonly producing capsules. **P. cuspidatum**
1b.   Leaves serrate from apex to base, to almost or quite entire; leaves commonly rounded at the apex . . . . (2)

2a.   Leaves not decurrent; leaf cells elongate, 20–35 x 40–80 μm; margin weakly serrate, with relatively few arrested teeth. **P. ellipticum**
2b.   Leaves decurrent; leaf cells slightly elongate, with corner thickening; teeth sharply serrate from base to apex. **P. medium**

P. cuspidatum. In this species the leaves are serrate from apex to mid-leaf, narrowed at the base, and long decurrent; the apex is sharply acute. The teeth are one-celled, sharp, and longest near the apex. The leaf cells are small (15–20 μm) compared to the other species. Synoicous, commonly fruiting. Frequent along streams in the foothills. Our specimens are from the Eastern Slope.

P. ellipticum. Most keys state that this species has sharp marginal teeth. In Colorado this is a rare occurrence. Koponen (1980) stated to the contrary that "entire leaves, or leaves with arrested teeth are more common than in any other European species of *Plagiomnium*, so that I have seen specimens of it identified as *Rhizomnium punctatum*." Our plants are mostly if not all male. The stem apex forms a rosette with massed paraphyses surrounding many antheridia. The leaves, when dry, are strongly contorted and have a fleshy look. It is mostly a subalpine to alpine wetland species.

P. medium. Here the marginal teeth are always present and numerous, one-celled, short-triangular, from base to apex, the leaves longer than wide and narrowly decurrent. (In *P. ellipticum* the leaves are not decurrent.) The leaf cells are larger than those of the other woodland species. Synoicous. A common foothills species along small streamlets in forests.

Rhizomnium [for the macronemata]

Coarse mosses with stout stems and broadly spreading oval or orbicular leaves commonly forming terminal rosettes. The usually branched stems are stout and clothed with reddish brown rhizoids (macronemata). The leaves are bordered with a bistratose series of elongate cells and are never toothed. The plants are large, conspicuous, and occur in conspicuous masses of stems, easily recognizable with the naked eye. The leaves are regularly arranged around the stem and arching stolons are not produced. Found in moist or wet forest floors, often near running water.

1a.   Larger plants; leaves elliptic-obovate, up to 1 cm long; costa percurrent or ending shortly below the apex; capsule ovate, outer peristome teeth yellowish, rarely brownish; peristome lamellae more than 20; dioicous. **R. magnifolium**
1b.   Smaller plants; leaves up to 7 mm long, nearly circular to broadly obovate; costa rarely percurrent, mostly vanishing below the apex; capsule globose; outer peristome teeth brown; peristome lamellae fewer than 20; synoicous. **R. pseudopunctatum**

R. magnifolium. A magnificent moss characteristic of the most mesic and often virgin spruce-fir forests of the interior parks and valleys. It and *R. pseudopunctatum* are similar. Koponen (1980) distinguishes them as follows:

"*R. pseudopunctatum* shares the leaf and rhizoid characters of *R. magnifolium*. However, the leaves are shorter in *R. pseudopunctatum*, and in most cases also more definitely obovate or more broadly elliptic than in

R. magnifolium, in which they may even be oblong. The shape of the marginal cells at the apex differs in being often quadrate or shortly rectangular in *R. pseudopunctatum*, but elongated and rectangular in *R. magnifolium*."

R. pseudopunctatum. Occurring in similar habitats, but not enough field observations have been made to determine how the requirements of each differ. Both are evidently restricted to the subalpine forested zone.

Excluded Taxa

Mnium spinulosum Bruch & Schimper is attributed to Colorado by Crum & Anderson (1981). This is a synoicous species, hence is one that frequently bears capsules. The peristome teeth are dark brown as in *M. spinosum* (which we have only sterile). The other *Mnium* species have yellow peristome teeth. It has the abaxial surface of the costa smooth or with small teeth; in *M. hornum, M. spinosum,* and *M. thomsonii* the costal teeth are sharp. It is similar to *M. thomsonii* but the leaves are broadly ovate, the teeth are much longer, and the plant is synoicous or paroicous. We have no authentic specimens, while *M. thomsonii* is not uncommon.

Plagiomnium rostratum (Schrader) Koponen. Crum and Anderson (1981) attribute the species to Colorado. We find no specimens from Colorado in the herbarium. *Kiener 4946* from LR is *P. ellipticum. P. rostratum* is synoicous, commonly fruiting, and the operculum is rostrate (the only such species).

NECKERACEAE (NCK)

Pinnately branched mosses with complanate stems, and leaves that are usually, but not always transversely undulate. They are delicate plants that grow in deeply shaded overhangs of cliffs and sometimes talus slopes under the loosely piled rocks. They appear to avoid sites that are really wet, but take advantage of mist from low-hanging cloud veils. In our species the costa is lacking or short and faint. The leaf cells are short, unlike other complanate mosses of our area.

Neckera [for N. J. de Necker]

N. oligocarpa. Frequent in deep shaded crevices of rock outcrops and ledges of granitic rocks, ranging widely in altitude from the lower foothills to the subalpine. Surprisingly, it sometimes occurs in what appear to be barren talus slopes. Here it may be found on the undersides of the talus blocks where small 'grottoes' occur in spaces between rocks.

In our material, we find a marked difference in size, length, shape, and undulation of the stem leaves *vis a vis* the branch leaves. In the field it almost appears as if two species are mixed. However, upon dissection of the mat, it is clear that the stem leaves are broadly oblong and truncate, at least three or four times the size of the branch leaves. The branch leaves, on the other hand, are narrowed from base to apex, slightly if at all undulate, those at the base of the branch more so than the gradually smaller leaves of the branch tip. This led us into the error of thinking we had *Neckera complanata*, in which all of the leaves, including the stem leaves, are not undulate. The various manuals do not mention any dimorphism of the stem and branch leaves of *Neckera*, except that some state that the branches may sometimes be flagelliform and probably serve to propagate the plant.

Neckera capsules are surrounded by a long rolled up sheath of perichaetial leaves. Our plants have been called *N. pennata*, but in that species the capsule is completely enclosed. In *N. oligocarpa* the capsule protrudes distinctly from the sheath apex or out through the side. Our specimens are mostly sterile, but a persistent search often reveals some capsules.

Excluded Taxa

N. douglasii Hooker. Listed for Colorado by Grout (1939) but not by Lawton (1971). The report is based upon a specimen at the New York Botanical Garden from the herbarium of E. A. Rau. Two specimens are indicated on the label; in fact there are two pieces, one on each side of the packet. "No.1. Hab. wet rocks, Colorado Springs, *Brandegee*; 2. Hab. ground, Colorado, *Mrs. Spence*." Both pieces (*vidi*) are *N. oligocarpa*.

ORTHOTRICHACEAE (ORT)

1a.   Leaves linear to linear-lanceolate; distal leaf cells roughened by small, elliptic papillae positioned continuously over cell lumina and walls; capsules terminating short setae, with 8 deep ribs and no peristome; forming dense patches in crevices of cliffs. **Amphidium**

1b.   Leaves lanceolate or broader; distal leaf cells with papillae on cell lumina only, or smooth; capsules terminating short or long setae, smooth or 8-ribbed; peristome present; on trees or boulders, usually forming rather loose tufts, widely distributed. **Orthotrichum**

Amphidium [little amphora, a vase]

Recent research places *Amphidium* in the Rhabdoweisiaceae. Strong morphological evidence for this was presented by Norris and Koponen (1999). Supporting molecular evidence by was presented by Stech (1999).

1a.   Leaf margins plane or nearly so; distal leaf cells with large, conspicuous papillae; autoicous, commonly fruiting; perichaetial bracts plainly differentiated, sheathing, the cells smooth. **A. lapponicum**

1b.   Leaf margins usually plainly recurved; papillae on distal leaf cells less conspicuous; dioicous, often sterile; perichaetial bracts not strongly differentiated, the cells papillose. **A. mougeotii**

A. lapponicum [from Lappland]. Common on shaded alpine and subalpine cliffs, easily recognized by the characteristic capsules which are almost always found if one looks hard enough. *Amphidium* is almost always likely to be found intermixed with *Grimmia torquata*, which has much shorter, more strongly crisped and spirally twisted leaves.

A. mougeotii [for J. B. Mougeot]. In this species the leaves are much more laxly organized than in the last, brighter green, and the margins are distinctly recurved. *A. mougeotii* is common on vertical cliffs at low altitudes in the outer foothills. BL, DA, LR.

Orthotrichum [straight hairs (of calyptra)]

It is difficult, without much experience, to identify specimens lacking capsules. Most of the critical characters are in the capsule, peristome and calyptra. The nature of the stomata on the capsule is critical, whether superficial or immersed. There may be only one or two stomata per capsule, and the capsule not only needs to be split and viewed from the dorsal side, but the spores should be washed away so that the tissue is clear. On a strictly local basis it is possible to learn with experience to recognize the species after one has a thorough understanding of them, but some are almost identical vegetatively.

la.   Upper leaves with long serrate hair-points as in *Grimmia*; rarely fruiting; on oak and juniper in southwestern Colorado. **O. diaphanum**

lb.   Leaves lacking hyaline hair-points . . . . (2)

2a.   Leaves thick and glaucous; plants of calcareous cliffs. **O. pellucidum**

2b.   Plants not as above . . . . (3)

3a.   Leaf-margins plane and involute throughout; gemmae scattered on the adaxial surface; minute plants with immersed capsules; on cottonwood and oak at middle altitudes in southwestern Colorado. **O. obtusifolium**

3b.   Leaf-margins revolute for at least a portion of their length; larger plants on rocks or conifers . . . . (4)

4a.   Capsules completely exserted, the seta evident . . . . (5)

4b.   Capsules immersed or only emergent (seta not visible) . . . . (6)

5a.   Capsule long cylindric, smooth; occupying various substrates. **O. laevigatum**
5b.   Capsule ovoid, smooth or ridged; restricted to limestone. **O. anomalum**

6a.   Basal leaf cells elongate, more or less thick-walled and nodose; stomata superficial (the accessory cells on the same plane with the guard cells) . . . . (7)
6b.   Basal leaf cells rectangular, thin-walled and not nodose; stomata immersed (the guard cells situated under the thick-walled accessory cells, whose margins usually conspicuously overlap and hide them, except in *0. pallens* in which the overlap is slight) . . . . (8)

7a.   Capsule smooth or only faintly ribbed; exostome teeth erect or somewhat spreading; endostome never present. **O. rupestre**
7b.   Capsule ribbed the entire length; exostome teeth reflexed; endostome often present. **O. affine**

8a.   Leaves with conspicuous, sharp-pointed simple or forked papillae. **O. alpestre**
8b.   Leaves with low papillae, or papillae lacking . . . . (9)

9a   Plants relatively large (over 1 cm); on rocks; calyptra hairy . . . . (10)
9b.   Plants minute (less than 5 mm); on bark; calyptra naked . . . . (11)

10a.   Capsules with 16 ribs, immersed or slightly emergent; leaves ovate-lanceolate, acute, unistratose or rarely with bistratose streaks. **O. cupulatum**
10b.   Capsules with 8 ribs, more or less emergent, oblong or oblong-ovate; leaves lanceolate to ligulate, obtuse or blunt, bistratose distally. **O. hallii**

11a.   Antheridia inside the perichaetium, not in special lateral shoots just below the perichaetium. **O. pallens**
11b.   Antheridia in small, few-leaved lateral shoots just below the terminal perichaetium that bears the capsule. **O. pumilum**

O. affine [akin to]. This species is attributed to Colorado by Vitt in the BFNA, but we have no Colorado specimens. It approaches closely in southern Wyoming. Evidently it is similar to *O. rupestre* but grows on the bark of deciduous trees, which, if true, would limit its likelihood of occurring here.

O. alpestre [of high mountains]. One of the most common and wide-ranging species, occurring on a variety of rock types and occasionally on trees. It resembles *O. pumilum*, but is somewhat larger, the capsule is almost exserted, and the leaves have sharp pointed conical papillae.

O. anomalum. An infrequent species of the lower outer foothills, restricted to limestone. The capsule is deeply furrowed and exserted on a long seta. *O. anomalum* is also known by its twisted seta, shortly exserted capsules with 8 long ribs alternating with 8 shorter ones. The leaves dark or blackish, rather broad, bluntly acute, rigid, with strongly revolute margins. See discussion under *O. laevigatum*. EP: Williams Canyon, Manitou Springs, 6500 ft, *Weber, Wittmann & Kelso B-112019.*

O. cupulatum. A species of the outer foothills and canyon country, probably restricted to sedimentary rocks. The capsule, while distinctive in shape, is inconspicuous because it is erect or nearly so on a short seta. Other species such as *O. anomalum* are sometimes confused with this but the long and twisted seta and more elongate capsule should distinguish them. BL: NE base of Eldorado Mountain, 6000 ft, *W&W B-111740.* Vitt (1973) cites a specimen from LP.

O. diaphanum [colorless and transparent]. The hair-point of this species would suggest *Grimmia*, which, however, does not occur on tree trunks. The plant is small, like *O. pumilum*. The leaves usually have gemmae. The capsule is minute, immersed, cylindrical, and ribbed, and the calyptra is naked. *O. diaphanum* is common in the southern counties, on junipers and oaks. FM: Phantom Canyon, on *Quercus*, 6000 ft, *W&W B-112147.*

O. hallii [for Elihu Hall]. Probably the most common species on rocks in the outer foothills. It is less so in the montane or subalpine, and occurs commonly with *O. rupestre* and *O. laevigatum* at lower altitudes. With some experience it is fairly easy to recognize in the field. The plants are small compared to *O. rupestre*, the leaves are short, blunt, and the capsules are immersed to emergent, narrow and constricted below the mouth when dry, with eight vertical ridges. The leaves are distally bistratose; most or all of our other species are unistratose. Until recently this was thought to be endemic to the southern Rocky Mountains, but it has now been reported (Tan et al., 1995)from China!

O. laevigatum [smooth]. A common species on boulders in the outer foothills. It can be recognized in the field by its slender, pale, cylindric, smooth or faintly ribbed, exserted capsules as contrasted with the oblong, strongly ribbed ones of *O. anomalum*, which is restricted to limestone, and the lighter green leaves. Most of our collections are from rocks, but it also occurs on trunks of scrub oak in the San Juans. GA: Head of Willow Creek, 9500 ft, *Weber, Wittmann, & Tidball B-112488*.

O. obtusifolium. The most distinctive of all our *Orthotrichum* species. Some authors prefer to segregate it as the genus *Nyholmiella*. It occurs on the bark of *Populus* and scrub oaks. In Colorado, relative humidity is usually too low to support mosses on tree trunks. GF: Rifle Falls State Park, on *Populus, Weber B-115882*; OR: Ouray, road to Camp Bird Mine, on *Populus, W&W B-111901;* SJ: Junction Creek C. G. on old *Quercus, Weber & Jamieson B-113430*.

O. pallens. A rare or infrequently collected species on trunks of deciduous trees in the outer foothill canyons and ravines of outwash fans. A tiny species occurring on bark of small trees in the outermost foothills. Similar to, but somewhat larger than *O. pumilum*, and has a naked calyptra, but the fertile main stem has small, few-leaved antheridial shoots situated just below the perichaetial stem. BL: Rocky Flats, 7 mi S of Boulder, on base of *Crataegus, Kunkel & Shultz B-46172*; DA: Devils Head, *Flowers 9792* (!Vitt).

O. pellucidum [translucent]. A glaucous plant of seeping limestone or limey conglomerate cliffs. The leaves are often encrusted with calcium carbonate. GN: Cement Creek Canyon, *Weber, Wittmann, & Lehr B-113687*; MZ: Mesa Verde; MN: 2 mi below Buckeye Reservoir, *Weber B-8967*.

O. pumilum [small]. An infrequently collected but easily recognizable species restricted to the lower trunks of deciduous trees, particularly *Negundo* and *Quercus*, in the outer Front Range foothills, and frequent in the Western Slope canyon-bottoms. The leaves and capsules are minute (2 mm)! The plants are small, with short, broadly acute or apiculate leaves. The calyptra is naked; the capsules are slender cylindrical, strangulate and 8-ribbed below the apex, emergent (the short seta is hidden in the leaves), and the peristome reflexed. In *O. pumilum* the antheridia are within the terminal perichaetium. See discussion under *O. pallens*. PE: Beulah, 6750 ft, *W&W B-112061*.

O. rupestre [of rocks]. The largest and most abundant species, on large granitic boulders and outcrops from the outer foothills through the subalpine. It is able to tolerate considerable aridity, but the sites in which it occurs are probably indicative of local high temporary humidity rather than liquid moisture.

PLAGIOTHECIACEAE (PGT)

The Plagiotheciaceae in our area may be recognized without much difficulty. We have the single genus, *Plagiothecium*. The stems creep over wood or soil or occasionally ascend when growing in boggy places. The leaves are usually distinctly complanate, often quite asymmetric (one side broader than the other). The costa is absent or short and double, and the leaf cells are elongate-linear. These are forest mosses except for the open ground forms of *Plagiothecium denticulatum* discussed below. The leaves are almost or quite entire.

Plagiothecium [oblique capsules]

1a.   Leaves symmetric or nearly so, typically cochleariform, not complanate or loosely and irregularly so; leaves 1.5–3.0 mm long, often recurved at the tips. **P. cavifolium**
1b.   Leaves asymmetric, flat, distinctly complanate .... (2)

2a. Medial leaf cells long-rhomboid, 10–13 μm wide; decurrent portion of the leaf often auriculate, oval in outline, composed of many inflated, spherical cells in 2–8 vertical rows; capsules striate or wrinkled. **P. denticulatum**

2b. Median leaf cells linear-flexuous, 5–7 mm wide; decurrent portion of the leaf never auriculate, but tapering and triangular in outline, composed of mostly rectangular cells in 1–5 vertical rows; capsules smooth, rarely striate or wrinkled. **P. laetum**

P. cavifolium [cochleariform leaves]. Infrequent on moist rocks and on wet soil along brooks, subalpine forests. The capsules are erect and symmetric, but rarely produced here. BL: Gorge of Como Creek, 9000 ft, on cliff, *W&W B-110628*; LR: Rocky Mountain National Park, Cow Creek Valley, 8200 ft, *Rolston & Hermann 85170 (B-92649)*.

P. denticulatum. Frequent on cliffs, the bases of trees and over rotting logs in the forests from the lower foothills up into the subalpine. It is easily recognized as a *Plagiothecium* by the complanate, strongly asymmetric, leaves that curve outward and often somewhat back.

The capsules are strongly inclined to horizontal, curved and asymmetric. In the alpine zone a robust form occurs in wet tundra and boggy places that is less complanate, with the leaves more crowded and overlapping.

P. laetum [attractive]. Similar to the last, but much smaller, differing distinctly by the characters given in the key. The capsules, when present, are erect symmetric, usually found to be fruiting abundantly. It is relatively rare and seems to be restricted to the spruce forests.

POLYTRICHACEAE (PLT)

Note: From 6a onward, sporophytes and leaf cross-sections are desirable or necessary

1a. Leaves with a white hair-point. **Polytrichum piliferum**
1b. Leaves without a white hair-point .... (2)

2a. Leaves with the margins folded in over the lamellae .... (3)
2b. Leaves with the margins plane, not infolded .... (4)

3a. Lower stems covered with a white tomentum; plants of high mountain wetlands. **P. strictum**
3b. Lower stems naked; plants of various habitats. **P. juniperinum**

4a. Leaves thin, margin double-toothed (as in *Mnium*), the costa narrow and the unistratose laminal surface broad, with only 2-6 surface lamellae; calyptra smooth; capsules elongate-cylindric. **Atrichum**
4b. Leaves thick, the costa wide, the surface lamellae over 20, otherwise not as above .... (5)

5a. Stems fastigiate, with a few branches arising from the base; each stem may bear a sporophyte. **Meiotrichum**
5b. Stems simple; sporophytes single .... (6)

6a. Capsule smooth, perfectly round in cross-section or with up to six indistinct ridges, never clearly 4-angled. **Pogonatum**
6b. Capsule clearly quadrangular, equally so or with the two upper angles closer together than the others . (7)

7a. Capsule perfectly round and smooth, or only indistinctly angular. **Polytrichastrum**
7b. Capsule equally quadrangular, the base with a constriction between the urn and the button-like apophysis; calyptra densely hairy. **Polytrichum**

Atrichum [(calyptra) without hairs]

A. selwynii [Mount Selwyn, Canada]. Without capsules, this might be mistaken for *Timmia*. In the fresh state the broad, thin leaves have that general appearance, and one might miss the few low lamellae, which are rather inconspicuous. When dry the leaves curl inward and are undulate-margined, looking more like an *Encalypta* than one of the Polytrichaceae. Although it has a wide altitudinal range, occurring from the cool foothill canyons under Douglas-fir, to the moister spruce forests of the subalpine, and sparingly in the moister alpine areas, *Atrichum* seems to be rather rigidly restricted to specific micro-sites probably connected with high local humidity.

Mnium hornum is similar, and has a double-toothed margin, but lacks lamellae. The stems are uniformly leafy, the leaves are thin, becoming transparent, with a prominent narrow costa, not strongly crisped when dry. The stems of *Atrichum* are comose, denuded below, the leaves are broader and strongly crisped when dry, dark green and opaque.

Meiotrichum [fewer (calyptra) hairs]

Meiotrichum is common in the dry subalpine forest. Unlike the other genera, the stems are commonly fasciculately branched below. The leaves are more widely spreading than those of most *Polytrichum*, and the capsule, instead of being square-sided, has two of the four ribs closer together. The calyptra is only sparsely hairy.

M. lyallii [for David Lyall]. A common species of the forest floor, in moist slopes under Douglas-fir in the foothills, and along rivulets in open sites in the upper subalpine and alpine. The large plants with long leaves, widely spreading when moist, are distinct and will remind the eastern bryologist of *Polytrichum commune*, which is quite rare here and has unbranched stems. When dry the leaves are more appressed to the stem, with the long tips curving outward.

Pogonatum [bearded]

P. urnigerum [carrying urns]. In Colorado this species ranges from alpine tundra, often occurring in dry shelves under boulders, to montane-subalpine where it may be found infrequently on the forest floor. Whereas *Polytrichastrum alpinum* forms dense sods on peaty soil, *Pogonatum urnigerum* grows scattered, only a few stems together, on more sandy substrates. At Summit Lake it is minute, the leafy portion less than a cm high. The characters of the lamellae may be used, as a last resort, to distinguish them. *Polytrichastrum alpinum* fruits commonly, while we have never found *Pogonatum urnigerum* in fruit here.

Polytrichastrum [resembling *Polytrichum*]

1a.   Terminal cells of lamellae coarsely papillose; capsules terete, long-cylindric to ovate cylindric to suborbicular. **P. alpinum**

1b.   Terminal cells of lamellae not papillose; capsules weakly 4-angled . . . . (2)

2a.   Marginal area of lamina 5–9(–20) cells wide, somewhat incurved when dry; median cells of the leaf sheath rectangular (3–5:1). **P. longisetum**

2b.   Marginal area of lamina 2–5 cells wide, plane; median cells of leaf sheath elongate (5–9:1). **P. formosum**

P. alpinum. A characteristic, usually dwarf, plant of the wet tundra, especially in late snow areas and on frost-push hummocks. It commonly fruits and has short, broad, light-colored capsules. Leaf length seems to vary a great deal depending on the seasonal snow-cover, and striking variation can be seen from one margin of a patch to the center. It differs from *P. longisetum* in that the terminal cells of the lamellae are papillose; in *P. longisetum* they are smooth and rounded.

P. formosum [handsome]. Similar to the next, but an uncommon plant of wet forests rather than open fens. GL: Tolland to Apex, *Grout;* HN: Wager's Iron Fen below the ghost town of Carson, 11300 ft, *W&W B-111973.*

P. longisetum. Restricted to subalpine fens. Fortunately, it frequently occurs in fruit, for vegetatively it is somewhat difficult to recognize in the field. This species shows remarkable environmental plasticity, since when the plants are in wet sites or even seasonally submerged in melt-water, the lamellae may be few in number or even lacking, and the lamina thin and broad. The unistratose margin of the leaf has strong marginal teeth and extends only 4–8 cells beyond the lamellae. The terminal cells of the lamella in face view are perfectly flat and smooth, and the walls are not thickened. In cross section they are higher than wide and rounded on top. In *P. formosum* the terminal cells of the lamella are more or less incrassate.

Polytrichum [many hairs]

1a.   Leaves with a white hair-point. **P. piliferum**

1b.   Leaves lacking a white hair-point . . . . (2)

2a.   Margin of lamina sharply folded, covering the lamellae . . . . (3)

2b.   Margin of lamina not folded . . . . (4)

3a.   Stems densely white-tomentose **P. strictum**

3b.   Stems brown-tomentose only at the base **P. juniperinum**

4a.   Leaves straight or weakly recurved when moist, the blade caducous (tending to be brittle), margins entire or finely serrulate; capsule short-rectangular to cuboidal. **P. jensenii**

4b.   Leaves with a squarrose-recurved blade, not caducous; margins strongly toothed; capsule rectangular. **P. commune**

P. commune. In the absence of fruit, the shape of the terminal lamellar cell is distinctive although it is shared with *P. jensenii*. The species is uncommon in Colorado. BL, CC, GA.

P. jensenii [for C. Jensen]. In iron fens in the San Juan Mountains. This is a fairly short plant with long or short, easily broken leaves. The leaf margin has shorter teeth than those of *P. commune*, and the marginal cells of the lamellae are often less furrowed. In wet parts of the fen they are straight and appressed, but slightly curved

and with spreading tips on higher ground. Capsule-bearing plants are rare, but when they occur, they are short-rectangular. SJ: Chattanooga Fen, S of Red Mountain Pass, *W&W B-111028, B-111029.* This species is known from the Arctic and, heretofore, only from Wyoming in the 'lower 48'.

P. juniperinum. Abundant and widely distributed from the foothills through the alpine. In contrast to the next species, which is equally abundant, *P. juniperinum* occurs in heavier soils with more humus, tends to avoid excessively dry or exposed sites, and tolerates more shade. With a hand-lens, one can see that the margins of the leaves of both this and *P. piliferum* are have their margins strongly infolded. The reddish leaf-tip of *P. juniperinum* is diagnostic. The terminal cells of the lamella, in cross-sectional view, are suddenly narrowed into a thick-walled 'nipple'. The male sex organs are formed in terminal 'flowers'—rosettes of reddish or yellowish perigonial leaves. These are often conspicuous, and in optimum condition the antheridia can be teased or squeezed out of the antheridial head, and when placed on a slide with water the free-swimming sperm can be easily examined under the microscope.

P. piliferum. Abundant on gravelly open ridges and windswept summits, fell-fields and ledges of siliceous gravel, *P. piliferum* is one of the first mosses learned by a beginner. When it forms dense sods with short stems it can be mistaken for *Selaginella densa*, with which it often occurs. Despite its abundance, *P. piliferum* seems to have rather rigid ecological requirements or tolerances.

P. strictum. Subalpine fens, closely related to *P. juniperinum*, but easily distinguished by the dense white tomentum of the stem, and the wetland habitat.

POTTIACEAE (PTT)

1a.   Leaves with superficial outgrowths (lamellae, plates, or chlorophyllose filaments) arising from the adaxial surface of the costa . . . . (2)

1b.   Leaves lacking any superficial outgrowths except papillae . . . . (4)

2a.   Leaves with several erect sheet-like lamellae arising from the costa near the leaf apex. **Pterygoneurum**

2b.   Leaves lacking lamellae but with a dense cluster of erect, green filaments on the upper lamina . . . . (3)

3a.   Leaves thick and rigid, with an obtuse, cucullate apex, hair-point lacking. **Aloina**

3b.   Leaves with a long, hyaline hair-point, the lamina thin-textured, not cucullate or rigid. **Crossidium**

4a.   Capsule cleistocarpous (the lid not differentiated and the capsule rupturing irregularly, always present; ephemeral plants of early spring and summer. **Tortula acaulon**

4b.   Capsule with a differentiated operculum, or the plants lacking sporophytes . . . . (5)

5a.   Leaves broadly ovate, cochleariform, in a tight cabbage-like cluster, the cells not papillose; leaves with or without a hair-point; peristome of 16 undivided teeth; minute alpine mosses. **Stegonia** and **Tortula systylia** (see **Tortula** key)

5b.   Leaves oblanceolate, lingulate or spatulate; peristome divided into filiform prongs, or elongate and spirally twisted . . . . (6)

6a.   Leaves with a hyaline, golden or reddish hair-point often half as long as the leaf or longer . . . . (7)

6b.   Leaves lacking a distinct hair-point, but sometimes short-pointed . . . . (9)

7a.   Larger plants; leaves usually over 2 mm long, usually with well-developed inner basal hyaline 'windows'; cells usually papillose. **Syntrichia**

7b.   Minute species with small (up to 1.5 mm) leaves, without distinct basal hyaline 'windows' . . . . (8)

8a.   Leaves keeled, deeply channeled adaxially at the costa (cross-section V-shaped); costa with one stereid band. **Anoectangium**

8b.   Leaves ventrally broadly cochleariform, plane to only shallowly grooved (cross-section U-shaped). **Molendoa**

9a.   Leaves broad, oblanceolate, oblong or spatulate, often broadest above the middle . . . . (10)

9b.   Leaves narrowed from base to apex, sometimes from a relatively broad base . . . . (14)

10a.   Leaves large (2–6 mm), with inner basal hyaline 'windows'. **Syntrichia**
10b.   Leaves small (up to 2 mm), lacking distinct basal hyaline 'windows', although leaf base may consist generally of rectangular hyaline cells not framed by shorter, green, marginal cells . . . . (11)

11a.   Leaves sheathing at the base, usually coarsely toothed near the apex, sharply mucronate at the apex; lamina bistratose. **Rhexophyllum**
11b.   Leaves not sheathing at the base, entire, not mucronate; lamina unistratose . . . . (12)

12a.   Leaves minute (less than 1 mm long); weedy plants of disturbed sites on the plains; sterile. **Barbula**
12b.   Leaves larger (2 mm or more long); native plants of the mountains; commonly fruiting . . . . (13)

13a.   Leaves red in KOH; peristome lacking; operculum systylious. **Hennediella**
13b.   Leaves yellow in KOH; peristome present or rarely absent; operculum not systylious. **Tortula**

14a.   Leaf margin recurved or revolute . . . . (15)
14b.   Leaf margin plane or involute . . . . (17)

15a.   Leave margins tightly rolled, the inside of the rolled margin packed with larger cells; rare, recently found in Las Animas County. **Pseudocrossidium**
15b.   Leave margins merely recurved; common mosses . . . . (16)

16a.   Minute, densely tufted mosses of calcareous travertine cliffs, the leaves often encrusted with lime; capsules usually present, the urn short and wide; operculum with a long, bent beak. **Hymenostylium**
16b.   Minute or larger mosses not associated with travertine, in loose tufts, usually without capsules; capsules elongate, cylindric; growing in various habitats (see key to **Didymodon** and its relatives, including **Bryoerythrophyllum**)

17a.   Leaves with a basal area of rectangular, transparent cells encroaching upwards along the margin, thus forming a V-shaped contact line between these and the inner and upper chlorophyllose cells. **Tortella**
17b.   Leaves without a distinct V-shaped contact zone . . . . (18)

18a.   Leaf cells with a single high, sharply conical mammilla on upper and lower surfaces giving the costa and margin a toothed appearance, the lamina in cross-section accordion-pleated in appearance due to the mammillae (see **Dichodontium**, in Dicranaceae)
18b.   Leaf cells smooth or only low-papillose . . . . (19)

19a.   Leaves with a few distinct marginal teeth at the 'shoulders' (the junction of the lower elongate cell zone and the upper quadrate cell area). **Eucladium**
19 b. Leaves lacking distinct marginal teeth at the shoulders . . . . (20)

20a.   Leaves obtuse, ribbon-like; densely tufted plants of calcareous or travertine-forming cliff faces. **Gymnostomum**
20b.   Leaves acute or acuminate . . . . (21)

21a.   Leaves entire, neither undulate nor fragile, narrowly involute from base to apex; the basal portion not clearly differentiated into a semi-sheathing base; leaves forming a short basal rosette, the stem never elongate; commonly fruiting; plants of moist rock crevices in mostly sedimentary rocks. **Weissia**
21b.   Leaves irregularly dentate near apex, the lamina undulate and fragile; the basal portion forming a somewhat sheathing base; stems elongate, in erect tufts; sterile; two species, one alpine and one desert-steppe. **Trichostomum**

Aloina [resembling the genus *Aloe*]

1a.   Leaf with a long hair-point; marginal basal cells quadrate, firm-walled. **A. bifrons**
1b.   Leaf lacking a hair-point . . . . (2)

2a.   Leaf base with the marginal cells quadrate, not hyaline or thin-walled. **A. hamulus**
2b.   Leaf base with a few marginal rows of elongate, thin-walled cells. **A. rigida**

A. bifrons [two brows, the allusion unclear]. One locality known: MN: Intimately mixed with lichens on steep slopes of bare gypsum domes, Paradox Valley just NE of Bedrock, with *Didymodon nevadensis, Weber, Kunkel, & LaFarge B-43543.* Not reported for Colorado in the Flora of North America (2007).

A. hamulus [little hook]. A small, brown species with short, broad leaves. Its range in North America is the high plains, but it also occurs in Mexico and Central America. Our only collection is from YM: Bluff overlooking the south bank of Arikaree River, *Denham 81395 (B-80420)* (!Delgadillo).

A. rigida. The minute leaves are thick, rounded-oblong and involute with the edges rolled over the cushion of photosynthetic filaments. When fresh the leaf looks like a wiener bun. It occurs in two widely separated places in specific habitats on arid, sandy bluffs on the eastern High Plains: WL: Pawnee National Grassland, 5000 ft, *Weber & Hermann B-37491,* on bad-land mud-stone colluvium at the base of the Pawnee Buttes, in sites that are only moist for short periods. Usually there are only scattered individual stems intermixed with *Gemmabryum cf. klinggraeffii* and *Pterygoneurum subsessile.* We have tried to find it several times since, without success. Evidently it is an annual that only appears in seasons that achieve some critical moisture balance. In May, 2003, the site showed absolutely no evidence of mosses. It is also abundant locally just at timberline, ST: Gravelly limestone pavements, base of Blue Lake Dam, 11000 ft, *Weber, Wittmann, & Spribille B-111104.*

Anoectangium [open or wide-mouthed capsule]

Anoectangium is a moss with minute leaves and stems (less than 0.5 mm wide) that forms tightly packed erect sods in which the stems may be extremely short. The leaves are costate, narrowly triangular, slightly spreading when moist but not at all curved. The leaf cells are rounded and coarsely papillose; the papillae are so conspicuous along the leaf margins as to make the leaf crenulate. The leaf apex, however, contains three or four cells with clear lumina and no papillae. The leaves immediately turn orange-red with the application of a drop of KOH.

Technically, the genus is distinguished from the similar genera of the subfamily Pleuroweisiae: *Eucladium, Hymenostylium, Gymnostomum,* and *Molendoa,* in having the female gametangia on short lateral branches. In the other genera they are terminal on the main stem. It has not been found fruiting here.

1a.   Stems short; leaves bright blue-green, forming a thin but confertate mat on friable soil in crevices of desert-steppe rimrock. **A. handelii**
1b.   Stems elongate, slender and tightly packed; on limestone cliffs in the upper subalpine. **A. aestivum**

A. aestivum [of summer]. Known in Colorado from one locality: ST: Blue Lake Dam; on limestone cliff above several low terraces, with *Amphidium, W&W B-111183* (!Zander).

A. handelii [for H. Handel-Mazzetti].This is undoubtedly the smallest of all our Pottiaceae, even smaller than *Weissia controversa.* It was described from Kurdistan. Thus far, it has been found at a few stations in the western Hemisphere, in Colorado (Zander & Weber, 2005) and Nevada. BL: Hall Ranch Open Space, 5600 ft, *Weber, Wittmann, & Lehr B-114035;* W slope Steamboat Mountain, 6200 ft, May, 1971, *Hermann* *23589, 23593*; LR: Buckhorn Creek, Stove Prairie road, *Hermann 27016.*

Having rediscovered the species recently we can give a description of its ecology. In May, 2004, Colorado was suffering a severe drought. A lucky few days of heavy rains prompted us to visit a recently-designated Open Space area on the west side of a cuesta in the outermost foothills of the Front Range east of Lyons. Here the slope was less steep than on Steamboat Mountain, where small scraps of the species (then thought to be *Molendoa*) had been collected years ago in cracks of the rimrock. The slopes are covered by *Cercocarpus montanus* and *Opuntia phaeacantha,* with scattered trees of *Sabina virginiana*. A variety of calcareous and non-calcareous sandstones form ledges of red and tan layers.

We discovered that the moisture had made visible a variety of mosses, including *Syntrichia caninervis, Weissia ligulifolia, Encalypta vulgaris, Jaffueliobryum wrightii, Grimmia anodon, Tortula obtusifolia,* and *Crossidium squamiferum*. Here also, occurring adjacent to tufts of *Didymodon rigidulus,* were scattered patches of *D. anserinocapitatus,* which we had found only once before in southern Colorado.

After analyzing most of the collections in the laboratory, we turned to the last one. This was what appeared to be a minute, nondescript sterile pottiaceous moss that formed a dense mat with erect blue-green leaves suggesting a miniature golf putting green. This resisted analysis, but in the American flora it seemed to resemble *Gyroweisia tenuis*. This would have been a remarkable find. We needed more. Ron said that he knew just where it was found, so we returned to the site a week later, where we followed up one of the larger gulches to an area of white rimrock.

The layers of rock were almost horizontal, and separated by deep fissures. The moss grew in the deepest parts of the recesses. In order to reach it, Ron had to extend his arm all the way to the shoulder, holding his knife, and scrape the mat out. We learned a few days later that there were rattlesnakes in the area, a risk we should have appreciated, but we had been concerned only with the possibility of meeting mountain lions, since warning signs were posted nearby.

The critical feature of the microhabitat appears to be the direction of exposure of the rocks, which was such that the opening of the crevices faced southwest. This provides protection from soil erosion and a guarantee of some small amount of moisture throughout the year, permitting the entry of sunlight during the winter when the sun lies low on the horizon. We had noticed this feature in southwestern Colorado earlier, in the Four Corners desert where the soil was always too dry to support *Weissia ligulifolia* and *Entosthodon* elsewhere.

Barbula [a beard, alluding to the peristome]

Our species of *Barbula* are minute mosses growing in sand or gravel in disturbed areas. The stems are short, almost lacking. The leaves are less than 1 mm long, and when dry they shrivel up and are hardly visible. When moist they are apple-green, oblong, arranged in rosettes, and they spread out widely. The two species are relatively easily separated by the color of their setae, but in our area we have not yet found capsules. The key given here is a combination of that provided by Crum and Anderson (1981) and by Nyholm (1954-1969). Nyholm does not distinguish the two leaf shapes. In fact, the European specimens we have examined of *B. convoluta* clearly show similar apiculate leaf apices. Nyholm mentions that the tufts (stems) of *B. convoluta* are tomentose (with rhizoid masses) whereas in *B. unguiculata* the tufts have only few rhizoids. All of our material has well-developed rhizoidal masses.

1a.   Seta yellow; perichaetial leaves larger than the vegetative ones, sheathing, truncate to rounded or broadly pointed; vegetative leaves with broad to rounded-obtuse tips, the margins slightly revolute toward the base; stem with few rhizoids. **B. convoluta**

1b.   Seta reddish-purple; perichaetial leaves not sheathing; leaves terminating in an apiculus, the margins usually more conspicuously revolute in the lower half or more; stem with copious rhizoids. **B. unguiculata**

B. convoluta. Probably common, a weedy species rarely seen fertile. BL: Wall of old Business School, U. of Colorado (fertile), *W&W B-114735*; LO: Moist, shaded depression under cottonwoods on south bank of South Platte River, 3700 ft, 1.5 mi S of Crook, *Hermann 29088* (!Zander).

B. unguiculata [clawed]. A somewhat weedy species of compact soils of flood plains at low altitudes, volunteering in lawns. The leaves usually have a slight apiculus and the leaf tip curves slightly inward. The upper and median cells are quadrate and multipapillose, and the cells of the basal part are rectangular and clear.

Bryoerythrophyllum [red-leaved moss]

1a.   Plants forming a confertate turf, with multicellular gemmae attached to the rhizoids; never bearing sporophytes; alleged calciphile in the alpine tundra. **B. ferruginascens**

1b.   Plants forming loose tufts, lacking rhizoidal gemmae; commonly bearing sporophytes; widespread, but most common in the foothill canyons. **B. recurvirostrum**

B. ferruginascens [becoming rusty]. A rare species of the high alpine tundra, occurring only sterile and so nondescript as to be overlooked except by specialists. The vital characteristic for certain identification is the presence of multicellular gemmae attached to the rhizoids. There are several slender turf-forming Pottiaceae in the Blue Lake area of Summit County. Superficially they look much alike, all having minute (<1mm), broadly oblong-lanceolate leaves with essentially quadrate cells. *Bryoerythrophyllum* has distinctly multi-papillose cells and a strong, well-differentiated red costa reaching the apex. *Molendoa* has the same leaf shape but the cells are indistinctly papillose (one papilla per cell), this broad and low, and the costa is neither red nor is it very distinct. In *Anoectangium* the leaf cells are rounded and coarsely papillose; the papillae are so conspicuous along the leaf margins as to create a crenulate effect. The leaf apex, however, contains three or four cells with clear lumina and no papillae. The leaves turn orange-red with the application of a drop of KOH. ST: Blue Lake, 3000 m, *W&W B-111194.*

B. recurvirostrum. Unlike the other *Barbula* relatives, this species tends to grow commonly on the forest floor, especially along damp gullies in the outer foothills. The plants often reach 1 cm high and are rusty-tinged, all but the youngest leaves being reddish. The leaves are lanceolate from a sub-sheathing base, and the lamina curves out from the base so that the leaf cannot lie flat. The blades are irregularly curled, with a prominent costa and revolute margins. The leaf apex usually has a few coarsely-toothed hyaline cells. The upper leaf cells are quadrate, papillose with C-shaped papillae, contrasting with the elongate-rectangular hyaline cells of the sheath. Plants commonly bear elongate, erect, straight, reddish capsules resembling those of *Tortula*.

Crossidium [a small fringe or tassel]

1a.   Distal leaf cells not incrassate or differentiated from the median; laminal filaments low, hardly visible without making a cross-section; terminal cell of filaments sub-spherical; leaf cells (often faintly) papillose. **C. aberrans**
1b.   Distal leaf cells incrassate, with narrow or obsolete lumina; filamentous brush well-developed; terminal cells of filaments cylindrical or conical, thick-walled and with a few blunt papillae; leaf cells usually smooth. **C. squamiferum**

C. aberrans. Uncommon on *Sarcobatus* alkali flats in the western tier of counties. This is a difficult identification to make because the 'filaments' are usually only one cell high, and hardly create a brush-like distal costa. One first has the impression of a richly fruited *Tortula*, but the leaves are small, ovate, convex and incurved, and the cells practically smooth (somewhat mammillose in cross-section), arranged in a 'cabbage-head'. Observation of the abaxial side of the costa in the distal end of the leaf will show some protruding spherical cells, but the costa is little enlarged by these filaments. Besides, their occurrence and development is variable. The peristome is definitely that of a *Tortula*, as is the slender cylindric capsule. It grows intermixed with masses of *Pterygoneurum subsessile, Syntrichia caninervis,* and *Tortula guepinii.* ME: Gunnison River Valley 3 mi S of Whitewater, *Weber, LaFarge, & Kunkel B-43370.* This was thought to be an American endemic. It is now known from Asia (Bai 2002).

C. squamiferum var. **pottioideum**. BL: Sandstone ledge, Steamboat Mountain, 2 mi NW of Lyons, 6000 ft, *Weber B-5520*; MZ: Jct. of Yellowjacket and McElmo Creeks, 4800 ft, *Pursell 3230 (B-65337).* Elsewhere in America it has been found in Arizona and California.

Didymodon and relatives

1a.   Leaves, the lower at least, with a distinctly reddish color . . . . (2)
1b.   Leaves green, brown, or blackish . . . . (4)

2a.   Leaves semi-sheathing, bent, and not lying flat on the slide; apex usually denticulate; upper leaf cells strongly papillose with C-shaped papillae . . . . (3)

2b. Leaves not sheathing, lying flat when dissected; apex smooth; leaves mammillose but never strongly papillose, with beautifully golden cell walls; leaves spreading widely when moistened. **Didymodon asperifolius**

3a. Common and usually fertile; plants of forest floors and streamsides in the middle altitudes, forming loose tufts with distinctly reddish lower leaves; gemmae lacking. **Bryoerythrophyllum recurvirostrum**

3b. Rare, always sterile; minute plants of tundra slopes, forming dense tufts; multicellular gemmae occurring on the rhizoids. **B. ferruginascens**

4a. Leaves oblong, with rounded, blunt, or merely apiculate apex. **Barbula**

4b. Leaves with pointed, often acuminate tips, or if blunt, then with an ovate outline, broadest at the base. **Didymodon**

Didymodon [divided teeth]

Most of our species of *Didymodon* can be nicely sorted out by habitat. Alpine wetlands, on calcareous rock and travertine: *D. subandreaeoides*; on granite gravel, alpine: *D. asperifolius;* granite cliffs, montane: *D. vinealis, D. tectorum;* vertical cliffs by waterfalls: *D. tophaceus*; sandstone rimrock: *D. anserinocapitatus, D. brachyphyllus, D. rigidulus*; gypsum deposits: *D. nevadense.*

1a. Leaves blunt or merely acute, ligulate; plants of saturated places, especially on calcareous mud or travertine. **D. tophaceus**

1b. Leaves acute or narrower; plants of dry places or seeping cliffs .... (2)

2a. Leaf apices caducous or fragile; plants of dry, usually sedimentary, rocks .... (3)

2b. Leaf apices intact or only occasionally broken; plants of wet areas or seeping cliffs .... (4)

3a. Leaf apices swollen, falling as a propagulum. **D. anserinocapitatus**

3b. Leaf apices not swollen, usually evenly narrowing; small, few-celled gemmae commonly in the leaf axils. **D. rigidulus**

4a. Plants reddish to black; leaves ovate-lanceolate, 1–3 mm long with a distinct broad hyaline base, the lamina recurved when moist but appressed when dry, the leaf cells incrassate, the upper ones reddish, with rounded lumina. Wet gravels in the alpine tundra. **D. asperifolius**

4b. Plants usually green, if black, then with small leaves and on dry terrain .... (5)

5a. Plants black, resembling a species of *Andreaea*; leaves minute, not keeled, not recurved, the margins finely crenulate by bulging cell walls. Rare, on irrigated limestone terraces, mostly at high altitudes. **D. subandreaeoides**

5b. Plants not resembling *Andreaea;* leaves larger, sometimes keeled and/or highly recurved, the margins usually entire or dentate but not minutely crenulate .... (6)

6a. Leaf cells sharply papillose with forked papillae. **D. ferrugineus**

6b. Leaf cells with no or only low, rounded papillae .... (7)

7a. Leaves long-lanceolate, the leaves when wet disposed like a pinwheel; leaf base generally short and gradually narrowing to the apex, weakly recurved to revolute below and plane to narrowly recurved above to near the apex; propagula almost always absent. Seeping granite cliffs. **D. vinealis**

7b. Leaves deltoid to short-lanceolate or ovate, margins recurved or revolute to near apex; propagula sometimes present. Plants mostly of arid ground .... (8)

8a. Costal section showing thin-walled adaxial epidermal cells, remainder of costa thick-walled; costa blunt apically, wider at mid-leaf than below, with bulging adaxial surface forming a long-elliptic unistratose pad of cells; guide cells in 2(–3) layers; leaf margins loosely revolute; tubers occasional on the larger rhizoids. Restricted to gypsum soils. **D. nevadense**

8b.   All cells of costal section about equally thickened; costa often with an apical conic cell or costa short-excurrent, gradually narrowing distally, the adaxial surface nearly flat and not forming a wide pad of cells (but costa occasionally thickened and bulging adaxially); guide cells usually in one layer; leaf margins recurved to tightly revolute; small spherical gemmae often present in leaf axils .... (9)

9a.   Leaves deltoid to deltoid-lanceolate, 1.0–1.5 mm long; base oblong; apex straight or somewhat reflexed; costa excurrent from obtuse apex as several-celled mucro. On seeping granite cliffs. **D. tectorum**

9b.   Leaves ovate or ovate-lanceolate, 0.7–1.0(–1.5) mm long; base ovate or weakly differentiated; apex cucullate or weakly cochleariform; costa percurrent or just sub-percurrent in the obtuse or broadly acute apex, often apiculate by 1–3 smooth cells. On dry soils, plains and upper subalpine. **D. brachyphyllus**

D. anserinocapitatus [goose head]. Recently described from western China. Resembling *D. rigidulus* except for the swollen and easily broken leaf apex. Until recently the only other American collection is from New Mexico: SM: Along the Pecos River, 21 Oct. 1939, *Richards & Drouet 456*. See Zander & Weber (1997). Very rare in Colorado. FN: On sandstone cliffs, mouth of Phantom Canyon, *Weber, Hermann, Feddema B-3752*. After the first collection, several subsequent unsuccessful visits were made to Phantom Canyon. However, we since have discovered a 'mother lode', BL: Hall Ranch Open Space, SW of Lyons, rimrock of Lyons Formation, *W&W B-114031*. It is evidently always associated with *D. rigidulus* and may be much more common and widely distributed in the West than supposed.

   D. anserinocapitatus often occurs side by side with *D. rigidulus* and can be recognized in the field by its more tousled appearance, the leaves sticking out at odd angles to show the swollen apices. It also has a greasy-black color, the stems are slender, weak, and short. The clumps also tend to separate into small blocks, *D. rigidulus* tends to have the leaves more strict and appressed to the stem, and its color is a matte black.

D. asperifolius. A characteristic species of snow melt basins in the alpine tundra, occurring in loose mats, loosely attached to sandy gravels in periodically inundated meltwater rills. It is recognized by its handsome reddish coloration and its leaves which are recurved from the stem when moist; when dry the leaves are appressed to the stem. Under the microscope the leaves are particularly beautiful. The cells are incrassate, the upper quadrate with rounded lumina, the lower more rectangular but with rounded edges. The cells are chlorophyllose, often with an accumulation of oil drops. The combination of copper-gold and green, and clearly visible cell structure is most attractive. The leaves are narrowly revolute, ovate at the base.

D. brachyphyllus [short-leaved]. Distinguishable from *D. vinealis* by its ovate leaf shape, margins strongly recurved or revolute to near the apex, and reproduction by small axillary, spherical gemmae when sterile. It differs from *D. rigidulus* in having leaves not at all attenuate but triangular-ovate, margins recurved from base to apex; there is a deep adaxial groove, and the basal cells are mostly clear, not green. Altitudinal distribution ranges from the plains to the subalpine, and substrates from granite to sandstone. BL, WL, MZ, ST.

D. ferrugineus [rust-colored]. An infrequent species, smaller than *D. vinealis* but with leaves appressed when dry, out-curved when moist but not in pinwheel fashion. The papillae on this species are sharply pointed. JF: Clear Creek Canyon, on calcareous schist, *W&W B-110836*; SJ: Near Silverton, *Hermann 23292*; LR: Rocky Mountain National Park, Cub Lake Trail, *Weber & Evenson B-112424*. For further discussion see Crum & Hall (1995).

D. nevadense. A recently described species restricted to gypsum-salt domes. Zander's (1995) (not in bibliography) discussion follows. "*Didymodon nevadense* differs from its apparent closest relative, *D. brachyphyllus*, which is also in sect. Vineales, by the costa usually lacking an apical conical cell, the costa wider at mid-leaf than below with a strongly bulging ventral surface forming a long-elliptical unistratose pad of cells, guide cells in 2–3 layers, ventral superficial costal cells usually thicker through than wide (as seen in section), and the leaf margins loosely revolute."

   This species forms dense patches only 1–2 mm high. The leaves are densely packed and appressed, brown or reddish, broad and, short, with a broad costa. Thus far, we have found it only in Paradox Valley, where it grows among crustose lichens with sparse stems of *Aloina*. MN: Just NE of Bedrock, *Weber B-43543, B-43544*.

D. rigidulus. Common on boulders and ledges in the lower altitudes throughout Colorado, forming loose mats on periodically irrigated rocks. The leaves are long and narrow, from a base that is not much wider, and semi-sheathing. On ledges of calcareous sandstone in the outer foothill hogbacks it is a blackish plant. Elsewhere it is usually dull green. From *Didymodon asperifolius* it differs in never having a red coloration, never occurring in the snow-flush habitat, and in having smooth, not mammillose or papillose cells. From *B. vinealis*, it has generally narrower leaves and non-papillose cells. Forms with short, broad leaves may be mistaken for *D. brachyphyllus*, but there is no ventral groove in the leaf, the basal cells are green and quadrate and the leaves are not revolute from base to apex.

D. subandreaeoides [like *Andreaea*]. Forming black mats with the aspect of a species of *Andreaea*, it can be minute, only about 5 mm high. The species forms minute clumps on drying travertine, but on irrigated limestone slopes it forms extensive black mats. Known in Colorado only from a few localities: GN: Cement Creek travertine cliffs, *W&W B-113672*; SJ: Cunningham Gulch, *Jamieson 10922*; ST: Blue Lake, 11700 ft, *Weber & Anderson B-34249, W&W B-111141, B-111183, Spribille 9908*.

D. tectorum [of roofs]. Zander & Ochyra (2001) write: "*Didymodon tectorum* is an East Asian-North American species. The stems are extremely slender and the leaves short and broad, appressed when dry and merely spreading somewhat when dry. In common with *D. nevadense*, the leaves are short and broadly ovate. Its habitat is on vertical granite cliffs. The main center of its occurrence is on mainland Asia in eastern China."

BL: Gregory Gulch between Flagstaff and Green Mt., *Weber B-10580* (!Zander). A specimen from YM: 3.2 km S of Bonny Reservoir *(Hermann 23631)*, cited by Zander, appears to belong to *D. brachyphyllus*.

D. tophaceus [of tufa or travertine]. Common on wet faces of small check-dams and other irrigated calcareous sites, and in the spray area of waterfalls, generally at low elevations on sedimentary rocks in the canyons. The leaves are rather distant on the stem and are bowed out and curved in toward the stem when dry. The broad, blunt leaves with recurved margins and cells irregularly quadrate from base to apex covered with small, blunt papillae, are diagnostic.

D. vinealis [of vine-lands]. An abundant moss on N-facing cliffs. From the other *Didymodon* species it is distinguished by the bright green leaves which, when moist, curve around, suggesting a pinwheel. Zander (*in litt.*) writes: "The laminal cells have reddish walls in KOH under high power and there is a light-colored short groove right up near the leaf apex on the ventral side of the leaf."

Excluded Taxa

Didymodon australasiae (Hooker & Greville) Zander. This is reported for Colorado in the Flora of North America (2007). The basis is evidently a single specimen, LR: W of Fort Collins, *Hermann 25668, B-45249*, (! Lawton). Zander (*in litt.*, 1976) examined this material in depth and decided that it is *D. rigidulus*.

Eucladium [with straight stems]

E. verticillatum [whorled]. An uncommon species of limestone seeps or calcareous sandstone cliffs in the plateau and canyon country of western Colorado. It forms rounded tufts along bedding planes and may become encrusted with lime. It is usually sterile but is recognized by the distinct teeth on the leaf margins at the leaf shoulders, and where the basal hyaline area meets the distal half of the leaf, the cells are quadrate and papillose. ME: Colorado National Monument: Head of Serpent Trail, on sandstone cliff, *Weber & Khanna B-12300*; MZ: Sand Canyon, 1 mi N of McElmo Canyon, alcove of Navajo sandstone, *Colyer B-100790*.

Gymnostomum [naked, lacking a peristome]

G. aeruginosum [coppery blue-green]. Common on cliffs in the western foothill and canyon country on calcareous sedimentary or metamorphic rock, forming dense tufts that become encrusted with lime. Stems are slender and tightly packed together; leaf margins are plane or slightly incurved; leaves are narrow and ribbon-like with a well-differentiated, papillose distal half of quadrate cells about 6-8 μm diameter, and a base with elongate-rectangular cells. *Gymnostomum* can be confused with *Eucladium* and *Hymenostylium*, which grow in similar sites. *Eucladium* always shows a few distinct marginal teeth at the leaf shoulders and has narrower leaves with the costa occupying almost a third of the width. The upper leaf cells of *Eucladium* are more irregular, with mixed rectangular and quadrate cells. The leaves of *Gymnostomum* are fragile, flat, and irregularly crisped or crumpled when dry, while those of *Hymenostylium* are strongly involute, longer and more rigid, stiffly curving over each other at the stem tip; on careful observation, the margins of moist leaves are distinctly recurved. *Gymnostomum* is usually found in fruit, while *Eucladium* does not fruit here.

Hennediella [for R. Hennedy]

H. heimii [for E. L. Heim]. This is keyed under *Tortula*, but it is quite distinct although rather difficult to place in a key. The leaves are broadly lanceolate, dark green (in *Tortula* the leaves are usually glaucous-green), becoming straw-colored in age, coiled, not strongly tapering to the apex, the margins are inrolled. The proximal cells are rectangular and clear, the distal ones quadrate and coarsely papillose; there is a narrow border of clear or less papillose cells distally and the leaf apex is slightly irregular, with one or two clear terminal cells. The capsule is most diagnostic, being long-cylindric, lacking a peristome, and with a systylious operculum. The leaves turn red in KOH (yellow in *Tortula* and *Syntrichia*). We have a single collection, verified by Zander. GF: At the base of travertine cliffs at Rifle Falls, *Weber et al. B-46266* (!Zander).

Hymenostylium [slender style (beak)]

H. recurvirostrum. On calcareous or travertine cliffs in the middle altitudes of the canyon country. The plants often form tight tufts impregnated with lime. The leaves are linear-lanceolate with slightly recurved margins. Usually found fruiting and easily recognizable by the short seta (2 mm), short, broad urn (1 mm) with flat operculum provided with a slender beak, bent to the side. The leaves, when dry, are curved over the imaginary stem axis in a characteristic way. The recurved leaf margin will distinguish sterile material from *Gymnostomum*, which grows in similar habitats.

Molendoa [for L. Molendo]

M. sendtneriana [for O. Sendtner]. A tiny species making low confertate tufts on calcareous sandstone or conglomerate rock, in desert-steppe rimrock and alpine cliffs. When wet, it has a somewhat glaucous aspect, resembling a tiny *Weissia*, but the leaves are shorter and broader, about 0.5 mm long; in drying they are appressed to the stem rather than coiled, and the margins are distinctly revolute. The leaf is unistratose, and the median cells rounded-hexagonal, 6–8 μm. The costa is rather stout and sub-percurrent. The leaves have 1–2 low papillae, but the cells are not obscured.

We have this from a few widely separated sites: BL: Steamboat Mt., NW of Lyons, on sandstone rimrock, 6000 ft, *W&W B-110664*, DA: Castlewood State Park, *W&W B-114928;* SJ: Silverton, *Hermann 24408*; ST: Blue Lake, on limestone terraces, 10000 ft, *W&W B-111191*.

Pseudocrossidium

P. replicatum [turned back upon itself]. Plants forming dense turfs on desert soils. The stems are about 1 cm long, the leaves lanceolate, loosely contorted, dull. The tightly revolute margins are easily seen with the hand lens. In cross-section the leaves resemble the capitals of ionic columns; they are unistratose, low-papillose, and the costa has a very prominent abaxial steriod layer. No other genus can be confused with it. July 4, 2007The range extends southwards through New Mexico, Mexico, Central America, and via the Andes, South America. It also occurs on the Galápagos Islands. LA: North-facing lip of Tobe Canyon, in a sandstone cliff. Sporophytes abundant, but capsules immature, June 2007, *Dina Clark B-116472*. Not reported for Colorado in the Flora of North America (2007).

Pterygoneurum [winged vein]

Like *Crossidium*, *Pterygoneurum* is easily recognized by the superficial outgrowths on the leaves. These, in *Pterygoneurum*, are a few chlorophyllose lamellae arising from the costa. They are narrow at the base of the leaf, only one cell high, increasing toward the leaf apex so as to resemble extra leaf-laminae. These lamellae are toothed and irregular distally, and sometimes have protonema-like projections. *Pterygoneurum* is an inconspicuous moss of semidesert areas, but it occurs on favorable sites on the Great Plains as well. The plants form bulb-like shoots, the leaves being broad, brown or pale, deeply cochleariform, and usually provided with long hair-points which so dominate the plant as to hide the leaves from view. Its active season is in the winter and early spring. Ecologically, *Pterygoneurum* seems to be sensitive to soil instability and is absent from sites where there is much soil movement.

1a.   Sporophyte immersed to emergent; calyptra small, mitrulate, shorter than the capsule and splayed out on the sides; leaves with long and slender, serrulate hair-points, concealing the leaves. **P. subsessile**

1b.   Sporophyte exserted on a seta longer than the capsule; calyptra cucullate, often longer than the capsule; hair-points shorter and stouter, smooth, the leaves usually evident. **P. ovatum**

P. ovatum. Colonizing loose soils derived from sandstone or limestone, desert-steppe areas on the cuestas of the outer foothills of the Front Range, and abundant in the canyon country. BL, MR, MZ

P. subsessile. Abundant at low altitudes on both the Eastern and Western slopes. Usually the hair-points dominate the appearance, the leaves often imbedded between the surrounding sand grains.

Rhexophyllum [fragile leaves]

R. subnigrum. Superficially this has the appearance of a *Syntrichia*, but without a hair-point. Leaves dark green, involute and incurved when dry, recurved when moist, with a reddish sheathing base; margin plane, coarsely serrate distally; costa with a few red-walled adaxial stereids, few if any abaxially; coarse low papillae mostly evident on adaxial surface. A monotypic genus of Mexico, South America, Arizona, New Mexico, and Texas. Rhexophyllum reaches its northern limit in North America in southern Colorado. EP: Williams Canyon, on vertical faces of a granite tor at head of trail, with *Grimmia, Porella, Leptogium,* etc., *Weber, Wittmann, & Kelso B-112129* (!Zander).

Stegonia [roof or cover]

S. latifolia. A tiny plant common on alpine tundra, mostly on dry, gravelly sites. Easily recognized by its broadly ovate, cochleariform leaves that are aggregated into a cabbage-like head. Otherwise the plant resembles a small *Tortula*. The common form has rounded or only short-apiculate leaves, but a variant, forma *pilifera*, has long hair-points. This might be confused with *Tortula systylia*, and is keyed under *Tortula*.

Syntrichia [united teeth]

1a.   Gemmae present, either near the costa or in the axils of the upper leaves . . . . (2)

1a.   Gemmae absent . . . . (3)

2a.   Large pale fusiform gemmae present in a dense cluster at the stem apex, surrounded by the terminal rosette of leaves; costa smooth abaxially, usually with a shiny surface. **S. pagorum**

2b.   Gemmae present, forming masses of minute spherical, few-celled bodies attached to the ventral face of the costa; leaf cells with a single central papilla on the abaxial face. **S. papillosa**

3a.   Costa percurrent, or excurrent only as a short mucro or apiculus, never a conspicuous awn . . . . (4)

3b.   Costa excurrent as a long or short awn . . . . (7)

4a.   Leaves large (3–5 mm long), distinctly reddish-sheathing, the leaves not fragile, the sheaths persistent and conspicuous on the stem. **S. norvegica**, awnless form.

4b.   Leaves smaller, hardly sheathing; apex muticous or nearly so . . . . (5)

5a.   Costa not reaching the leaf apex; leaves 1.5–2.0 mm long, with a 'pinched' and reflexed apex when dry. Rare plants of limey sandstone crevices. **S. cainii**

5b.   Costa reaching the apex or shortly excurrent; leaves over 2 mm long . . . . (6)

7a.   Hair-point short and straight, red; leaves plane, not squarrose; plants forming neat hemispherical patches on boulders, especially of limestone. **S. sinensis**

7b.   Hair-point long, often wavy, hyaline; leaves recurved or revolute, erect or squarrose . . . . (8)

8a.   Leaves short, broadly ovate, bistratose; stereid layer hardly differentiated, all dorsal cells of the costa with large lumina and relatively thin walls. **S. caninervis**

8b.   Leaves longer, ovate-lanceolate or lance-oblong, unistratose; stereids always well-developed . . . . (9)

9a.   Leaves erect to spreading, not squarrose-recurved when moist, usually not over 2.5 mm long; hair-point fine and fragile, easily detached . . . . (10)

9b.   Leaves distinctly squarrose-recurved when moist, over 3–4 mm long . . . . (11)

10a.   Leaves small, 1.5–2.0 mm long excluding the hair-point; one leaf margin slightly puckered in upper third; costa pale; cross-section showing only 1–2(3) layers of stereids. **S. virescens**

10b.   Leaves larger; leaf margin not puckered; costa red; stereids in more than 3 layers, forming a strongly bulging mass in cross-section. **S. calcicola**

11a.   Papillae low, C-shaped . . . . (12)

11b.   Papillae high, branched, in section (upper leaf cells) the papillae arising from a conical hollow extension of the cell, branched as small antlers, equaling the cell lumen in height. **S. papillosissima**

12a.   Hair-point reddish at base; leaves more or less appressed to the stem when dry, showing the stem usually reddish from the overlapping reddish leaf sheaths; leaves weakly squarrose when moist, with margin recurved from the base to half the leaf length, plane in the upper part. **S. norvegica**

12b.   Hair-point hyaline; stems not conspicuously reddish, the leaves densely clothing the stem and the sheaths not usually visible; leaves strongly squarrose when moist, the margin recurved along the entire length. **S. ruralis**

S. cainii [for R. F. Cain]. This is the smallest of our species of *Syntrichia*. Although Crum and Anderson (1981) describe the plant as being 1–2 cm high, ours have stems only a few mm high. As the authors say, it totally lacks hair-points, and the costa disappears below the leaf apex. The abruptly narrowed, pinched, and reflexed leaf-tips are diagnostic. This can be seen well even in dry specimens. The small leaves are more or less folded and curved around an imaginary axis. The costa is distinct as a light colored midline.

Up until now *S. caini* was only known from a few collections in Ontario, where it grows on limestone. This rare species is known in Colorado from two collections: BL: S-facing slopes of Steamboat Mountain (calcareous Lyons sandstone), in horizontal crevices, *W&W B-110664* (!Zander); and Lake Mesa, south edge of Marshall, in crevices of Laramie sandstone rimrock, 5600 ft, *W&W B-113850*. Mishler (2007) suggests that Colorado records of *S. cainii* are misidentifications; however, he has not examined our material.

S. calcicola [calciphile]. This species is smaller than *S. ruralis*. The leaves are 2.0–2.5 mm long, little if at all recurved when moist, the leaf apex is truncate, broadly rounded, or emarginate, and the hair-point is similar to that of *S. ruralis* but slender and fragile. It has been called, incorrectly, *S. intermedia*. We are following Gallego (2006) in her treatment of the *Syntrichia* species of Spain. We have found it best developed on accumulations of silt covering granitic or metamorphic rocks in the oak belt of southern Colorado and on subalpine slopes in the San Juans. It forms dense and neatly compact mounds.

S. caninervis [hoary costa]. A typical species of the desert-steppe, most abundant in the western counties. It occurs commonly at the base of shrubs, where the snow tends to lie the longest in winter. A few records come from the Eastern Slope on sandstones at the base of the Front Range. The tufts are small and almost black. The bistratose leaves and prominent, coarsely papillose, pruinose costa are diagnostic features. First described from Tibet, it occurs throughout southwest Asia.

S. norvegica. Common on moist rock outcrops in the foothills, usually on granite, as well as saturated ground under willows in subalpine fens. While *S. ruralis* may occur in the habitat of *S. norvegica*, the latter never occurs on the desert soils in which *S. ruralis* is often abundant. In *S. norvegica* the leaves are less tightly grouped on the stem, and the pinkish leaf-sheaths are obvious. The hair-point is typically reddish quite far up on the hair-point, but in *S. ruralis* only the base of the hair-point is often distinctly reddish. A muticous form is frequent on cliff habitats.

Mishler writes (*in litt.*), "*S. norvegica* is much more mesic than *S. ruralis*, which does overlap in altitudinal range, but when you see *ruralis* at those higher elevations it is on rock out in the sun. And the plants are really quite different. *S. norvegica* is usually smaller and always softer than *ruralis*. The most obvious difference under the hand lens is that, while *ruralis* leaves fold up and twist around the stem evenly, *norvegica* leaves twist at the base but sort of crumple up (like a fried rasher of bacon) at the top, because the supportive stereid layer disappears there. The red hair-point is a good character for *norvegica* but not infallible because *ruralis* has statistically a red hair-point too. Under the microscope *S. norvegica* has statistically larger cells, and margin not recurved in the upper part of the leaf."

S. pagorum [of a town or district]. Frequent on moist cliff faces in the foothills of the Front Range. Even when not obviously gemmiparous the key can be used to distinguish the species satisfactorily. When gemmiparous the gemmae are large and visible with a hand lens.

S. papillosa. Frequent on moist cliff faces in the outer foothills of the Front Range, and on bark of oak saplings in the canyons of the Arkansas Divide. This minute species forms loose colonies, the short stems comose rather than uniformly leafy. In *S. papillosa* the leaves are dark, not revolute, sometimes constricted below the middle, rounded at the apex, and abruptly short hair-pointed. Spherical 3–10-celled gemmae are always present in large numbers on the adaxial leaf surface near the leaf apex along the costa, and when dry the costa is distinctly grayish and papillose even under the hand lens. The papillae are solitary but are not easily visible except in profile view along the leaf margin. *S. papillosa* and *S. pagorum* occur almost exclusively on trees in most other parts of the world, but in the exceedingly low humidity of the central Rockies they occur in the compensating environment of cliff faces with a humid microclimate. At high altitudes in the desert ranges of New Mexico the humidity is high due to Gulf air influence; there *S. papillosa* again occurs on trees. *S. papillosa* reproduces exclusively by gemmae except in

Australia where it produces sporophytes abundantly.

S. papillosissima. A species of deep soils along the rims of arroyos and under desert shrubs in western Colorado and Utah. It is widely distributed in southeast Asia and Mediterranean Europe. The species is taller than *S. ruralis*. The branched, antler-like papillae are unique in the genus.

S. ruralis. Probably one of the most abundant and ubiquitous mosses in Colorado, occurring from soils in the arid steppe up through the mountains to the alpine tundra, usually on soil but also on rocks. Its beautifully recurved leaves (when moistened) earn it the local name of 'star moss'.

S. sinensis [for China]. Frequent in the southern outer foothill canyons of the Front Range, especially on limestone. This was first collected in Colorado in EP: Williams Canyon, *Jewett* in 1913 (!Mishler). The plant was rediscovered by *Tass Kelso (B-111807)* along the old abandoned road to Cave of the Winds, where it is common. We now have it from BL: Bear Canyon, *Weber, Wittmann, & Mazurek B-112180*; FN: Phantom Canyon, *Weber & Kelso B-106231*); JF: Deer Creek, *Carmer B-43447*.

The following notes on this species were made from examination of fresh specimens: Leaves spreading widely when moist but not recurved, congested in the dry state. Evidently the older leaves are so fragile that only a terminal rosette of dark green mature leaves remains, and a cluster of small light green leaves is visible at the center. Plants are about 5 mm high, densely foliate; leaves somewhat constricted in the middle, revolute proximally; costa percurrent or slightly excurrent into a stout, short, red or brown point; leaf apex rounded or truncate; lamina fragile; median cells 10–15 μm, multi-papillose with C-shaped papillae. Leaf base fairly short, not well defined.

S. sinensis, first described from China, occurs in southern Europe, the Russian Altai, Japan and Afghanistan. Morphologically, *S. sinensis* and *S. bartramii* appear to be very similar, but according to Gallego (2006), *S. sinensis* is autoicous and *S. bartramii* appears to be dioicous and grows on trees. However, we have one tuft of the Williams Canyon material that has one mature sporophyte and several old broken ones.

S. virescens [greenish]. The characters separating *S. calcicola* and *S. virescens* are relatively minor but seem to be consistent—small plants with small leaves, narrow and weak when dry, with a 'puckered' (termed 'constricted' by Gallego (2006)) margin, and a weakly developed stereid layer in the costa. The clumps of this species are discrete, and when dry the leaves appear twisted and narrow, distinctly greener than *S. calcicola* and slightly glaucous. The leaves of *S. calcicola* seem to be consistently browner, thick, and with a prominent red costa.

     BL: Flagstaff Mountain, Tenderfoot Trail, 6700 ft, on gravel in mixed *Pinus ponderosa-Pseudotsuga* forest, forming small, rounded patches in open ground, with *Bryum calobryoides Rosulabryum laevifilum,* and *Ceratodon purpureus, W&W B-113374*; OT: Devil's Canyon N of Delhi, on sandstone rimrock (with *Tortula muralis*), *Weber B-36799* (!W. Kramer); CC: on gravelly tundra, 13500 ft, Mount Evans, *W&W B-115156*. (!Gallego).

Tortella [little *Tortula*]

1a.   Leaves not fragile .... (2)
1a.   Leaves fragile, many of the leaf tips breaking off .... (3)

2a.   Plants appearing black below, densely foliose, the leaf bases hidden; stems with the tomentum hidden in the bases of the branch innovations; leaves often squarrose-recurved when wet; restricted to alpine tundra. **T. arctica**
2b.   Plants appearing brown below, loosely foliose, some leaf bases exposed, the margins ruffled; stems conspicuously tomentose; leaves erect or erect-spreading when wet; plants of lower altitudes. **T. tortuosa**

3a.   Stems 1–5 cm tall, coarsely tomentose; central strand absent; distal leaves to 7 mm, densely crowded, rigid; with patches of elongated, non-papillose cells on distal margins of young leaves at the stem apex; leaf cells 10–12 μm long; lamina below subula bistratose; apical propagula falling in a single rigid unit; subulate limb bistratose to multistratose. **T. fragilis**
3b.   Stems 0.5–1.5 cm tall, scarcely or not tomentose; central strand present; distal leaves 1.5–2.0 mm long, sparse, soft; leaf tips without differentiated marginal cells; leaf cells 14 μm; lamina unistratose throughout the leaf; propaguloid leaves and apical propagula articulated by periodic constrictions, falling in several pieces; leaves bistratose only in patches or along costa distally. **T. alpicola**

T. alpicola [of high mountains]. Like a small edition of *T. fragilis*. "This is the smallest of the North American *Tortellae*. The plants are noted by their small size, absence of tomentum on typical (sterile) stems, broken tips on most of the leaves, vividly shining white leaf bases on the stem in contrast with the deep or bright green of the limb, and large leaf cells (to 14 μm wide)." (Eckel 1998). The species is widely scattered in western North America, northern South America, India, Hawaii, and Antarctica. In Colorado, at least, it is not an alpine, but a desert-steppe species. BL, FN, GF, MN, OR, SJ, SM.

T. arctica. Common in moist tundra, often on solifluction terraces. The sods formed by this species tend to be more compact than in the lowland *T. tortuosa* and darker in color. The species is similar otherwise and is distributed across arctic America and Asia, China, and disjunct in British Columbia and Colorado.

T. fragilis. Although this species reaches its best development on seeping granite ledges in the foothill canyons, where it forms extensive yellowish sods over rocks, it also occurs on limestone in the outer hogbacks, sandstone cliffs in the western canyons, and on (probably calcareous) screes. OR: Ouray; rock outcrops on trail to Box Canyon Falls, 2350 m, *W&W B-111012*.

T. tortuosa. Locally abundant on cliffs and ravines, usually of a granitic or sandstone parent rock in the outer foothill canyons. The tufts are large and loose compared to *T. arctica*. *T. tortuosa* is a good example of a species that, in the more oceanic climates of western Europe, behaves as a distinct calciphile. The species is widely distributed across the Northern Hemisphere.

Tortula [twisted, alluding to the peristome]

1a.   Operculum absent or, if present, not dehiscent; minute ephemeral mosses .... (2)
1b.   Not as above .... (3)

2a. Capsule sessile, globose, truly cleistocarpous. Common in early spring on open, grassy sites on outwash slopes of outer foothills, eastern Colorado. **T. acaulon**

2b. Capsule on a short seta, spindle-shaped, with a peristome, but this only revealed by dissecting the capsule. Very rare, in desert-steppe rimrock crevices, western Colorado. **T. protobryoides**

3a. Leaves oblong or lingulate, obtuse; leaf cells clear, papillae absent or only present distally; capsules long and narrow, often somewhat curved .... (4)

3b. Leaves ovate-lanceolate, acute, usually broadest at the base; leaf cells usually papillose .... (6)

4a. Peristome present (almost always fruiting); common low altitude species. **T. mucronifolia**

4b. Peristome lacking .... (5)

5a. Capsule systylious; rare plant of travertine deposits. **Hennediella heimii**

5b. Capsule not systylious; rare plant of saline soils. **Tortula nevadense**

6a. Costa (in cross-section) with a massive adaxial pad (a row of bulging cells). **T. atrovirens**

6b. Costa not as above .... (7)

7a. Capsule systylious, the operculum remaining attached to the columella and not falling away .... (8)

7b. Capsule not systylious .... (9)

8a. Leaves not bordered; costa long-excurrent; leaves not papillose; peristome well-developed (capsules necessary for identification); alpine tundra. **T. systylia**

8b. Leaves acute or acuminate, bordered; costa percurrent; distal cells papillose; peristome none; calciphile of low altitudes. **Hennediella heimii**

9a. Leaves broadly ovate .... (10)

9b. Not as above .... (11)

10a. Leaves 2 mm or more long, pale green, in a cabbage-like head; leaves awned or awnless. Alpine tundra. **Stegonia**

10b. Leaves minute, up to 1.5 mm long, dark green or with reddish tints, appressed when dry, always hair-pointed. Alkali flats. **Tortula guepinii**

11a. Leaves with a long smooth hair-point often as long as the leaf .... (12)

11b. Leaves lacking a hair-point or hair-point not so long (or a stout mucro) .... (13)

12a. Distal leaf cells about 8 μm wide; papillae C-shaped or circular. **T. plinthobia**

12b. Distal leaf cells about 9–12 μm wide; papillae irregularly warty. Weedy moss of walls and sidewalk cracks, forming indeterminate tufts. **T. muralis**

13a. Leaves longitudinally folded and flattened, usually with a strong mucro and stout costa, the group forming a strongly curved horizontal 'pinwheel' (more like a *Syntrichia* than a *Tortula*). **T. inermis**

13b. Leaves not as above .... (14)

14a. Leaf cells densely papillose throughout; seta contorted and capsule pendulous. **T. laureri**

14b. Leaf cells smooth or faintly and minutely papillose distally, not enough to obscure the cell detail; capsule erect or cernuous .... (15)

15a. Capsule cernuous or pendulous; distal leaf cells smooth or minutely papillose, not enough to obscure the cell detail .... (16)

15b. Capsule erect, straight; distal leaf cells papillose .... (17)

16a.   Capsule minute, ovoid; seta nearly straight. **T. cernua**
16b.   Capsule larger, narrowly elliptical; the seta strongly curved. **T. laureri**

17a.   Upper leaf cells less than 10 μm diam.; leaves rounded at the apex, with or without an excurrent hair-point or costa; wide-ranging plants from the plains to the alpine. **T. obtusifolia**
17b.   Upper leaf cells 11–30 μm diam.; leaves various; plants of soil, foothills and subalpine . . . . (18)

18a.   Leaves 2–3 mm long, oblong or spatulate, with or without a hair-point, the upper leaf cells 15–21 μm diam. **T. hoppeana**
18b.   Leaves minute (less than 1 mm long), gradually tapering from the middle or below, slender-pointed or more or less acute; upper cells 11–16 μm diameter. **T. leucostoma**

T. acaulon [stemless]. A minute, almost stemless species that one hardly ever encounters except in early spring or late fall when the ground is moist and the plants are fruiting. After a few weeks the soil is baked dry and the mosses go into a summer dormancy and disappear from view. Its habitat is primarily the open grassland or grassy slopes in the lower outer foothills. Fruiting material is easily recognized because the capsule is sessile, spherical, has no differentiated operculum (cleistocarpous) and of course no peristome. The spores are densely papillose and about 25 μm diam. The leaves form a cochleariform cluster surrounding the reddish-brown capsule, are broadly oblong-ovate and abruptly narrowed to the apex. The costa is excurrent into a slender, smooth, golden hair-point.

Vegetatively the plant resembles a small *Bryum* more than a small *Tortula*, but the quadrate or rectangular cells mark it as pottiaceous, whereas the leaf cells of *Bryum* are usually hexagonal. The leaf cells may or may not be distinctly papillose.

T. acaulon demonstrates the desirability of making accurate observations on habitat and occurrence. In March, 2004, in the midst of one of the most long-enduring droughts in Colorado's history, we made an excursion of a few hours to Flagstaff Mountain, near Boulder, to show a newcomer some of the common species. We walked along a well-trodden trail around the west side of the summit. The soil and vegetation was tinder-dry. Except for a few species of *Grimmia* and *Schistidium* on the rocks, and the ever-present *Ceratodon purpureus* on the packed soil of the trailside, there were no mosses to be seen. We were passing by a fairly steep, gravelly slope strewn with larger boulders when something green caught my eye. It was a small, bright green moss tuft. The slope had a scattering of plant debris (pine needles, broken leaves and twigs) that were slipping down-slope. In the alpine we would call these solifluction lobes. These slowly sliding sheets of dry vegetation created, at their lower ends, small overhangs of an inch or two, creating miniature caves, a few of which were visible, but most could be seen only by brushing away the overhanging debris. In any of these tiny caves were small tufts of *Tortula acaulon*, recognizable on sight by the bright green, shiny leaves and long golden hair-point, and the spherical green or brown capsule nestled in the rosette of leaves.

The summer of 2004, on the other hand, was unusually wet in the eastern foothills. On November 6, Wittmann and Lehr were walking on the colluvial outwash fan south of Boulder and discovered acres of *Tortula acaulon* on bare ground exposed after the fall dieback of herbaceous cover. Capsules were present but immature. Evidently these plants remain dormant under the snow, ready to explode when warm weather returns in the spring. Thus, a rare species becomes common once one knows its life history!

T. atrovirens [dark green]. The broad, irregularly contorted, muticous, ovate, revolute leaves with bulging cells on the costa are characteristic. Probably not uncommon, but we have only two collection: BL: NW of Lyons, soil on talus of sandstone bluff along St. Vrain Creek, 24 Oct. 1970, *Hermann 23497 (B-93567)*; University of Colorado campus, shaded retaining wall below Macky Auditorium and Varsity Pond, 5 March 2005, *W&W B-114745*.

T. cernua [slightly drooping]. A small plant 3–10 mm high that is probably so inconspicuous that it is only collected in the fertile condition. Its chief characteristic, in contrast to other *Tortula* species, is the short and broad capsule, only about twice as long as broad, that is inclined at a 45 degree angle rather than erect. The leaf cells are almost smooth, with a few scattered C-shaped papillae. This might well be classed as a weed moss. The ecology is not known, but since one came from a railroad right-of-way, and the other from a ditch, these were probably alkaline areas at low altitudes. Our two collections come from LR: Railroad between Fort Collins and La Porte, 1895, *Crandall 11 (B-18058)*; CF: On log by ditch, 0.5 mi NW of Buena Vista, 8000 ft, *Conard 40-795 (B-18057)*.

T. guepinii [for J.B.P. Guépin]. A minute moss with hardly any stem, and leaves only 1.0–1.5 mm long, broadly ovate and rounded or even truncate at the apex, with a long, smooth hair-point. The hair-points are almost the only visible part of the plant. In this respect it resembles a species of *Pterygoneurum* except that it lacks any lamellae. We have one collection: DT: SE of Delta, on a low alkaline flat, under older *Sarcobatus* shrubs, associated with *Crossidium,* 1500 m, *W&W B-111875.* It is a rarely collected moss found in Arizona, California, New Mexico, and Texas.

T. hoppeana [for D. H. Hoppe]. A common species in the subalpine and alpine. In the tundra it is often the only moss growing on dry sites of fell fields. It fruits abundantly. The hair-point is variably developed. Without a hair-point it is called var. *mutica.*

T. inermis. Easily recognized by the oblong, muticous or stoutly mucronate leaves that are folded, the sides flat, forming a horizontally sickle-shaped rosette, curved around the axis of the stem, never crumpled. FN: Phantom Canyon, on red sandstone ledges, *Weber & Kelso B-102631*; ME: Colorado National Monument, on soil in Kodel Canyon, *Shushan & Weber 26315*; 7 mi SW of Whitewater, on benches above East Creek, *Weber et al B-43495*; MZ: Mesa Verde, Rock Canyon, *Erdman B-3694.*

This species has been moved back and forth between *Syntrichia* and *Tortula.* Zander & Eckel (Flora of North America, 2007) say: "*Tortula inermis* has the aspect of a *Syntrichia* with its ligulate, apiculate leaves and strong costa, but the plant is yellow or orange in KOH solution, and the costal section reveals a rounded stereid band. It is related to *T. subulata* and *T. mucronifolia* but the lack of a strong mucro and the narrow but nearly complete recurving of the leaf margins are diagnostic." Also, the short stems with a single terminal rosette of leaves does not fit with *Syntrichia.*

T. laureri [for J. F. Laurer]. Even when sterile this species can be recognized by the tiny dispersed papillae and the large leaf cells (up to 60 x 20 μm). The pendulous capsule makes the plant look somewhat like a small *Bryum.* A rarely collected species thus far found only in along streams in the montane sagebrush and the mossy alpine tundra. CC: Saddle between Mt. Evans and Mt. Epaulet, 13200 ft, *W&W B-115161*; SH: Along the vertical banks of a small streamlet in sagebrush hills 9500 ft along Los Pinos Creek, *W&W B-112938.*

T. leucostoma [white-mouthed]. Probably frequent in the subalpine and alpine, although we have relatively few collections. The leaves are oblong, up to 2 mm long, broadest above the middle, not tapered to the apex, suddenly drawn out into a short point with elongate, incrassate, clear cells; distal leaf cells rounded-quadrate, 10–15 μm long, with few rounded or C-shaped papillae; proximal leaf cells broadly rectangular, clear, smooth. Capsule ovoid, erect, up to 2 mm long. Peristome red, short, densely coarse-papillose, the units with variable length narrow gaps but essentially not split until near the apex. CC: Guanella Pass, willow fen, *Weber, Wittmann, & Tidball B-112507*; GA: Middle Park, *Weber & Tidball B-112443.*

T. mucronifolia. Common in shaded sites, never far from running water or seepage water, on rock cliffs, retaining walls, edges of watercourses in ravines, etc., from the outer foothills up to the subalpine. Without capsules it would be difficult to place this in *Tortula* because of the non-papillose leaves, but fortunately capsules are almost always present.

T. muralis. Abundant in mortar-filled cracks in old walls. It appears to prefer drier sites than *T. mucronifolia,* which occupies similar substrates in more shaded, well-watered sites. The setae are much shorter in this species than in *T. mucronifolia.* Steere, in Grout (1936, p. 231), wrote: "This common and variable species is most apt to be confused, when sterile, with *T. plinthobia,* but that species has smaller leaf cells, no differentiated cells at the leaf border, and crescent-shaped or circular papillae which are not irregularly verruculose as in *T. muralis.*" The leaves are oblong, about 1–1.5 mm long, narrow, rounded to emarginate at the apex, with a long, smooth hair-point. BL, ME, OT.

T. nevadensis. Evidently a rare plant of saline soils and clay, differing from *T. mucronifolia* in lacking or having a rudimentary peristome. CS: South Colony Creek, gravelly soil among rocks, 11700 ft, *Kiener 10255* (MO).

T. obtusifolia. A ubiquitous moss on thin soil over sedimentary rocks from the plains to the dry alpine tundra. The leaves are oblong, the lower part with clear cells, the distal half of the leaf with small, densely papillose cells difficult to distinguish as units.

T. plinthobia [on bases of columns]. Infrequent on sedimentary outcrops on the plains and outer foothills. FM: Phantom Canyon between Florence and Victor, on steep canyon wall, 6000 ft, *W&W B-112157.*

T. protobryoides. Uncommon or rare in Colorado. The leaves are ovate, folded and curved when dry, with a strong excurrent costa. The upper leaf cells are multipapillose. Technically it cannot be distinguished from *T. acaulon* without dissecting the capsule to prove the existence of a peristome, which adheres to the inside of the operculum. Unlike *T. acaulon*, however, there is a short seta up to 4 mm long, and the capsule is pointed at each end. We have only one record of this species, from ME: 7 mi SW of Whitewater, on arid benches above East Creek N of Gibler Gulch, on calcareous soil in a deep recess under an overhanging sandstone block, *Weber, Kunkel & LaFarge B-43494*.

T. systylia. The epithet 'systylia' alludes to the operculum which remains attached to the columella after separating from the capsule rim, so that it stands like an umbrella over the open capsule mouth for a long time. This is not unique to *T. systylia*, however, and may occur in *T. hoppeana* as well as in *Stegonia*. *T. systylia* is frequently confused with *Stegonia*, and both are tundra plants.

Trichostomum [hairy mouth]

Trichostomum has the narrow leaves of a *Barbula*, but with the margins incurved, rather than recurved. The basal third is hyaline and sheaths the stem much in the manner of a *Tortella*. However, the hyaline cells are internal and usually are bounded on the margin by shorter, chlorophyllose cells. In *Tortella*, the entire base is hyaline, with the hyaline cells extending diagonally up the margin. The upper leaf cells of *Trichostomum* are multipapillose, and the margin somewhat crenulate, with here and there a few larger cells protruding as coarse teeth.

1a.   Minute, *Weissia*-like plants of desert-steppe, but leaves not as involute. **T. planifolium**
1b.   Larger plants of alpine tundra, resembling *Tortella* but without the V-shaped basal area of clear cells. **T. tenuirostre**

T. planifolium [flat-leaved]. A plant of desert-steppe and foothills sandstone habitats. LR: Big Thompson Canyon, 8 mi W of Loveland, Hermann 27636.5 (B-10103),; Greyrock Mt., 10 mi W of La Porte, *Hermann 23651 (B-48554)*; ME: 5 km S of Fruita, *Shushan & Weber B-26865*; 7 mi SW of Whitewater, *Weber et al. B-43497*. It is abundant in Utah.

T. tenuirostre [slender beak]. Rare or infrequent, alpine tundra, and at compensating environments on north faces of lower canyon walls. Resembling somewhat *Tortella arctica*, but lacking the shining abaxial costa surface. BL: Cirque below Devil's Armchair, on saturated soil on boulders along torrential runoff area, 10500 ft, *Weber & LaFarge B-43731*; 2 mi S of Ward, on granite wall of old mine-shaft, seepage area, *Shultz B-43340*; ravine, Boulder Canyon 3 mi W of Boulder, 7000 ft, *Weber B-10562*; CC: Summit Lake, Mt. Evans, 12500 ft, *Weber, Porsild, & Holmen B-4360*.

Weissia [for F. W. Weiss]

1a.   Leaves narrowly linear, hardly broadened at the base; foliage dark green; costa inconspicuous; papillae simple and low. **W. controversa**
1b.   Leaves broader and shorter than linear, oblong, or ligulate, or ovate; foliage apple-green or chartreuse when moist, somewhat glaucous when dry; costa usually quite obvious; papillae forked or branched . . . . (2)

2a.   Leaves deltoid-ovate, the much broader base up to half the length of the lamina, with definite shoulders. **W. condensa**
2b.   Leaves ligulate, little broader at the base, the basal portion less than half as long as the lamina. **W. ligulifolia**

W. condensa. The ovate-deltoid leaves are distinctive; when moist, they spread like short, yellow-green flower petals. When dry, this and *W. ligulifolia* have apple-green leaves, more like small garden-hoses than narrow wires or ribbons. Its distribution is spotty, in southern Arizona, New Mexico Utah, and Texas. Distribution in Colorado evidently limited to the arid steppe area of the outer foothills of the Front Range and outcrops on the eastern plains, where it is locally common on calcareous sandstones. Where we have found this species, it usually associates with *Didymodon anserinocapitatus, D. rigidulus,* and *Grimmia anodon*. Sporophytes are infrequent. *W. condensa* has been thought to be exclusively European, but Flowers (1973, plate 11, fig. 9–14) illustrates his single collection from Utah as *W. tortilis*, a synonym of *W. condensa*. Flowers at one point planned to describe this plant: *Flowers 2771*, from Arizona: Mohave Co., Pipe Spring, as *Weissia obtusifolia sp. nov. TYPE*, and wrote in pencil above, "*tortilis?*" It is interesting that he was aware of this morphological expression. Stoneburner (1985) also recognized the presence of *W. condensa* in western North America. Thus far in Colorado we have not found *W. condensa* on the Western Slope.

BL: Steamboat Mt., Lyons, *W&W B-110667*; FN: 20 mi E of Canyon City, *Hermann 23200a (B-37946)*; LR: Buckhorn Creek, WNW of Masonville, *Hermann 18963 (B-80232); Kiener 5671 (B-85265);* YM: Bonny Reservoir, *Weber & Hermann B-37465.*

W. controversa. The plants of *W. controversa* are really rather dark green, without any glaucous or purplish tints. When dry, the leaves of *W. controversa* are so tightly rolled that they resemble slender wires. Those of *W. ligulifolia* and *W. condensa* are shorter and wider; only the upper parts are clearly convolute. Sporophytes are usually present. The species is common, but all of the material we have from Colorado comes from granitic soil on cliffs and ledges in the foothills.

W. ligulifolia. This is the common species in xeric desert-steppe habitats of western Colorado, Utah, New Mexico, Texas, and Arizona. It consistently has leaves intermediate between the broadly triangular ovate ones of *W. condensa* and the ribbon-like ones of *W. controversa*. In southern New Mexico and Texas it is often richly fruiting. Stoneburner (1985) noted, and we agree, that "*W. condensa* is more frequently found in small, weathered pockets of crevices in boulders, while *W. ligulifolia* occurs on soil around the bases of boulders."

Excluded Taxa

Crumia latifolia (Kindberg ex Macoun) Schofield. This was reported for Colorado, as *Merceya latifolia*, by Grout (1939) in Vol. 1: 247. 1939. Flowers (1973) repeated the report. However, there is no specimen in the Flowers herbarium at COLO. Schofield (1966) did not show it on his distribution map or list it in his citation of specimens. Neither does the BFNA list this for Colorado. The species might be expected to occur in waterfalls in the canyon country of western Colorado.

Didymodon revolutus (Cardot) Williams. We have no Colorado specimens. Zander (*in litt.*) has verified one from El Paso, Texas (*Worthington 6848*). Although this is said to be closely related to *D. rigidulus*, in aspect it is nothing like that species. The entire leafy stem is bulb-like, no more than 1 mm high, with short, broad, blunt leaves; the plants are densely packed together and bound by sand grains into a close, hard mat. Zander (*in litt.*) points out that the leaf has a somewhat spurred costa and unicellular gemmae produced in the leaf axils. The single published report of this is from SJ: Deadwood Gulch, San Juan Mountains, 3.2 km SSW of Silverton, 2600 m, *Hermann 23287*. This is a strange site for a desert-steppe species. In Arizona, Bartram collected it on the calcareous banks of dry washes; this is where one should look for it. For a discussion of the species, see Zander (1981). A fine illustration is found in Sharp et al. (1994, p. 310, Fig. 228).

Trichostomum alpinum Kindberg. Zander has provided us with a few stems from the type collection: Mt. Carbon [Gunnison Co.], *N.L.T. Nelson 2462 (S)*. This proves to be *Dicranoweisia crispula!*

Trichostomum coloradense Austin is considered a synonym of *Timmiella crassinervis* (Hampe) L. Koch, a species known from British Columbia, Washington, Oregon, and California. It has, in fact, never been found in Colorado, despite the epithet. Lesquereux & James (1884, p. 413) wrote: "As remarked by Watson, Bot. Calif. 2:367, this species is based upon specimens without fruit, and the genus therefore indeterminable. The name of the collector is also uncertain, the specific name is a misnomer, and the moss is not known from Colorado. Austin states that it was collected in Yosemite Valley by a Mr. James, probably B. W. James."

PTERIGYNANDRACEAE (PTR)

1a.   Leaf cells obviously prorulate, the leaf margins serrulate; plants forming abundant patches, the stems directed downward, on tree trunks and boulders. **Pterigynandrum**

1b. Leaf cells indistinctly prorulate, the leaf margins entire; plants slender and never forming patches; usually confined to protected crevices of granitic rocks. **Leptopterigynandrum**

Leptopterigynandrum [*lepto*-, slender]

L. austroalpinum [of South America]. A small, creeping moss like *Pseudoleskeella tectorum*, forming thin mats on cliffs in the foothills, occasionally mixed with *Neckera*. It is not blackish but glaucous yellow-green when fresh. It differs in not having incurved leaves; the leaf tips are distinctly spreading, and the costa is better developed than in *P. tectorum*, and forked. The median leaf cells are oblong rather than rhomboid, the distal ones incrassate and slightly prorulose, sharply differentiated from the numerous quadrate alar cells, and under oil immersion, the cells appear to be dotted with minute papillae. This is disjunct species occurring in Alaska, South America, and the Himalaya, but it is relatively frequent in the southern Rocky Mountains. See Weber (2000).

Pterigynandrum [wing, + female, + male]

P. filiforme. A small, creeping moss with elongate, julaceous, hardly pinnate, leafy stems and broadly ovate, acute leaves, characterized by rhombic cells throughout, except for a large area on each alar region, of smaller quadrate cells, costa short and double. The leaves are small, only 0.5 mm long, broadly orbicular-ovate, distally with serrulate margins, and strongly prorulose abaxially. The genus takes its name from the distinctly lateral male and female branchlets. It is a common forest species especially on sloping boulders where the colony tends to grow downward as a smooth carpet.

RHYTIDIACEAE (RHY)

Rhytidium [wrinkled]

R. rugosum. A large, coarse and shaggy brown moss with strongly transversely wrinkled leaves, that is easily recognized. It forms deep but fairly loose cushions. The stems are irregularly pinnate, with ovate, falcate leaves. The leaves are costate to above the middle, with a distinct area of small incrassate, quadrate or rounded alar cells, linear or oblong-linear median cells with strongly incrassate and pitted walls, strongly prorulate on the dorsal side. It occurs from protected north slopes in the foothills canyons up onto the alpine tundra.

SCOULERIACEAE (SCL)

Scouleria [for J. Scouler]

S. aquatica. An unmistakable semi-aquatic moss that has been collected close to the Colorado line in the Laramie Range of southern Wyoming. It grows attached to boulders in fast-flowing shallow streams and is seasonally covered with water. The plant is quite black except for the dark green younger leaves. Capsules are depressed-globose at the branch apices. The leaves usually bear tufts of rhizoids on the proximal abaxial leaf surface. This unique moss was long included in the Grimmiaceae. A search for this in the northern Park Range of Colorado was unsuccessful, yet it is to be expected, for the Northern Rocky Mountain flora has an isolated outlier in the Park Range, with *Trillium, Azaleastrum, Mimulus lewisii, Erocallis triphylla*, and other marker species.

SELIGERIACEAE (SLG)

1a.   Plants exceedingly minute (1–3 mm high), frequently fruiting. On sandstone or limestone, often protected by overhanging rock. **Seligeria**

1b.   Plants not minute; always sterile in our area. On siliceous rock (seepage lines) or wet gravel. **Blindia**

Blindia [for J. J. Blind]

B. acuta. *Blindia* ranges from the montane to the tundra, and forms thin turf-like mats less than one cm high, often tightly attached to the granitic or calcareous substrate, on sites where water percolates seasonally over the surface. It can be recognized by its peculiar copper-brown sheen, greener when wet. Our collections are sterile, but recognition in vegetative condition is quite easy. The plant resembles a small *Dicranum*, with leaves swept somewhat to one side and tapered to a cylindrical tip. The leaf apex, however, is not sharp; under the microscope it is actually rounded. The costa is broad, and the entire distal half of the leaf is composed of the costa alone, the proper lamina narrowing and fading out a little above the middle. The leaf base has enlarged, thick-walled, orange, alar and basal cells. BL: Boulder Canyon at Castle Rock, *Weber B-35829*; ST: Blue Lake, in calcareous schist and limestone, 11700 ft, *W& W B-111202, B-115172*.

Seligeria [for I. Seliger]

Seligeria is perhaps the smallest of our mosses and the most difficult to see, for it is only 1–2 mm tall and often grows upside down in the dark on the overhanging lips of horizontal cliffs. All of our collections were made by sharp-eyed David Jamieson in the San Juan Mts. (Jamieson 1986b).

1a.   Peristome lacking. **S. donniana**

1b.   Peristome present . . . . (2)

2a.   Capsule systylious, the columella exserted and the operculum tardily deciduous. **S. tristichoides**

2b.   Capsule not systylious, the columella not exserted and the operculum readily deciduous. **S. campylopoda**

S. campylopoda [bent foot]. The plants are dull brown to dark green with narrow, curved leaves. The sporophyte has a yellow seta and a sub-cylindric urn about 0.5 mm long.

S. donniana [for George Don]. SJ: along Cascade Creek at Engine Creek Falls, *Jamieson 13123, 13124*.

S. tristichoides. [somewhat three-ranked]. SJ: Engine Creek Falls, 9500 ft. *(Jamieson 13145, 13717, 13732, 13735, 13736, 13750)*. The species occurs on limestone near waterfalls.

SPHAGNACEAE (SPH)

Sphagnum is hardly a moss in the usual sense, since its characteristics set it apart clearly as a separate class of plants, the Sphagnobrya. *Sphagnum* is one of the only mosses that is deliberately and extensively used by man. The fuel, peat, built up by the dead plants in northern bogs, has made life possible for human survival, notably in Ireland. The sterile nature and absorbent quality of sphagnum moss was used in World War I as a dressing for wounds.

Sphagnum

Colorado has a poor *Sphagnum* flora because the elevated landscape provides mineral-rich aquatic habitats; we have no areas in which the wetlands are dependent upon rain-water alone. Our fens are not very acid, hence many of the northern *Sphagnum* species are excluded from this area. Richard Andrus is our expert on the Colorado species this difficult genus. Quotations are from the late sphagnologist Cyrus McQueen (1990).

1a.   Branch leaves squarrose or nearly squarrose (the leaf is not merely spreading, but the distal half appears to be bent out at an angle) .... (2)

1b.   Branch leaves imbricate or at most only recurved near apex, especially when dry .... (3)

2a.   Branch leaves strongly squarrose throughout plant; stem leaves long triangular with truncate apex; pale green. **S. squarrosum**

2b.   Branch leaves imbricate in upper part of plant and slightly squarrose in lower portion; stem leaves tongue-shaped with rounded or flat apex (apex may have 3–4 low, broad teeth); plants brown in open areas, pale green in shade. **S. teres**

3a.   Hanging and spreading branches similar, about the same length, or plants with few or no branches ... (4)

3b.   Hanging branches usually longer and more slender than spreading branches, the distinction between hanging and spreading branches pronounced .... (5)

4a.   Stem leaves up to 1.2 mm long; branches of head curved to one side. **S. contortum**

4b.   Stem leaves over 1.2 mm long; branches not curved to one side. **S. platyphyllum**

5a.   Stem leaves much smaller than branch leaves and usually hanging downwards on the stem; plants various shades of green, yellow, or brown, never red (but sometimes branches and stems pinkish); on plants with stellate heads, paired branches or fascicles present between rays of branches (aquatic plants usually lack this feature); plants of wet mineral-rich depressions, submerged or near the water level .... (6)

5b.   Stem leaves nearly the same size as branch leaves or larger and usually upright on the stem; plants various shades of green, brown, or red; on plants with stellate heads, branches of fascicles single or absent between rays of branches; plants of drier mineral-poor acid habitats, not normally submerged, usually growing above water level (Acutifolia) .... (7)

6a.   Apex of stem leaf apiculate (the spreading branch leaves are indistinctly 5-ranked). **S. balticum**

6b.   Apex of stem leaf acute or rounded **S. angustifolium**

7a.   Stem leaves fan-shaped or tongue-shaped, and fringed at least at apex .... (8)

7b.   Stem leaves variously shaped, the margins entire, never fimbriate, with rounded or pointed apex. .... (10)

8a.   Stem-leaves flabelliform (fan-shaped) and fringed from apex down the sides nearly to base; terminal bud large. **S. fimbriatum**

8b.   Stem-leaves lingulate (tongue-shaped), fringed only at apex or with a distinct notch; terminal bud lacking or poorly developed .... (9)

9a.   Stem leaves fringed across apex; plants stiff and wiry, with long, slender, spreading branches perpendicular to the stem; plants green, never red. **S. girgensohnii**

9b.   Stem leaves with distinct notch at apex; plants weak-stemmed, not especially wiry or stiff; plants green, but usually splotched with red. **S. russowii**

10a. Heads flat, stellate; branch leaves in five rows; plants of spruce-fir swamps and rich minerotrophic fens; plants green, but stems usually purplish-red, rarely red throughout. **S. warnstorfii**

10b. Heads flat, rounded, or hemispherical; branch leaves imbricate, not in five rows. **S. fuscum**

S. angustifolium. "Widely distributed in loose lawns or depressions of hummocky, shrub-covered peatlands (*S. recurvum*) . . . The plants are green or, in the sun, brownish-yellow, and as drying progresses during the summer, the plants or at least their branch tips become whitish. The capitulum is 5-radiate, and round pendent branches, as seen between rays of the capitulum, appear to be paired. The stem tips and bases of spreading branches are pink-flushed. The stem cortex is scarcely differentiated. The broadly lingulate stem leaves are more or less eroded at a rounded-truncate apex. The leaves of young spreading branches in the capitulum are spiraled when moist. The branch leaves, when dry, are somewhat flattened out, spreading at the tips, and undulate at the margins. Their hyaline cells have few pores on the outer surface but show apical 'window pores' congruent with similar pores on the inner surface in addition to relatively large pores with thin margins in other corners." This is the most abundant and characteristic species of iron fens in southwestern Colorado.

S. balticum. Known from a single station, iron fens in the San Juan Mountains (Cooper et al., 2002). Emergent at the margins of shallow pools, 5–20 cm deep. Distinctive in the field with its long side branches and darkish capitulum.

S. contortum. Our two species in Section Subsecunda are characterized by having blunt, broadly ovate, strongly cochleariform leaves, especially in the capitulum. In *S. contortum* the capitulum branches are brownish and distinctly curved. In *S. platyphyllum* they are pale glaucous green. Collections: LR: Fen below Sandbeach Lake in Wild Basin, Rocky Mountain National Park, in the floating mat of a kettle pond with *Carex utriculata*, Cooper, 1688 and 1692 (herb. Andrus) see Cooper (1991a); DT: Grand Mesa, N of Cedaredge, Leon 1 Fen, Skinned Horse Fen, Safety No. 2 Fen, *G. Austin (B-114119, B-114121, B-114122).*

S. fimbriatum [fringed]. "Grows in loose, wide mounds, often supported by the lower branches of shrubs at bog margins and also in hardwood swamps. The plants are slender and pale green when moist but grayish-green when dry. The large, pointed terminal bud is grayish-cobwebby because of the greatly fringed stem leaves of which it is constructed. The branches of the capitulum are slender and noticeably flexuose. The cortical cells of the stem are porose. The fan-shaped stem leaves are lacerate across a broad apex and down the sides, except for a broad border at base embracing large cells in a median, triangular area; the hyaline cells

are mostly divided and largely resorbed on both surfaces. The species is widespread across the continent from the arctic southward to California, Colorado, Missouri, and Maryland."

S. fuscum [brownish]. "[Forms] compact masses at the dry tops of fen hummocks and also in low mounds in rich fens (often in association with separate mounds of *S. warnstorfii*). The densely interwoven plants are slender and brown or green mottled with brown at the surface and light brown below. The stems are also brown, and the stem leaves are long-lingulate with hyaline cells lacking fibrils and usually divided." Frequent in subalpine fens, particularly with *Betula glandulosa*. This species forms compact mounds on relatively high ground above the normal water table of the fen, and can be recognized in the field by the brown color of the leaves and stem, the slender stems and short branches.

S. girgensohnii [for G. C. Girgensohn]. In this species the stems are stout and crisp, breaking easily; the internodes are short so that the branching system is quite dense. The cluster of terminal branches forming the head is inconspicuous and short, with leaves smaller than those of the branches subtending it. The main branches are stiffly spreading at right angles to the stem, and the hanging branch is slender. The stem leaves are broad and truncate, with an extensive absorption area across the top.

Known from only one location, where it occurs on the margins of ponds in an iron fen. Instead of being out in the fen proper, as *S. russowii*, the present species grows luxuriantly between the tussocks of *Carex*, where it is not visible until one looks directly down on it. PA: Iron fen, Upper Geneva Park, 5.9 mi S of Guanella Pass, 11200 ft, *Cooper B-104998, Weber B-112037* (!Andrus).

S. platyphyllum [flat-leaved]. "Grows prostrate in shallow water, in ditches, for example, but becomes stranded as the summer progresses. The stems and branches have a tumid appearance, and there is no obvious capitulum. The terminal bud is large and the stem cortex 2–3-layered. The branches are 1–3 per fascicle and scarcely differentiated as to spreading and pendent types. The stem leaves are relatively large and ovate, much like branch leaves in size, shape, and structure. The hyaline cells of both stem and branch leaves have few to many commissural pores on the outer surface and few on the inner. [In addition to the characters given in the key, this species can be separated from the *S. warnstorfii-russowii* complex by the fact that the chlorophyllose cells of the branch leaves are more broadly exposed on the dorsal rather than the ventral leaf surface. The species has a scattered distribution in eastern North America and also in the west, from Alaska and the Yukon to Arizona." EA: Long Meadow, Cross Creek, Holy Cross Wilderness Area, *Cooper 1501, 1510, 1511, 1516* (!Andrus); ST: Above Boulder Lake, Arapahoe National Forest, 9700 ft, *Dougherty B-93227*.

S. russowii [for E. Russow]. Closely related to *S. warnstorfii*, the two species occur together in willow fens and on wet forest floors. The combination of large cell pores and porose cortical cells is diagnostic. "Dull green and often red-mottled (often because of highly colored antheridial inflorescences). The plants are generally cushion-forming and of medium-stature. At least some, and usually most, cortical cells are porose. The stem leaves are lingulate and somewhat eroded in the middle of a flat, rounded apex, and the border is abruptly broadened at the base. The hyaline cells are short and undivided near the leaf apex but longer, often divided, and membrane-pleated below. They generally lack fibrils and have outer surfaces intact and inner surfaces largely resorbed. The species is common across the continent at least as far southward as California, Colorado, Iowa, and North Carolina."

S. squarrosum. A large, coarse species, easily known by the stiffly spreading, involute leaf tips. It usually occurs in willow and peat fens along with *S. warnstorfii* and *S. russowii*, but it is less abundant and tends to inhabit the more saturated depressions.

S. teres [terete]. This is like a small edition of *S. squarrosum*, with leaves more of the magnitude of *S. russowii* and *S. warnstorfii*, but has, especially when dry, characteristic squarrose distal leaf tips. One is likely to mistake the out-curving leaves of the common *S. russowii* for *S. teres*, but in the latter, the leaves, though small, are distinctly bent outward above the base, not merely curved. We have this from only a few localities (Cumbres Pass, the Silver Lake area of the Boulder Watershed, and the Lake City area).

S. warnstorfii [for C. F. Warnstorf]. "A slender plant of the North, forming dense lawns in rich, open fens and in the shade of calcareous swamps. It is dark green in the shade but deep red to purplish-red in the sun. The hyaline cells of stem leaves are mostly divided and have no fibrils or only vestiges of them, and the outer surface is membrane-pleated, the inner resorbed as gaps. The branch leaves are noticeably 5-ranked with neatly spreading tips when dry. On the outer surface of hyaline cells, toward the leaf tips, are tiny, round, strongly ringed pores near the commissures and often contiguous with them (discernible as mere pepper dots at 100X magnification); farther down the pores become larger, elliptic, and less strongly ringed."

Study of a large series of specimens is needed for one to become convinced of the value of the pore size character, but once recognized, *S. warnstorfii* is really easily learned, although in the field I find nothing to surely distinguish it from *S. russowii*. Both species seem to be equally abundant in the subalpine willow fens along the beaver-dammed streams of the high country. Herbarium collections run about 20 to 1, *S. russowii* to *S. warnstorfii* (Cooper correspondence 1999).

Excluded Taxa

Sphagnum capillifolium (Ehrhart) Hedwig was allowed for Colorado by Andrews, according to Crum & Anderson (1981), but we believe the report is erroneous.

Sphagnum lindbergii Schimper has been reported for Colorado in the Flora of North America (2007). The specimen of record is actually from Yellowstone National Park, Wyoming.

SPLACHNACEAE (SPL)

The Splachnaceae are a peculiar group of mosses, most of our species occurring on dung, decaying animal remains, or other highly organic substrates. Several species have a highly swollen hypophysis (basal part of the capsule). In various species this structure may be colored pink, yellow or white. Such species are commonly called 'parasol mosses'. In the Colorado species the hypophysis is usually merely differentiated in some minor way from the capsule proper, being either broader or narrower than the urn.

The leaves of the Splachnaceae are usually loosely areolate, that is, the cells are large and easily visible with a hand-lens, suggestive of casement windows. The species, for the most part, are rare alpine or subalpine mosses growing in moist situations, especially willow fens and snow melt basins. Vegetatively they remind one of *Mnium* or *Funaria*, but the leaves are usually narrower and lingulate, and are never bordered. Vegetative specimens are virtually impossible to identify.

la.   Capsules cleistocarpous (irregularly dehiscent), long and narrow, tapering to the apex. **Voitia**
lb.   Capsules stegocarpous (regularly dehiscent), with an evident apophysis .... (2)

2a.   Apophysis short to elongate, as wide as, or narrower than the urn and similar to it in color; calyptra constricted above the base; stem with axillary club-shaped hairs. **Tayloria**
2b.   Apophysis clearly differentiated, elongate and somewhat to much broader than the urn; calyptra not constricted above the base .... (3)

3a   Apophysis narrowly pyriform, somewhat broader and usually longer then the urn, the same color or darker; peristome teeth at first joined in fours, later in twos, not chambered. **Tetraplodon**
3b.   Apophysis globose to turbinate; peristome teeth sometimes approximate or fused in pairs, chambered. **Splachnum**

Splachnum [a generic name used by Dioscorides]

S. sphaericum. Known from a single collection, from 'Our Moraine' at University Camp, 10500 ft, 1 Aug. 1938, *Sayre 666 (B-26137)*, det. P. Marino, 1996. Earlier, this specimen had been misidentified as *Haplodon* *wormskjoldii* and *Tetraplodon urceolatus (= T. mnioides)*. *Splachnum* species usually grow on animal dung. Perhaps with the introduction of Rocky Mountain goats more species may be found in the alpine tundra.

Tayloria [for T. Taylor]

1a.   Leaves acuminate, serrate; spores less than 20 μm diam. **T. acuminata**
1b.   Leaves obtuse, more or less entire; spores more than 20 μm diam .... (2)

2a.   Leaves spatulate, the apex rounded, lamina narrower at the base; club-shaped red hairs (probably gemmiferous) in the upper leaf axils; seta slender, 3-4 cm tall; peristome teeth at maturity 16, single; spores yellowish-green, granulate or smooth. **T. lingulata**
2b.   Leaves oblong, apex obtuse; axillary hairs absent; seta stout, ca. 1 cm tall; peristome teeth at maturity in pairs; spores brown, densely papillose .... (3)

3a.   Operculum systylious (not deciduous, remaining attached to the exserted columella); leaves clustered at the tip of the stem and branches. **T. hornschuchii**
3b.   Operculum deciduous; columella not protruding; leaves more or less uniformly distributed down the stem. **T. froelichiana**

T. acuminata. A rare or infrequent alpine species. GN: South of Emerald Lake, on quartzite cliffs, 10500 ft, *Khanna B-15336* (!Marino); CC, HF, SJ. The chromosome number of this specimen, determined by K. R. Khanna, is n=30.

T. froelichiana [for Friedrich Frölich]. This species is critically close to *T. lingulata*. According to Howard Crum (*in litt.*), in *T. froelichiana* the peristome has paired teeth and the leaves are rather erect and appressed (conspicuously contorted in *T. lingulata* when dry). It grows in seepage over soils at the base of cliffs in the upper subalpine or alpine. CC: NW of summit of Loveland Pass, 12500 ft, *Shushan B-97805* (!Marino);

GN: East River below Emerald Lake, on quartzite cliffs, 10500 ft, *Weber B-26110* (!Persson, !Crum, !Marino); CC, GN, PA, SJ.

T. hornschuchii [for C. F. Hornschuch]. This species typically grows on vertical sides of frost-heaved hummocks in the alpine. It is evidently uncommon because of the rarity of these optimum sites. Usually only a few plants are found in any one place. Even when sterile, it is recognizable by its broad, obtuse, loosely areolate leaves. ST: Blue Lake Dam, 11000 ft, *Weber & Anderson B-34238* (!Marino); CC, PA, SJ.

T. lingulata [tongue-shaped]. The red, club-shaped

hairs in the upper leaf axils and the transparent leaf areolation are distinctive. The hypophysis is narrow and the urn short and broad (wine-glass-shaped) with erect, gray-white peristome teeth. BL, CC, GN, LK, PA, SM, ST. This species occurs abundantly in a willow-sedge hummock area of a snow melt basin on the summit of Loveland Pass, where we have found it in association with *Scorpidium revolvens*. It does not seem to be associated with animal droppings, but forms dense turfs in depressions between the hummocks.

Voitia [for J. Voit]

V. nivalis. This is one of the rarest of our mosses. Except for Alaskan records, Colorado is the only state where it has been found, and then only twice, on Argentine Pass, in 1886, by William Trelease, and at Summit Lake and the Mount Epaulet saddle on Mount Evans, *W&W B-104081*. It is truly rare, occurring as occasional solitary tufts in moist tundra at 13000 ft. Fortunately every collection has been richly fruiting!

Excluded Taxa

Tetraplodon mnioides (Hedwig) Bruch & Schimper. There is one report of this species (as *T. urceolatus*) among the historical specimens collected by Major Downie at Twin lakes in 1868 (Lesquereux & James, 1884). We have not found any to justify its inclusion in the flora. The Downie collections should be at the Smithsonian, but his collections have not been located there or at Harvard. We have included *Tetraplodon* in the key because its occurrence in Colorado is considered likely.

TETRAPHIDACEAE (TTR)

This family is monotypic, containing only the single genus *Tetraphis*. It is a small, erect moss, immediately recognized, since it is almost always found in fruit, by the capsule, which has only four peristome teeth. No other moss has this number, and they can easily be seen and counted under the hand-lens. Vegetative shoots frequently develop elongate, almost leafless shoots terminated by a shallow cup of short, broad leaves in the center of which is a cluster of gemmae. Without sporophytes or gemmae, the species may be recognized on experience by the rather distantly arranged, almost perfectly elliptical leaves, somewhat prominently keeled, with translucent, rounded cells, which turn a beautiful reddish color when dry or old.

Tetraphis [four teeth]

T. pellucida [translucent]. Unmistakable because of its erect capsules with only four narrowly triangular peristome teeth. Frequent, particularly on well-rotted tree stumps in moist spruce-fir forests, and sometimes abundant on wet ground, almost always growing on old wet wood. Much more common than the few collections would indicate. It does require a moist forest environment that is not too common in Colorado.

THELIACEAE (THL)

Myurella [little mouse tail]

Myurella is easily recognized, being a minute, light blue-green pleurocarp with strongly cochleariform leaves usually strongly appressed to the axis (julaceous) producing a rope-like shoot. There is considerable variability in the leaves in their degree of cell papillosity. In most plants the leaves are smooth, but some collections have strongly papillose cells (dorsally), the papillae arising from the distal end of the cell. In these specimens the leaf is often abruptly apiculate with a recurved point consisting of a few cells.

1a. Stems julaceous, the leaves closely imbricate, obtuse or sometimes short-apiculate; the margins entire or weakly serrulate. **M. julacea**

1b. Stems not julaceous, the leaves distant, concave, their bases spreading, not concealing the base of the next leaf above, the upper half erect with spreading-recurved, acuminate apices, weakly serrulate. **M. tenerrima**

M. julacea. Fairly common but inconspicuous, easily recognizable by its pale green, julaceous stems with muticous, appressed leaves. *Myurella julacea* occurs in the subalpine and alpine, on moist cliffs and solifluction terraces, sometimes occurring in pure tufts but usually a few stems at a time intermixed with other mosses.

M. tenerrima. Evidently a calciphile; alpine, occurring mixed with *M. julacea* in seepage areas on sloping calcareous terraces. One collection: ST: Blue Lake, 11000 ft, *N. G. Miller B-115104.*

THUIDIACEAE (THD)

Abietinella [little fir tree]

A. abietina [like the fir tree, *Abies*]. A robust, pinnately branched, dark green plant forming massive but loose mats over rock outcrops on steep slopes in the foothills and on forest floors. It occasionally occurs up into the alpine tundra to 12500 ft, where it may grow in protected sites on the sides of grass hummocks. The stem leaves are broadly deltoid, much larger than the branch leaves, and the color is a deep dull green. When moist, the leaves become brighter green and spread widely, and the plant seems softer. The leaves are triangular ovate, and each cell has a single papilla. The stem has numerous paraphyllia, which may be simple or branched filaments. They are knobby, with squarish cells roughened by papillae.

The main difference between *Abietinella* and *Thuidium* is the simply pinnate branching of *Abietinella*. Even this is not always reliable, since *Abietinella* occasionally has bipinnate tendencies. The eastern American *Abietinella* was incorrectly reported by Sayre (1938) as *Thuidium [=Haplocladium) microphyllum* but the specimen she cited is typical *Abietinella*.

TIMMIACEAE (TMM)

Timmia [for J. C. Timm]

Timmiaceae is a monotypic family with the single genus, *Timmia*. The species, with the exception of *T. norvegica*, are large and coarse mosses with the general aspect of *Atrichum* or *Polytrichum*, having long straight leaves (when wet) with a distinctly sheathing base. However, there are no lamellae, and the leaves are unistratose. Fertile plants show a unique calyptra. The capsule, instead of carrying the calyptra on its ascending apex, escapes from the calyptra, leaving it standing on the seta by its involute base just below the base of the capsule. The combination bears a droll likeness to the head of an Indian brave with his single feather standing up behind.

   Timmia grows in a variety of situations, from the alpine tundra, where it occurs in loose soil in the lee of large boulders, to north-facing slopes of the foothill canyons in the *Pseudotsuga* zone where it probably indicates long-lasting snow-cover. Only *T. norvegica* occurs in really wet sites.

1a.   Leaf-sheath cells papillose; rare small plants, only a few cm long, of wet streamsides in the subalpine. **T. norvegica**

1b.   Leaf-sheath cells smooth; robust, common *Polytrichum*-like plants . . . . (2)

2a.   Leaf-sheath orange-brown, at least in part. **T. austriaca**

2b.   Leaf-sheath pale or hyaline. **T. megapolitana**

T. austriaca. Two forms of the species occur in Colorado, the typical form with leaves spreading and more or less crispate when dry, and forma **brevifolia**, with the leaves stiffly erect and incurved when dry. Forma *brevifolia* seems to be a high-sunlight ecological modification most commonly occurring in the alpine zone or on sunny canyonsides. The orange-brown sheath is a good field character. This is a species of forest floors and exposed rock faces.

T. megapolitana. A much smaller and weaker plant than *T. austriaca*. Evidently this has a narrower ecological and altitudinal range than *T. austriaca*, occurring in the foothill canyons and the Western Plateau canyons. We have found it locally abundant on moist soil in crevices and at the base of shaded overhangs. It is a delicate species, not forming dense tufts with stiff stems.

T. norvegica. An Arctic-subalpine species. *T. norvegica* is a small species, and might not be recognized as a *Timmia* immediately. A cross-section of the leaf shows the characteristic highly mammillose cells of the upper surface of the lamina. The Colorado records are the only ones for the United States outside of Alaska. In Eurasia it occurs both in the Arctic and in the southern mountains. In Colorado it occurs along wet streamsides and moist tundra slopes in the general area of Hoosier Pass. PA: Solifluction terraces, south side of Hoosier Ridge, 12000 ft, *Weber, Porsild & Holmen B-4407*; ST: Monte Cristo Creek, *Weber & Vogelman B-16252;* ST: Solifluction terraces, Hoosier Pass, 12000 ft, *Weber, Porsild, & Holmen B-4407*.

   I discovered this species accidentally while sitting on my porch cleaning debris from the roots of a fresh specimen of *Parnassia kotzebuei*, a plant new to Colorado. Hoosier Pass and adjacent South Park soon proved to be a mother lode for Colorado discoveries of vascular plants including *Armeria sibirica, Braya humilis, Cystopteris montana, Eutrema, Primula egaliksensis, Ptilagrostis, Salix calcicola* and *S. myrtifolia, Saussurea weberi,* and *Thellungiella salsuginosa*.

PLATES

Some Special Colorado Mosses
Drawings by **Patricia Eckel**

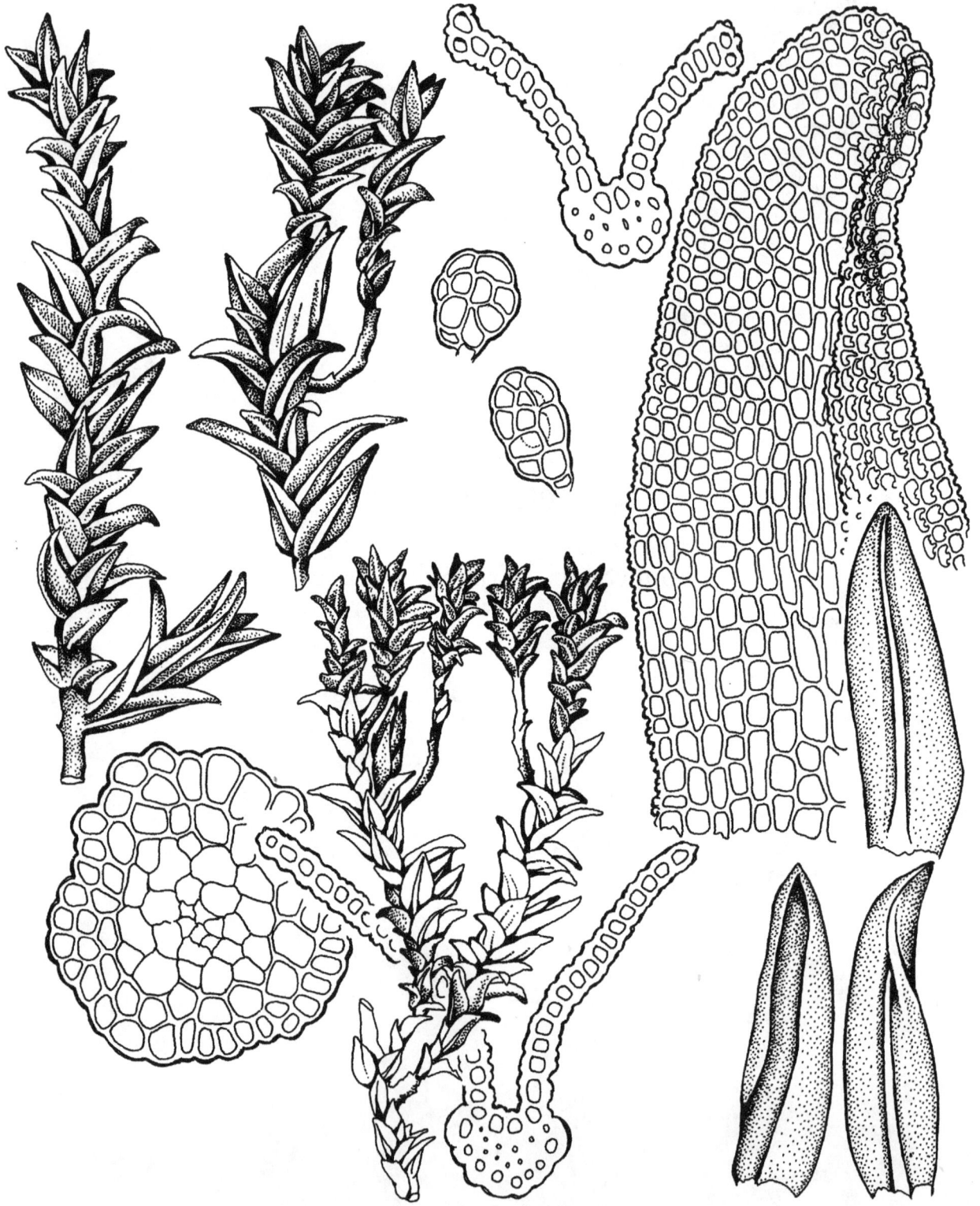

Anoectangium handelii

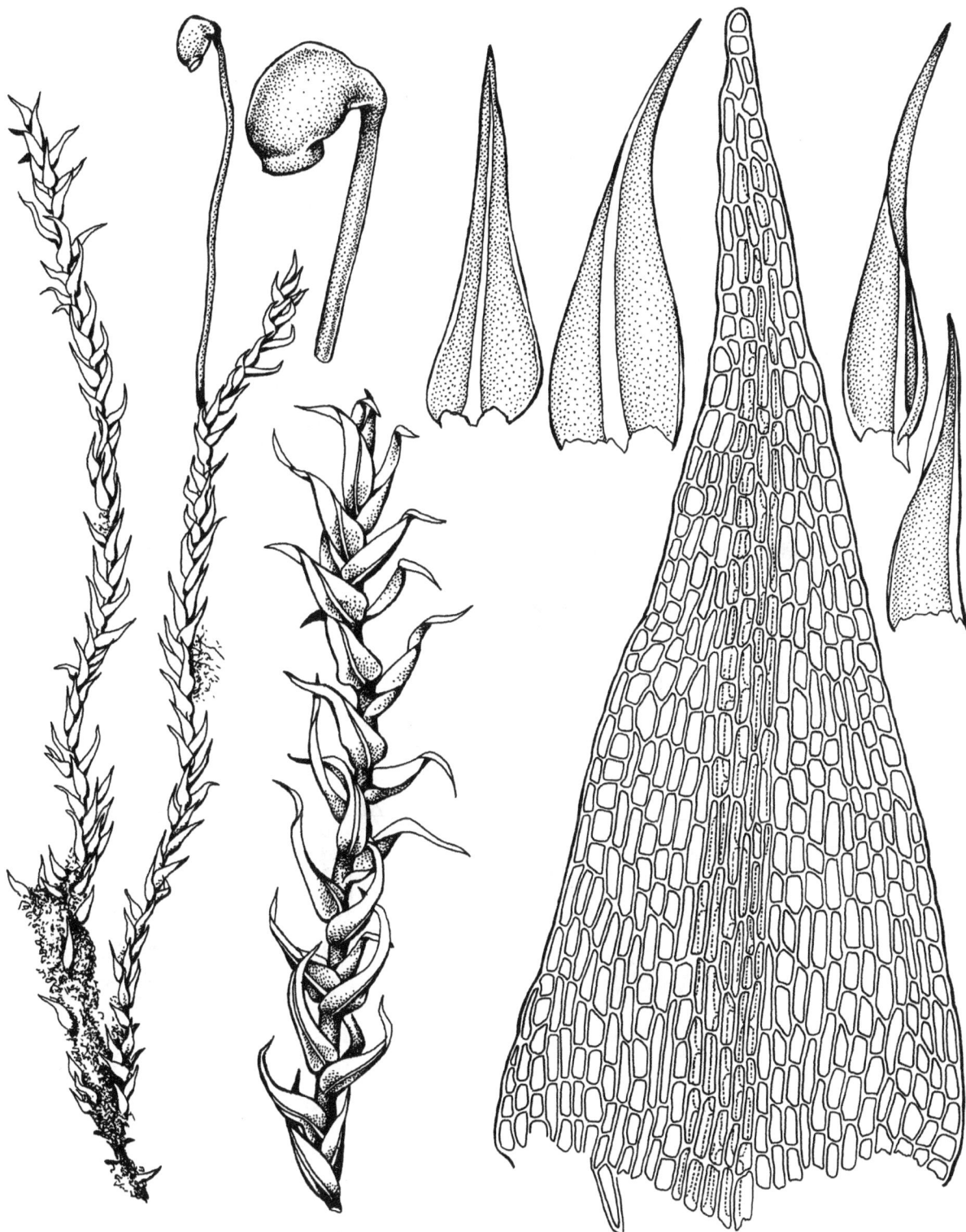

Catoscopium nigritum

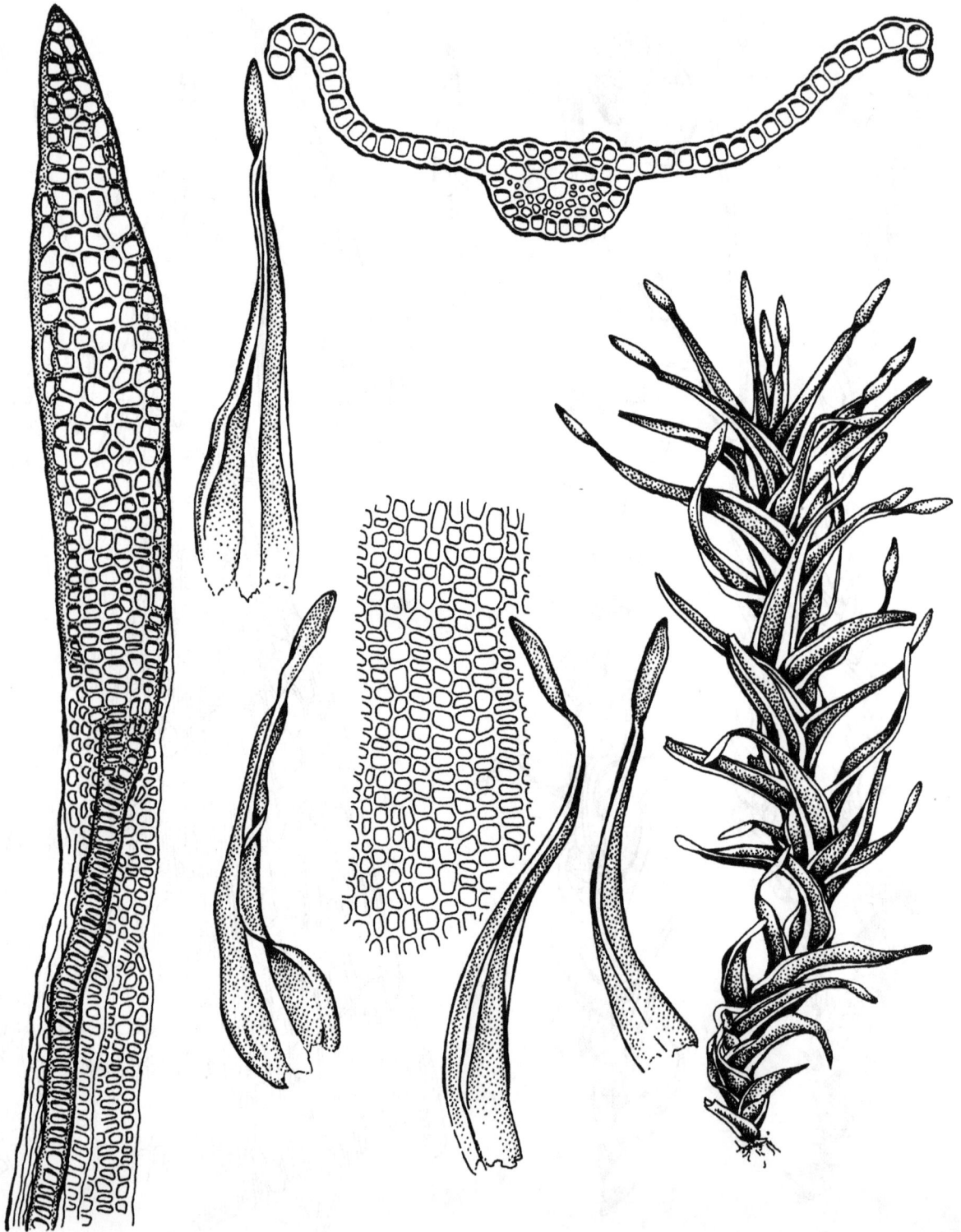

Didymodon anserinocapitatus

Leptodon smithii

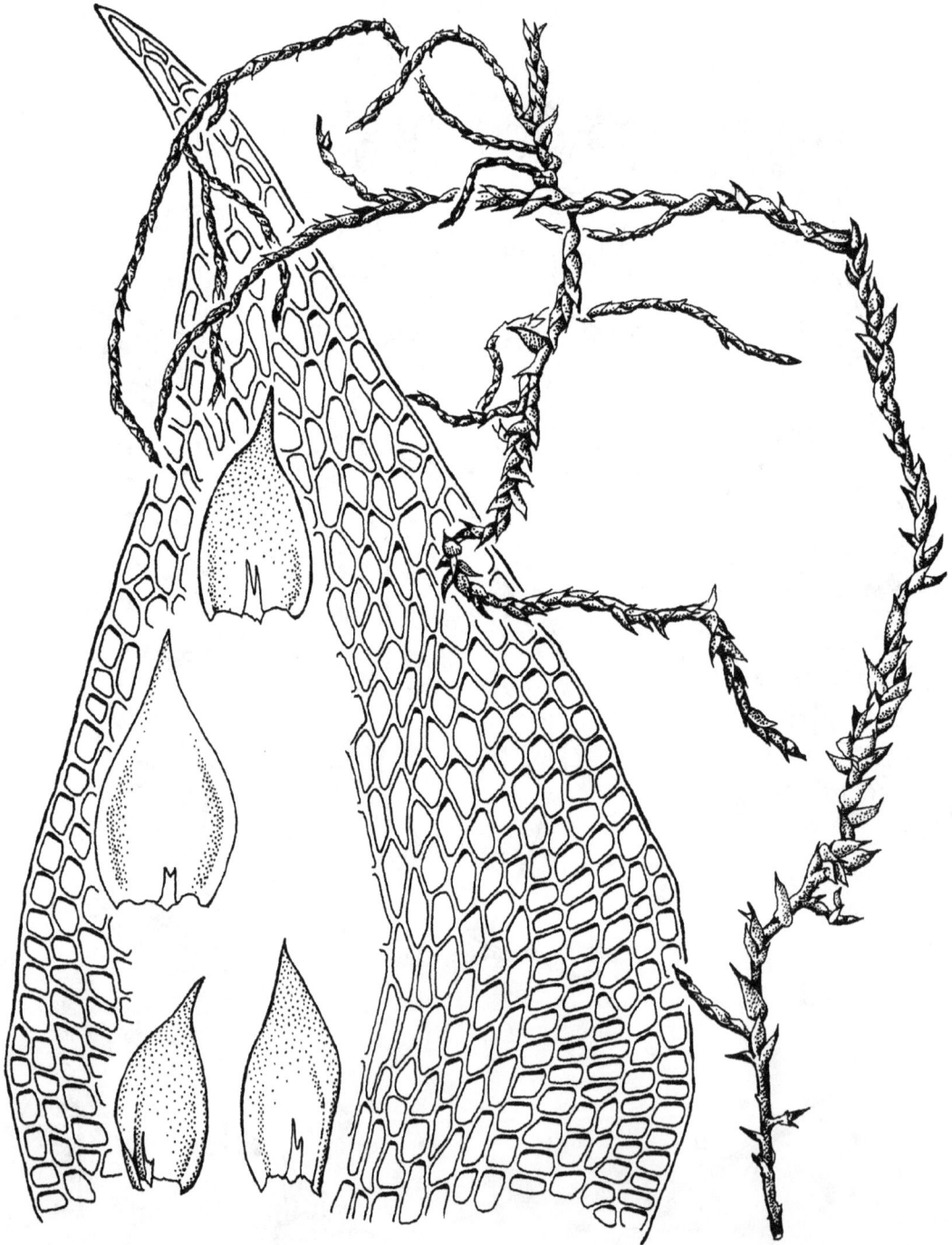

Leptopterigynandrum austroalpinum

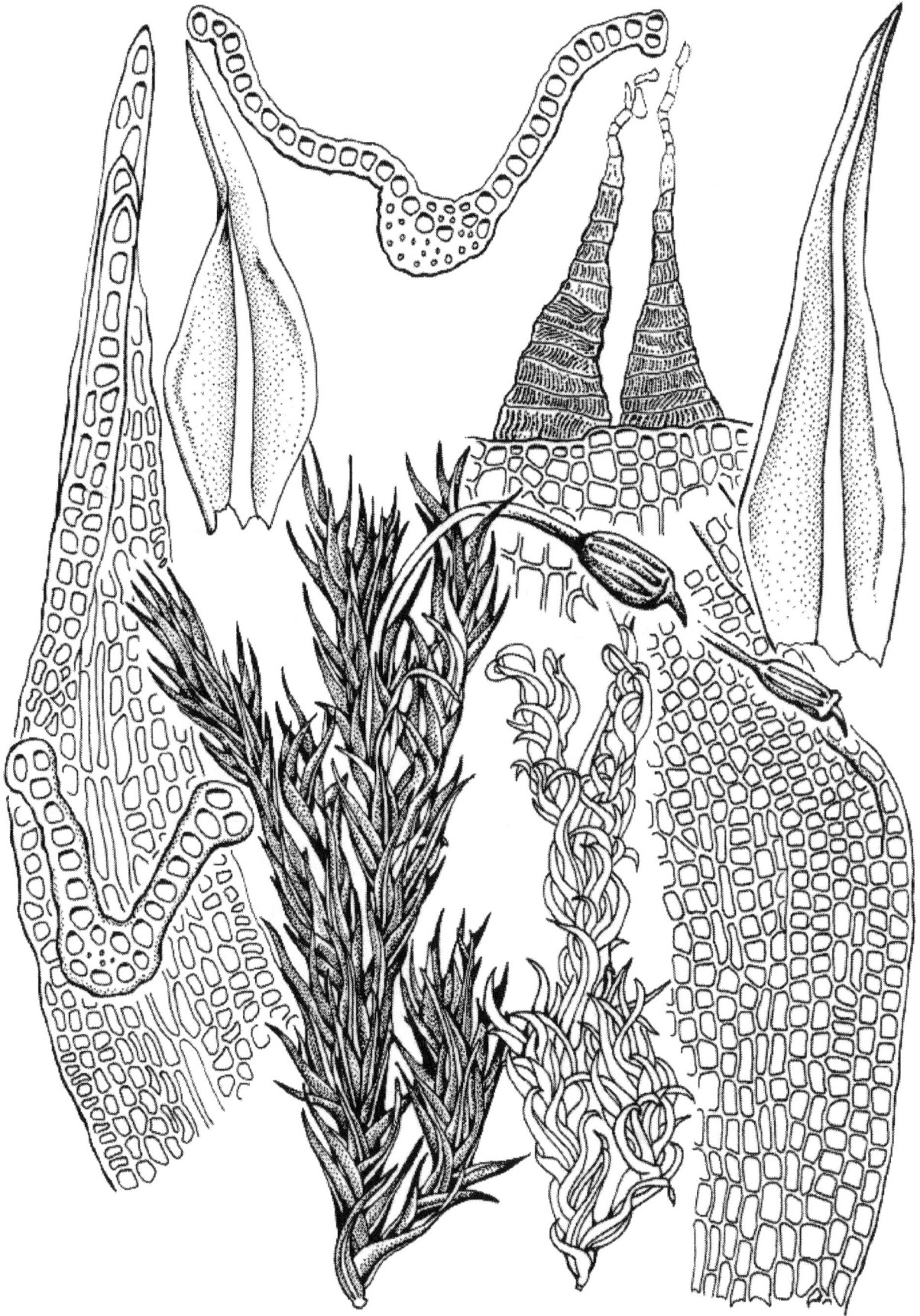

Oreas martiana

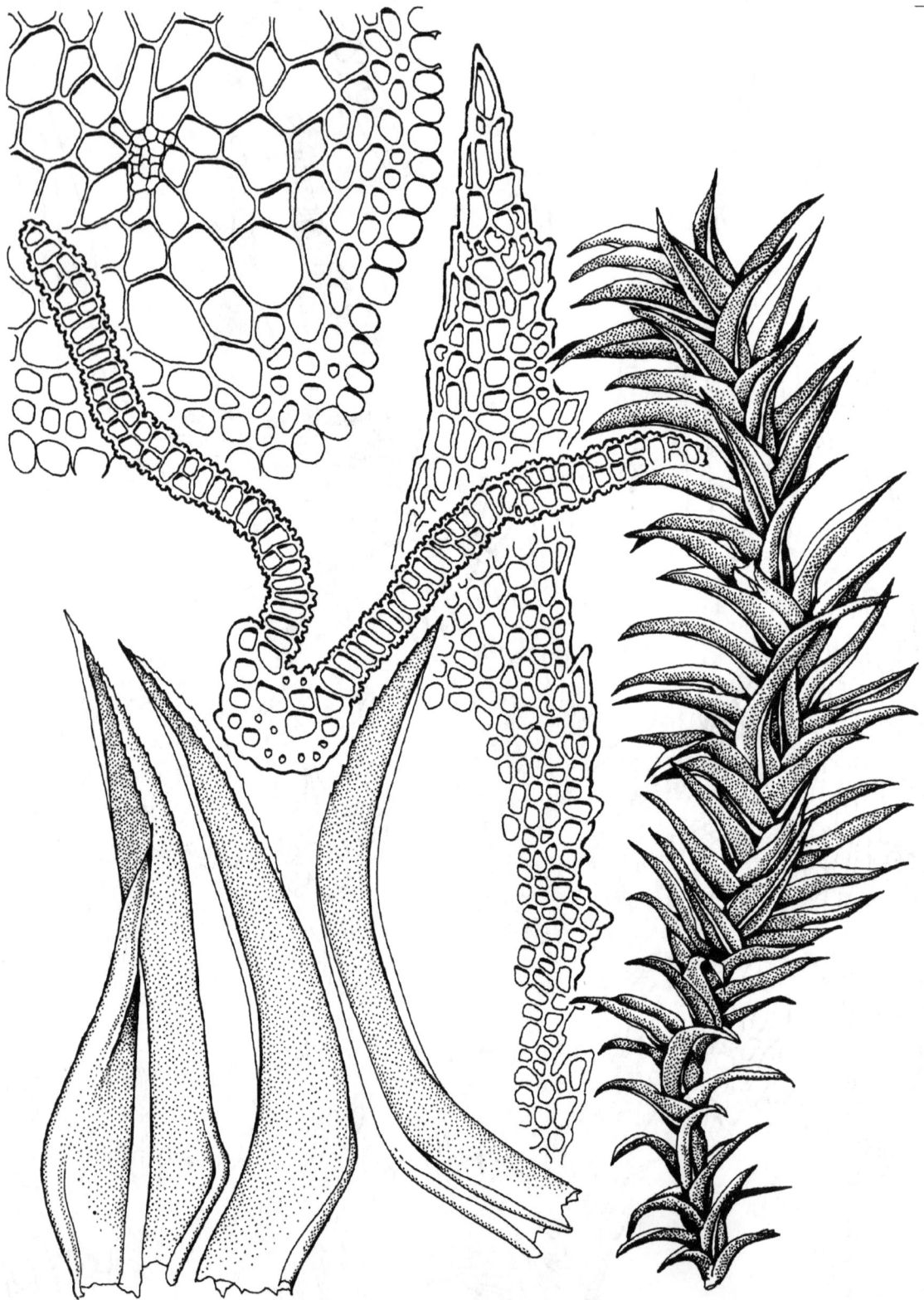

Rhexophyllum subnigrum

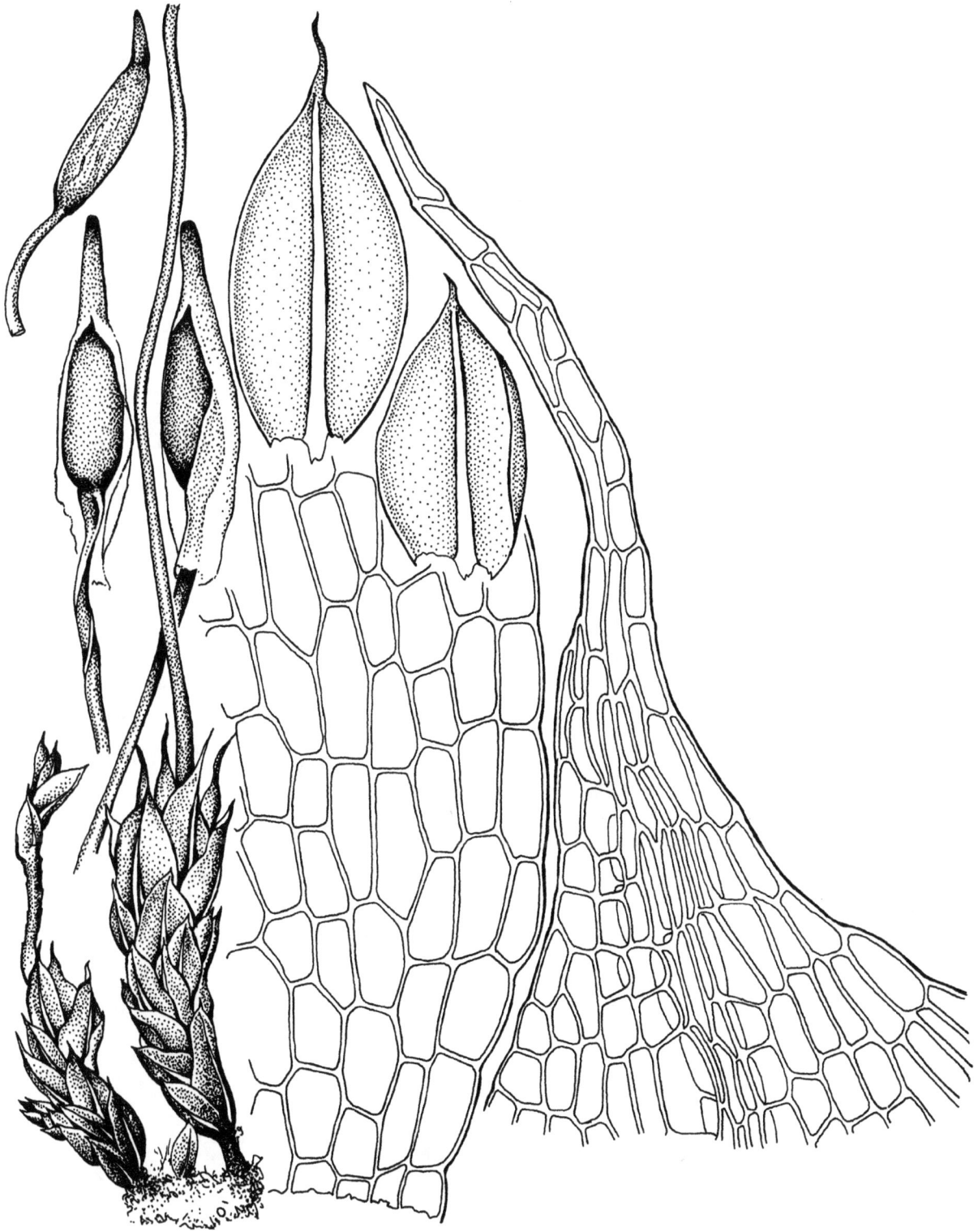

Voitia nivalis

PART 2. LIVERWORTS AND HORNWORTS

Taxonomic Treatments by Family

The liverworts of Colorado are still too poorly known to permit a definitive treatment. This would require the work of a specialist. Our knowledge of the species in the field still is very limited. The best we can do report what we have learned so far and to cite the reports of the species and improvise keys from the available literature. The most useful and fairly non-technical work available to us has been that of Schuster, *Boreal Hepaticae* (1953), Fortunately, the work of Damsholt covers practically all of the species occurring here, ands lacks treatment only of two very small genera: The hornwort *Notothylas* and the thalloid liverwort *Plagiochasma*. Here is a golden opportunity for a field specialist in this interesting but very difficult group of bryophytes.

What we commonly call liverworts comprise three different lines of evolution. Unlike the mosses, which are similarly complex, they are still recognizable by a certain unity which causes botanists to have no hesitation in continuing to call them by that common name On the other hand, the three groups of what we have been calling liverworts are so different from each other that we must recognize the hornworts, the thalloid liverworts, and the leafy liverworts as separate major groups. This requires a slight change of format.

The hornworts, represented by only two taxa in Colorado, as described below. The thalloid liverworts encompass plants that are not differentiated into stems and leaves, but consist of a dorsiventrally oriented tissue that is either broad or narrow, simple or lobed, and having a specialized mode of reproduction involving unique antheridial and archegonial structures. The leafy liverworts have leafy stems, but they characteristically have the leaves in three rows—two lateral and one underneath the stem.

The most useful book on the subject is that of Damsholt (2002), *Illustrated Flora of Nordic Liverworts and Hornworts*, which covers virtually all of our species in great detail and is beautifully illustrated by Annette Pagh. Won Shic Hong of the University of Great Falls, MT, has treated a number of genera in numerous publications (see Bibliography).

The three groups are treated sequentially, and within them the genera and species are arranged alphabetically: hornworts, p.157; thalloid liverworts, p. 158; leafy liverworts, p. 165.

Hornworts

The hornworts, as these are called, form a separate class from other liverworts, the Anthocerotae. The sporophyte, rather than having an ephemeral seta, is green and persists until the death of the thallus. The capsule is long and cylindric instead of round or oval. The capsule is two-valved rather than four. The cells of the thallus are thin-walled, never with corner thickenings (trigones), and they have a single, large saucer-like chloroplast.

1a.   Sporophytes tall, erect, cylindrical, projecting far beyond the perichaetia at maturity. **Anthocerotaceae**
1b.   Sporophytes small, nearly horizontal, surrounded until maturity by the perichaetia. **Notothyladaceae**

ANTHOCEROTACEAE (ANC)

Phaeoceros [brown horn]

P. laevis [smooth]. We have two collections: BL, Cliff of Laramie Formation, White Rocks, 5300 ft, 1908, *Bethel.* on moist vertical faces, 13 March 1961, *Weber 5585.* In *Phaeoceros* the spores are yellow. In *Anthoceros*, for which we have no reports, the spores are black. Fortunately the White Rocks locality is a protected reserve, and collecting is forbidden. Protection of the site is naturally reinforced by a dense cover of poison ivy and grasses.

NOTOTHYLADACEAE (NTT)

Notothylas [dorsal bag, the involucre]

N. orbicularis. Our only collection is from AA: East edge of Alamosa, on alluvial deposit, bank of Rio Grande River; scattered, with *Riccia frostii, McGregor 7465*. There has not been an attempt to rediscover this, and the town of Alamosa probably has grown to encompass and destroy the locality. The species forms small rosettes similar to those of *Riccia*, but the thallus is clear green and not appressed to the ground.

Thalloid Liverworts

Microscopic characters: For field botanists it will be impossible to follow some generic keys that deal with cell sizes and other microscopic characters. Until you have the necessary experience that you feel you are able to do careful microscopic work, you may have to be satisfied with recognizing the genus.

ANEURACEAE (ANR)

1a.   Thallus large, unbranched or nearly so; thallus thick, bright green and greasy in appearance, 10–15 cells thick and 4–10 mm wide; oil bodies absent; calyptra hairy. **Aneura**
1b.   Thallus small and rather delicate, thin, dull, 4–9 cells thick, less than 2 mm wide, regularly pinnately branched, the ultimate branches thin-edged about 2–3 cells wide and 1 cell thick. **Riccardia**

Aneura [lacking a vein]

A. pinguis [fatty)]. This plant consists of a simple, usually unbranched thallus up to 6 mm wide, opaque green, rather rigid and thickish, with a greasy appearance. There is no mid-rib. It is closest in appearance to *Pellia*, but that is thinner and has a mid-rib. Female plants have erect, club-shaped green calyptras. The fragile seta and black, oblong capsule stick up through the top of the calyptra. It is probably common in wet forests beside streamlets in the subalpine, but we have only a few records, from CF, GA, and ST.

Riccardia [attribution is unclear, according to Frye]

R. multifida [much-divided]. One record, LR: On soil, Chiquita Creek, Rocky Mountain National Park, *Hong* *78-921* (Hong 1980). We have no specimens at COLO.

AYTONIACEAE (AYT)

In this family, a critical character is the structure of the pores that open into the air-spaces of the thallus. Making thin sections showing the surface view or the cross-section of these is not an easy task, but it is worth while to explain below what the architecture of the surrounding tissue is like. Many times the determination of the genus depends on our knowledge of the pores. Also, sporophytes are usually necessary for identification. Count yourself lucky to find specimens with well-developed female receptacles. These are beautiful structures and vital in distinguishing the genera. Treat them gently so that they are not damaged by pressing.

1a.   Plants with female receptacles . . . . (2)
1b.   Plants lacking female receptacles . . . . (4)

2a.   Female receptacles (carpocephala) with a white pseudoperianth surrounding the capsule; stalk lacking scales at the base. **Asterella**
2b.   Female receptacles lacking a pseudoperianth . . . . (3)

3a.   Stalk short (3 mm or less),arising from the thallus surface, lacking a rhizoid furrow; the receptacle cap smooth; involucre 2-lipped (lengthwise). **Plagiochasma**
3b.   Stalk elongate, arising from the thallus notch, with a rhizoid furrow; receptacle cap with low coarse tubercles, not lipped. **Mannia**

4a.　Thallus pores bounded by 4–6 cells, not in several radial rows. **Plagiochasma**
4b.　Thallus pores bounded by several radial rows of cells . . . . (5)

5a.　Subalpine species. **Asterella**
5b.　Foothills species. **Mannia**

Asterella [little star]

Asterella is a thalloid, short ribbon-like liverwort with a bluish-green color. It is easily recognized when fruiting because it produces a stalked umbrella that is simply hemispherical, not lobed, and under the umbrella four sessile sporophytes that are enveloped by a hyaline pseudoperianth. This splits into a fringe of linear white segments or remains attached at the end, forming a 'Chinese lantern' cage. In *Mannia fragrans* there are white scales, but these are on the upper end of the stalk, not around the sporophytes. The thallus surface has small hexagonal markings (these lines indicate the places where the air chambers are separated). The ventral scales are small and do not protrude as a tuft at the thallus tip as in the similar *Mannia fragrans*.

1a.　Thallus about 2 mm wide. **A. gracilis**
1b.　Thallus 4–6 mm wide. **A. lindenbergiana**

A. gracilis [slender]. Evans (1915) reported this from BL, *Bethel 4*; GL: Near Tolland, *Young*. GN, LR, SJ. All of our specimens are from the subalpine, 10000–12000 ft.

A. lindenbergiana [for J.B.G. Lindenberg]. Infrequent in wet subalpine spruce forests. BL: Trail from 4th of July Canyon to Arapaho Glacier, 10–11000 ft, *Shushan B-15728*. ST: Mouth of Monte Cristo Creek, 2 mi N of Hoosier Pass, 11000 ft, *Weber & Holmen B-4428* (!M. Hicks).

Mannia [attribution unknown]

1a.　Stalk of sporophyte with slender, white scales along its length. **M. pilosa**
1b.　Stalk of sporophyte naked . . . . (2)

2a.　Thallus 2–4 mm wide, 1.5–2.0 cm long; appendages of ventral scales large, 2–3 in number, hyaline and forming a conspicuous cluster at the thallus apex; epidermis firm, not becoming lacunose; plants strongly aromatic. **M. fragrans**
2b.　Thallus 1.5–3.0 mm wide, 0.8–1.5 cm long; fertile plants with appendages of ventral scales 1–2, not forming a 'beard' at apex of thallus; epidermis fragile, becoming lacunose in age; plants without odor. **M. rupestris**

M. fragrans. Our most common species, mostly on ledges in the Front Range foothill canyons. BL, GN, LP, LR, MZ.

M. pilosa [with slender hairs]. Rimrock of the Mesa de Maya. BA: Big Hole Canyon, Carrizo Creek drainage, *W&W B-100956*; TL: Twin Rocks Valley, Florissant Fossil Beds, *Edwards et al. B-89815*.

M. rupestris [of rocks]. Front Range foothills canyons. BL: 4 mi NW of Lyons, 6500 ft, on north-facing ledges, *Weber B-27931*; LR: Moss-covered ledges of N-facing wall of Little Thompson Canyon 10 mi NW of Lyons, 7500 ft, *Weber B-6092*.

Plagiochasma [lateral opening]

Plagiochasma is a southern genus, appearing in only the southernmost counties, on protected ledges of sandstone rimrock. Unlike *Mannia*, which grows mostly as solitary thalli, *Plagiochasma* forms large colonies. The species are easy to distinguish when they are fresh. *Plagiochasma* shares with *Athalamia* (Cleveaceae) the character of the sporangiophore being produced on the surface rather than in a terminal notch of the thallus, in this instance with a fringe of white scales surrounding its base. The thallus rolls up to show a purple underside. The genus evidently does not occur in Europe but is found in China.

1a. Thallus true green; epidermal hexagons inconspicuous, not easily made out with the hand lens; pores usually surrounded by six radial rows of 2–3 cells. **P. wrightii**

1b Thallus glaucous, blue-green; epidermal hexagons easily visible with a hand lens; pores usually surrounded by one circle of 4–6 cells with no further radiate layers. **P. rupestre**

P. rupestre [on rocks]. Common on bases and crevices of rimrock cliffs of the Mesa de Maya. BA, LA.

P. wrightii [for Charles Wright]. BA: Canyonside, NE-exposure, Dodge Ranch SW of Utleyville, 4500 ft, *Shushan 166*; DA: Castlewood State Park, *W&W B-114927*.

BLASIACEAE (BLS)

We have the single genus, *Blasia*, a rather small, thin, green thalloid liverwort with distinctly crenate and ruffled margins. Scattered over the thallus are dark purplish, swollen spots which are actually symbiotic colonies of *Nostoc* living in the thallus. The plants commonly form rosettes. The branches have a more or less distinct mid-vein. The most outstanding feature is the production of gemmae in flask-shaped organs on the upper distal surface of the thallus. The flasks point forward in the direction of growth and are abruptly narrowed to an elongate neck through which the oval gemmae are discharged a few at a time. *Blasia* may also produce star-shaped gemmae directly on the thallus surface.

Blasia [for Blazius-Biagi]

B. pusilla [little]. *Blasia* occurs typically on raised grassy hummocks in wet grasslands. LR: Heavy soil on knoll in swale at base of bluff near Little South Cache La Poudre River, ca. 9000 ft, *Hermann 16980 (B-9120)*; SJ: South Mineral Creek, *Jamieson 7931*.

CLEVEACEAE (CLV)

The family consists of a single monotypic genus, *Athalamia*. The thallus surface has hexagonal lines marking the internal air spaces, as in *Marchantia*. The underside has hyaline or pale violet scales with tapering appendages protruding beyond the thallus margins. (Scale cells are large and iridescent.) Thallus color is pale crystalline green, the margins not very purplish (compare with *Mannia*). Carpocephala are dorsal on the thallus center (as in *Plagiochasma*), not from an apical notch. Pit-like depressions fringed with white scales mark where the carpocephala will arise. Pores have thickened radial (stellate) cell walls. Slender, white stalks emerge from sunken antheridia.

Athalamia [lacking internal chambers]

A. hyalina. A common thalloid liverwort of the alpine and subalpine (montane). BL, GN, HF, LR, PA, SH.

See *Preissia* for discussion of differences.

CONOCEPHALACEAE (CNC)

Our only genus, *Conocephalum*, is the giant among Colorado liverworts. It is larger than *Marchantia*, more yellow-green in color, more flabby than stiff, and has coarse polygonal areoles, each with an obvious white pore margin. It is said to have a fragrant odor (we find it more peppery). There are no gemma cups. The carpocephala are rarely produced in Colorado, but when present the stalked ones are only female and conical (whence the name *Conocephalum*!). Antheridial receptacles are sessile on the thallus surface.

Conocephalum [cone-head, for the disc of the carpocephalun]

C. salebrosum. A common species along small streams in the foothills and montane zones. *C. conicum* (the name most used for our plant) is exclusively European, while *C. salebrosum* is circumpolar. The two species differ in the shiny surface of the former and the matte surface of the latter, as well as fundamental differences in the structure of the pores.

LUNULARIACEAE (LNL)

The only genus and species in the family is *Lunularia*, a greenhouse weed unlikely to be found in nature in Colorado. It looks like a small *Marchantia*, but the gemma cups are crescentic, and the thallus has a glistening appearance

Lunularia [little moon, alluding to the gemma cups]

L. cruciata [tortured]. This commonly occurs on the soil of greenhouse benches in Boulder, *Weber B-7287*. It evidently does not survive here in nature.

MARCHANTIACEAE (MRC)

1a.   Ventral scales deep purplish, in two rows; gemmae absent; female receptacles shallowly 4–5-lobed on the margins, or hardly lobed; ventral scales without oil cells (or at most with only one or two). **Preissia**
1b.   Ventral scales colorless or pinkish, in 4 to 6 rows; gemmae in distinct, cup-like receptacles, female receptacles deeply divided into 5–9 lobes or segments; ventral scales with oil cells. **Marchantia**

Marchantia [for Nicolas Marchant]

1a.   Thallus thin, with a dark mid-line; stalk of male and female receptacles slender. **M. polymorpha**
1b.   Thallus succulent, brittle, without a dark mid-line, stalk of male and female receptacles short and stout. **M. alpestris**

M. alpestris. Common on the edges of streamlets and rills in the subalpine and alpine. The plants are lighter green than the next, and much puckered and raised, thick and rather brittle when fresh.

M. polymorpha. Common along streamsides from the plains to the montane. Replaced at higher altitudes by *M. alpestris*. This is the liverwort most usually taught in the schools as a standard example of a liverwort.

Preissia [for Ludwig Preiss]

P. quadrata [in fours, alluding to the four groups of cells surrounding the thallus pores]. Frequent on packed moist soil, especially along seeping cliffs. Much smaller than *Marchantia*, not forming an elongated and branched thallus. If one has difficulty in distinguishing between similar genera, no one has given us better advice than Schuster (1953): "The moderate size (usually between 6–10 mm wide and 2–4 cm long), and the dull green color, with the lateral thallus margins somewhat purplish, and the ventral scales always blackish-purple, give the species a rather characteristic appearance, even when sterile. Confusion is possible only with *Reboulia* and *Athalamia* (which are similar in size), while the smaller size (of *Mannia*) at once eliminates that genus. From all of these genera, and all other genera except *Marchantia*, *Preissia* differs at once in its elevated, whitish, rather prominent compound [barrel-shaped] pores. In *Reboulia* the pores are simple (that is, flanked by a single layer of cells), and the epidermal cells are distinctly collenchymatous (in *Preissia* always thin-walled and lacking trigones). In *Athalamia* the pores are stellate and simple, and different from those of *Preissia*. *Reboulia*, with which *Preissia* frequently occurs, can usually be separated from *Preissia* in the field by the smooth and often slightly more yellow-green thallus, with the pores inconspicuous, while the thalli of *Preissia* are a little rough because of the many elevated, whitish pores, and usually of a duller green color. The margins of the thalli of *Preissia* are also usually bleached—never so in *Reboulia*."

METZGERIACEAE (MTZ)

This family is characterized by its thin, ribbon-like thallus one cell thick and only a few mm wide, with a distinct midrib of several cell layers; the thallus margin and in some species the surface, have marginal hairs. Our single genus, *Apometzgeria* is characterized by having a dense cover of hairs above and below. The family gets its name from Johann Baptist Metzger, a friend of Raddi).

Apometzgeria [*apo*, upon, for the dorsal thallus hairs]

A. pubescens. This is a flat, branched, one-cell thick thallus with a thickened midrib. In contrast with *Aneura pinguis*, it is thin and delicate and grows on relatively dry old growth subalpine forest floors with other liverworts. It is instantly distinguishable from the genus *Metzgeria*, one or two of which might be expected here, because the thallus is densely pilose on the surface.

PELLIACEAE (PLL)

The thallus in *Pellia* is broad (4–8 mm wide), is somewhat similar to *Aneura pinguis* but not fleshy or greasy, dull green, flat, and thin, more so on the margins. There is a faint suggestion of a mid-vein (absent in *Aneura*). It rarely is found fruiting here. The antheridia are minute pimple-like elevations on the surface. The pseudoperianth is a low flap or tube beyond which the calyptra may protrude. The seta and capsule are fragile, similar to those of leafy liverworts and are erect.

Pellia [for L. Pelli-Fabbroni]

P. neesiana [for Nees von Esenbeck]. This is common, occurring in a variety of wet sites, including streamsides, willow carrs, and edges of pools in the subalpine. BL, GN, and PA.

REBOULIACEAE (RBL)

Reboulia [for E. de Reboul]

The salient features of *Reboulia* are: Larger than *Mannia* or *Asterella* (about the size of *Preissia*); relatively smooth and yellow-green upper surface with inconspicuous pores (*Preissia* is rough, bluish-green, with elevated pores; the older thallus portions never become bleached or white (that of *Preissia* does); the carpocephala are moderately 5–7-lobed; the pores are surrounded by 4–5 concentric rings of cells; the epidermal cells have distinct trigones; the stalk of the carpocephalum has a cluster of narrow filamentous scales (in *Mannia* these are lanceolate-linear).

R. hemisphaerica. Uncommon in sandy soil. Evans (1915) reports a Brandegee specimen, and one from GA: Granby, *Bethel 3*. Other collections are from BA: Sand Creek Canyon, 4400 ft, *Weber B-4635;* BL: Four-Mile Canyon between Salina and Crisman, *Weber B-15736*. It is evidently uncommon here.

RICCIACEAE (RCC)

1a.   Thalli forming neat rosettes not more than 1 cm diameter; on wet mud or moist soil; if aquatic, then totally submerged and ribbon-like. **Riccia**
1b.   Thallus floating on the water surface with the epidermis exposed to the air. **Ricciocarpus**

Riccia [for P. F. Ricci]

1a.   Thallus with photosynthetic, internal tissue loose, forming large air-chambers (in cross section the thallus is not compact in structure; surface of terrestrial forms often lacunose and spongy with age; mostly aquatic or mud-inhabiting species of the plains . . . . (2)
1b.   Thallus with compact photosynthetic tissue; mostly mountain or desert-steppe . . . . (4)

2a.   Thallus with the segments narrowly linear, mostly less than 0.7 mm wide; occurring free-floating under the water surface, or stranded. **R. fluitans**
2b.   Thalli with segments relatively broad, usually over 0.7 mm wide, always terrestrial . . . . (3)

3a.   Thallus glaucous; dorsal epidermis becoming spongy. **R. frostii**
3b.   Thallus bright green; dorsal epidermis actually degenerating and disappearing. **R. cavernosa**

4a.   Thallus with either white ventral scales or marginal cilia . . . . (5)
4b.   Thallus lacking white scales or marginal cilia. **R. sorocarpa**

5a.   Thallus with conspicuous white ventral scales visible on margins of dried thalli; desert-steppe species. **R. austinii**
5b.   Thallus margins with distinct cilia, stout, usually numerous, 75–300 μm long. **R. beyrichiana**

R. austinii [for C. F. Austin]. A desert-steppe species occurring on sandstone-derived soils on the Western Slope. It is easily recognized because of the conspicuous white ventral scales that become visible when the thallus dries and curls up. Our single collection is from MF: North base of Douglas Mountain 6 mi W of Greystone, 2300 m; in soil accumulating in depressions at the base of shelving sandstone rimrock, *Weber B-49823*.

R. beyrichiana Hampe [for H. K. Beyrich]. We have a single specimen: JA: Moraine pond between Jack Creek ranch and Teller City, 12 mi E of Rand, in muck at edge of pond, 9300 ft, *Douglass 61-580*. Evans (1915) cited a specimen from BL: Lake Eldora, as *R. lescuriana* Austin.

R. cavernosa [for the degenerative openings in the thallus]. A fairly common species on wet streambanks and pond borders, low and middle altitudes. Evans (1915) reported this as *R. crystallina*. BL, EP, GA.

R. fluitans. A low altitude species. We have one verified collection, from the eastern plains: MR: With *Ricciocarpus natans*. in small side stream out of the main current, South Platte River 8 mi W of Fort Morgan, *Mattoon 7978*. It is probably frequent and widespread in the irrigated areas of eastern Colorado.

R. frostii [for C. C. Frost, intermediary between the collector, S. Watson, and the describer, C. F. Austin]. BL, JF, GN, ME, MF. Two noteworthy collections are

BL: Bed of drying pond 9 mi E of Boulder, *Weber & Shushan 7980*. The muddy bottom filled with this and *R. cavernosa* in the 1940s but eventually filled with cat-tails and no longer supports any liverworts; and BR: Just W of Church's lake, on drying mud flat around small pond 1 mi S of Broomfield, 5500 ft, *Weber 8659*. No liverworts have been found in recent years in the ponds of the Colorado Piedmont. Their margins have been dredged to prevent the formation of mud flats, and the margins of the larger lakes like Boulder Reservoir have been heavily polluted by Canada Goose droppings and support no vegetation.

R. sorocarpa [*sorus*, a heap, + *carpa*, body]. A common species in protected places especially on well-vegetated tundra slopes, generally on the sides and bases of overhanging tussocks of *Kobresia*. The spores are large (60 μm). BL, CC, MF, PA, PT.

Ricciocarpus

R. natans. An aquatic species that floats on the surface of slow streams in the lowlands. The thalli consist of whole or partial rosettes, especially where flooding prevents the development of competitive vegetation. It is probably common throughout the irrigated areas at low altitudes. CF, LP, MR, MZ. A representative collection is MR: South Platte River, with *Riccia fluitans*, in small side stream outside of main current 8 mi W of Fort Morgan, *Mattoon 7979*.

Leafy Liverworts

ANTHELIACEAE (ANH)

Anthelia [little flower]

Anthelia is an exceedingly common plant of melting snow-beds in the alpine tundra. However, it is rarely recognized as a plant because it forms black, more or less amorphous patches. It is only upon close examination that it is seen to consist of minute stems. In the field *Anthelia* usually has a grayish color because of a covering of either fungal hyphae or terpenoid chemical compounds.

A. juratzkana [for J. Juratzka]. There are two species of *Anthelia*. The second one, *A. julacea,* occurs in areas with a more oceanic climate than that of Colorado. Fortunately we do not need to distinguish them here, since technically they differ in spore size (*A. julacea* having spores 14–15 μm, *A. juratzkana* 17–20 μm). We have not seen *Anthelia* fruiting here. We have specimens from BL, GA, LR, PA, PT, and SJ. It can be found in most snow-melt basins.

CALYPOGEIACEAE (CLY)

Calypogeia [cup, + hypogeous, underground]

Calypogeia, our only genus, is characterized by having simple, entire leaves that are distinctly incubously inserted. The underleaves are large and either bilobed or virtually entire. At their base is a distinct cushioned area formed of smaller cells, the rhizoid-initial region. There is no perianth, but a fleshy, subterranean hairy perigynium arises from a short branchlet coming from the axil of an underleaf. The plants are pale, rather transparent, bluish-green in color.

1a.   Underleaves deeply divided (over half of their length) into two lobes . . . . (2)
1b.   Underleaves undivided, or shallowly (less than 1/4 of their length), into two lobes . . . . (4)

2a.   Leaf cells small (less than 30 μm in the apex, 25–35 x 30–45 μm in middle); underleaves more than 2(3.2–3.5) times wider than the stem. **C. suecica**
2b.   Leaf cells large (over 30 μm in apex, 35–50 x 45–70 μm in middle); underleaves less than 2(1–2) times wider than the stem . . . . (4)

3a.   Leaves bidentate or sharply pointed at apex; underleaves ca. 1.5–2 times wider than long, 1.5–2 times wider than stem. **C. fissa**
3b.   Leaves entire or narrowly rounded at apex; underleaves 1–1.5 times wider than long, 2–2.5 times wider than stem. **C. muelleriana**

4a.   Leaf cells 35–45 x 40–60 μm in mid-leaf; underleaves broadly orbicular, ca 1.2–1.5 times as wide as long, 2–3 times wider than stem; rhizoid initial area distinct; border of elongate cells indistinct. **C. integristipula**
4b.   Leaf cells 30–40 x 35–45 μm in mid-leaf; underleaves orbicular, almost as long as wide, 2–2.5 times wider than stem; rhizoidal initial area indistinct; border of elongate cells distinct. **C. neesiana**

C. fissa [cleft]. It is really questionable whether this occurs in Colorado. Hong's map shows the distribution as being confined to the Pacific coast states. Hong (1990) neither cites specimens nor shows the position of any Colorado stations on his map, p. 315. Nevertheless he states its range as including Colorado.

C. integristipula. Leaves obtuse, a little less wide than long; leaf border not apparent; leaf cells lacking trigones, 40–50 μm; oil bodies oblong, 3–4 per cell, lumpy, not colored; underleaves 2–3 times wider than stem, wider than long, shallowly lobed. the rhizoidal initial layer with debris clinging to it, hence very distinct (*B-112439*). Hong cites collections from BL, GA, GN. LR.

C. muelleriana [for K. Müller Freiburg]. Hong cites no

specimens, but has two stations marked on his map, p. 325, evidently from RT and JA.

C. neesiana [for Nees von Esenbeck]. Hong cites no specimens, but indicates two Colorado localities on the map, p. 327. evidently from GA and LR.

C. suecica [from Sweden]. Hong cites no specimens, but indicates a single station, probably in Rocky Mountain National Park, on his map, p. 317.

CEPHALOZIACEAE (CPH)

Cephalozia is our only genus. Its plants are tiny, creeping or prostrate. The stems have large, pellucid, thin-walled cortical cells. The leaves are obliquely inserted, bilobed, with the lobes often pointing toward each other, without cell wall thickening. There are no underleaves. The plants occur on peat, humus, or decaying logs. They are usually somewhat larger than *Cephaloziella*, which has almost transversely inserted leaves.

Cephalozia [head + (short female) branch]

1a.   Leaves slightly decurrent, orbicular, horizontally or obliquely inserted, bilobed 0.2–0.5 their length, the lobes somewhat connivent. **C. pleniceps**
1b.   Leaves never distinctly decurrent, ovate (longer than wide, 1.1–1.4 times as long as broad), almost transversely inserted, bilobed for 0.5–0.7 their length; lobes never connivent. **C. bicuspidata**

C. bicuspidata. Montane forests. BL: Spruce-fir woods below Ouzel Falls, Rocky Mountain National Park, 9400 ft, *Hermann 27787* (as *C. pleniceps*); GN: Iron fen. Mt. Emmons, 9600 ft, *Weber B-92630* (!Hong).

C. pleniceps [many heads]. Hong (1980) cites one collection: LR: Rocky Mountain National Park, Roaring River, *Hong 78-885*.

CEPHALOZIELLACEAE (CZL)

Cephaloziella

These are the smallest known liverworts. If you thought *Cephalozia* was small! The plants are only 1–8 mm long, filiform, with the leaves only a little wider than the stems. The leaves are distant and transversely inserted, bilobed for 1/2–3/4 their length, with the lobes only 2–9 cells broad at base. The leaf cells never have trigones. Our species are found on moist rotten wood and among mosses in high subalpine or alpine situations. They are usually discovered while one is dissecting a larger bryophyte.

1a.   Leaves often dentate; underleaves present and distinct. **C. divaricata**
1b.   Leaves edentate; underleaves absent or minute if present . . . . (2)

2a.   Leaf lobes oval-lanceolate, 3–8 cells wide at base; cells thick-walled; plants red-brown. **C. rubella**
2b.   Leaf lobes ovate-triangular, 5–15 cells wide at the base; cells thin-walled . . . . (3)

3a.   Plants purple-black; leaves erect or sub-erect; leaf lobes blunt or rounded, 6–12 cells wide at base; gemmae absent; calciphile. **C. varians**

3b.   Plants green-brown; leaves distantly strongly spreading; leaf lobes acute, 5–8 cells wide at base; gemmae pale green; non-calciphile. **C. hampeana**

C. divaricata [spread apart]. A fairly common green species from the foothills to alpine. BL: Boulder Canyon, *Weber B-10554* is representative.

C. hampeana [for Ernst Hampe]. Especially common on burned stumps. Hong cited specimens from BL, EP, and LR. A fine representative collection is BL: Rocky Mountain National Park, Bear Lake tourist trail; forming a continuous carpet on soggy ground of inlet area, *W&W B-111256*.

C. rubella. Easily recognized by its reddish brown color, this species is common on north-facing cliffs in the foothill canyons of the Front Range.

C. varians. A distinctive black species of the upper subalpine or alpine. Hong cited a specimen from LR: Rocky Mountain National Park, Trail Ridge, *Hultén s.n.* (S). ST: Blue Lake, on seeping limestone ledges, 11000 ft, where it occurs with the equally black *Didymodon subandreaeoides*. *W&W B-111181* (!Hong, as *C. arctica*).

FRULLANIACEAE (FRL)

Frullania has complicate-bilobed leaves, one lobe dorsal, the other ventral, but quite different from that of *Scapania*. The dorsal lobe is typically round, and the insertion incubous. A curious ventral lobe is seen underneath, which is smaller and most often three-dimensional, helmet-shaped like an upside-down cup, connected by a small stalk to the upper lobe. In some species the cup is imperfectly formed and not closed along the side (explanate). The stem also has underleaves, in our species always bifid. All of the species are pioneer types rarely mixed with other mosses except where they occur as thin colonies on cliffs. In Colorado all of our species tend to be cliff-dwellers, while in more humid climates they are more common on tree-trunks.

Frullania [for Leonardo Frullani, a Florentine municipal official]

1a.   Ventral lobes small (less than half the size of the dorsal lobes), strongly explanate (evidently with no helmet-shaped lobules), lanceolate; perianth with five keels; underleaves bifid to a third their length. **F. inflata**

1b.   Ventral lobes strongly inflated (helmet-shaped), rarely weakly explanate, especially near the stem apex; perianth with 3–9 keels . . . . 2)

2a.   Ventral lobes mixed, helmet shaped on more mature stems, explanate on the younger ones; underleaves large (3– 4 times as wide as the stem), wider than long. **F. riparia**

2b.   Ventral lobes inflated uniformly, never explanate; underleaves small (up to 2 times as wide as the stem). **F. brittoniae**

F. brittoniae [for E. G. Britton]. Frequent in seepage zones on vertical cliffs, CC, GA, HN, MN, RG.

F. inflata. This species is usually found on bark. It has not yet been found in Colorado, but should be expected in the southern counties, since it comes as far north as Las Vegas, New Mexico. There may be small pockets in Front Range canyons where the humidity is high enough to support the species.

F. riparia [of streams]. Our only collection is from LA: Carrizo Creek, *Shushan B-6903*.

GEOCALYCACEAE (GCL)

This family contains genera that are not obviously similar. The leaves are succubous or obliquely inserted; the underleaves are bilobed; rhizoids are restricted to the bases of the underleaves or scattered over the underside of the stem; fertile branches are lateral or ventral; androecial branches are spicate, and many other contrasting characters. It is best to learn *Geocalyx, Lophocolea*, and *Chiloscyphus* in the field and forget about the family!

Engel & Schuster (1984) reduced *Lophocolea* to synonymy under *Chiloscyphus*. Their opinion is followed by Damsholt. Hong rejects this merger because of differences in the perianth, male inflorescence, leaves, underleaves, and bracts. However, for convenience we are maintaining the traditional nomenclature.

1a.   Perianth present, elongate, exserted, trigonous; perigynium or marsupium lacking; capsule oblong-ovoid, the wall 4–5-stratose; rhizoids restricted to the bases of the underleaves. . . . . (2)

1b.   Perianth absent, more or less spicate; female inflorescence terminal; leaves bilobed; rhizoids scattered over the ventral part of the stem . . . . (3)

2a.   Male inflorescence more or less spicate; female inflorescence terminal; leaves bilobed; underleaves with a subulate-acuminate tooth on each side. **Lophocolea**

2b.   Male inflorescence undifferentiated; female inflorescence on short lateral branches; underleaves with a spinose tooth on each side. **Chiloscyphus**

3a.   Autoicous; perianth absent, replaced by a subterranean rhizoidous perigynium (marsupium); underleaves deeply bilobed, with entire margins; gemmae absent; male bracts in 4–8 pairs. **Harpanthus** (to be expected)

3b.   Dioicous; perianth present, fused with shoot-calyptra; underleaves divided, subulate; gemmae present; male bracts in 2–5 pairs. **Geocalyx**

Chiloscyphus [lipped cup, alluding to a perianth characteristic]

1a.   Plant body transparent, pale whitish- to yellowish-green; cells at mid-leaf 35–40 x 45–60 μm; perianth lobes spinose-dentate. **C. pallescens**

1b.   Plant body dull to deep green or often blackish; cells at mid-leaf 25–30 μm; perianth lobes entire or undulate. **C. polyanthus**

C. pallescens. Common throughout the montane and subalpine forested areas. This is a close relative of the next, but differs in chromosome number (18).

C. polyanthus. Common through the montane and subalpine forested areas. This species has 9 chromosomes. A variant occurring in swift-flowing streams is dark green (var. *rivularis*).

Harpanthus [sickle-shaped 'flower']

H. flotovianus [for J. Flotow]. Hong reports a collection from southern Wyoming. It probably occurs in northern Colorado, in the Park Range of RT or JA.

Geocalyx [earth, + cup (marsupium)]

G. graveolens [strong-smelling]. Schuster says this is one of the easiest bilobed species to identify. The stems are prostrate, with shallowly notched yellow-green leaves with parallel sides, spreading flat and perpendicular to the stem. Under the microscope, the bifid, rather large, underleaves, with the linear-lanceolate divisions nearly parallel, entirely unarmed on the outer side, are absolutely characteristic. The trigones of the cells are small, and fresh material has from 6 to 10 minute oil bodies. *Lophocolea* may have similarly divided underleaves, but the lobes usually bear a sharp tooth on the outer side; in *Geocalyx*, also, rhizoids occur not only on the bases of the underleaves, but some occur scattered over the rest of the ventral side of the stem. In *Lophocolea* they are limited to a small area at the bases of the underleaves . . . On the ventral side of the stem the fleshy perigynium (marsupium) is usually evident; when young these are small, spherical, and look like small tubers; when mature they are cylindrical and prominent. No other species with bilobed leaves has such perigynia.

We have one Colorado record: LR: Rocky Mountain National Park, 1.2 mi N of Deer Creek junction on Hwy 34 to Estes Park. On ground, tree bases, and rotting wood in mesic *Abies-Populus-Alnus* woodland with abundant dead-fall and extensive seepy areas, also along the roadside, *Vitt 15255* (cited by Hong). This collection represents the southernmost station for the species in North America.

Lophocolea [crested sheath (the ridged perianth)]

1a. Plants pale whitish-green; leaves entire or sub-entire to bilobed; gemmae rarely developed. **L. heterophylla**
1b. Plants greenish-yellow; leaves bilobed 0.2–0.35 of leaf length; gemmae always abundantly developed along leaf margins. sometimes making the leaf ragged. **L. minor**

L. minor. Abundant in moist forests and streamsides throughout the mountains. Probably much more common than the next.

L. heterophylla. Abundant in moist forests throughout the mountains.

GYMNOMITRIACEAE (GYM)

This family consists of plants with tiny bilobed, transversely attached, non-decurrent leaves. They are not fen plants but usually found either in forests on rotting wood (*Marsupella*), or on alpine cliffs or rocks (*Gymnomitrion*). In our experience they are rare.

1a. Leaves densely imbricate, grayish-white or grayish green to pale green. The texture is like coral. **Gymnomitrion**
1b. Leaves spreading from the stem, distant in sterile shoots, mostly forming dark green or black tufts. **Marsupella**

Gymnomitrion [naked cup, absence of a perianth]

1a. Plants silvery-gray; leaves bilobed to 0.05–0.15 their length; leaf lobes rounded; cuticle smooth; marginal cells thin-walled, sloughing off and making the leaf erose. **G. corallioides**
1b. Plants greenish to reddish brown; leaves bilobed to 0.15–0.35 their length; leaf lobes apiculate to acute; cuticle smooth to papillose; marginal cells thick-walled. **G. concinnatum**

G. concinnatum. Hong cites a specimen: BL: Long's Peak, *Kiener 8971* (F). See discussion below. Kiener gave his two specimens different numbers, so perhaps he recognized that they did look different in the field.

G. corallioides. A rare plant of high altitudes, 13000 ft or higher. We have three collections: BL: Longs Peak, on siliceous rock ledges, 13500 ft, *Kiener B-8970*. CC: Rocks in the inlet of Summit Lake, Mt. Evans, with *Stereocaulon rivulorum*, 12500 ft, *Weber B-9014*; SJ: Eureka Gulch, *Jamieson 13804*. Both of the Kiener collections

appear to be the same species, and were so verified by Evans in 1955 and by Steere in 1956. Neither collection is silvery gray, but brown. The marginal cells are fragile and their degeneration makes the leaf erose. I believe it is entirely possible that we have only one species and that it is *G. corallioides*. However, since Hong claims they often grow together, perhaps the Kiener specimen that Hong examined at Harvard (F) is actually *G. concinnatum*, so we cannot rule it out. We really need more collections of this rare genus.

Marsupella [little money-bag]

1a.   Cortical of the stem (in cross-section) resembling the inner ones; usually on moist cliffs. **M. emarginata**
1b.   Cortical cells of the stem thin-walled and twice as large as the inner ones; forest plants. **M. sphacelata**

M. emarginata. The only report is that of Evans (1915), who cited this from Pikes Peak, 1896, *Holzinger.*

M. sphacelata [rotten]. We have two collections: CR: On ground and over rotting wood under spruces, South Colony Creek, 9500 ft, *Kiener B-10218*; LR: Rocky Mountain National Park, on soil SE of Haynach Lake, *Willard B-62167* (!Hong).

HAPLOMITRIACEAE (HPL)

Haplomitrium [naked cup, alluding to the calyptra]

H. hookeri. *Haplomitrium hookeri* is known in North America from Mount Katahdin, Maine, and Mount Washington, New Hampshire, and one locality in Colorado. It is a strange liverwort, consisting vegetatively only of a slender stem no more than a cm long, the lower part of which is a white and tenderly succulent rhizome. The leaves are remotely scattered along the stem, somewhat diamond-shaped, and shallowly and unequally lobed or completely entire. The cells are thin-walled, 25–40 x 30–50 µm. For those who would like information about the sexuality, see the detailed treatment of Schuster. One location, BL: Green Lakes Valley, on 'Haplomitrium Hill', 11800 ft, *N. G. Miller B-43801*, *Weber, Gradstein, & Amann B-57020.*

Schuster writes: "The chief problem with *H. hookeri* is not to identify it but to be able to locate it. It is a difficult plant to find in the field. The plant will not under any condition be confused with any other liverwort. It is quite likely, however, to be mistaken for a moss because of the erect growth, the lack of differentiation between lateral and dorsal leaves, and the form of the leaves ... The plant is exceedingly sensitive to drying, apparently lacking all toleration for intermittent moisture conditions."

The plant hardly appears more than a few mm above the ground and is lost among the welter of other low-growing plants. The single Colorado collection is the only one known in America south of the area of Continental Pleistocene glaciation. At Green lakes Valley it was accidentally found when Norton Miller dug up a patch of peaty soil and found the white, coralloid basal portion of the shoot system. Miller was already acquainted with it in the field in Scotland and knew just where to look. Everyone who has had the joy of collecting this has never found more than a few plants.

JUNGERMANNIACEAE (JNG)

Jungermannia [for Ludwig Jungermann]

The genus is characterized by its dioicous or paroicous gametangia, prostrate to erect stems, lateral branching, abundant rhizoids, succubous, obliquely inserted leaves, absence of underleaves, well-developed perianth (fusiform, pyriform, and cylindrical) usually with an abruptly constricted mouth and oval capsule with bistratose spiral elaters. *Plectocolea* and *Solenostoma* are now regarded as subgenera of *Jungermannia.*

1a.   Perianth abruptly narrowed distally in a beaked mouth . . . . (2)
1b.   Perianth gradually narrowed, not beaked . . . . (4)

2a.   Perianth tubular, smooth; leaves ovate-oblong, parallel-sided. **J. subulata**

2b.   Perianth cylindrical, ovoid to clavate; leaves reniform-orbicular to slightly ovate, erecto-patent (subgenus Solenostoma) . . . . (3)

3a.   Plants green; leaves orbicular to orbicular-cordate; leaf cells 25–30 x 35–50 μm in mid-leaf; rhizoids arising almost exclusively from ventral side of stem, not forming a bundle. Paroicous. **J. sphaerocarpa**

3b.   Plants red-purple, 1.0–1.5 mm wide, 0.8–2.0 cm long; leaf with a distinct border; cells 20–25 x 25–35 μm in mid-leaf, with small trigones, more than one oil body per cell. Dioicous. **J. rubra**

4a.   Perianth plicate above only, with elongated cells; perigynium present; rhizoids sometimes reddish-purple (Subgenus Plectocolea) . . . . (5)

4b.   Perianth plicate for almost the entire length, with isodiametric cells, perigynium lacking; rhizoids usually colorless (Subgenus Jungermannia) . . . . (6)

5a.   Plants usually dark green to purple; leaves oval-ovoid; oil bodies 1–4 per cell, mostly small, smooth; female bracts reflexed; rhizoids purplish or not; cuticle striate. Paroicous. **J. obovata**

5b.   Plants pale green to brownish-red; leaves circular, trigones distinct; rhizoids mostly colorless; oil bodies 4–5 per cell, verrucose, oval; cuticle smooth. Dioicous. **J. hyalina**

6a.   Perianth fusiform; leaves broadly elliptical, cells 20–25 x 25–30 μm in mid-leaf. Paroicous. **J. pumila**

6b.   Dioicous . . . . (7)

7a   Plants 0.8–1.0 mm wide, 0.2–3.0 cm long; perianth ovoid or oblong-ovoid; androecia terminal; male bracts in 4–12 pairs. **J. lanceolata**

7b.   Plants 0.5–4 mm wide, 0.3–12 cm long; perianth slender fusiform. **J. exsertifolia**

J. exsertifolia. In general aspect, this elongate, usually black liverwort resembles *Scapania irrigua*. It is common in wet areas of montane and subalpine forests.

J. hyalina. This species grows "on moist sandy or clayey soil, such as the sides of ditches and forest paths, more rarely on moist, siliceous rocks" (Arnell 1956). SJ: Coal Creek, 10,500 ft, *Jamieson 7976 (B-114401)*.

J. lanceolata. There is one report, not verified: SJ: Silverton, 1931, reported by Frye (1937).

J. obovata. This always grows on wet rocks and with us it is alpine. The dark color, purple rhizoids, reflexed female bracts, striate cuticle and paroicous inflorescence are diagnostic. We have one collection:

ST: Blue Lake Dam area, Monte Cristo Creek Valley, 3000 m, on limestone terraces irrigated with snow melt; in tufts of *Distichium capillaceum*, *W&W B-111214* (!Hong).

J. pumila. A fairly common species of moist sides of hummocks and rivulets in the subalpine. BL, CF, GA, LR, ML.

J. rubra. Dominant in fens and evidently restricted to the San Juan Mountains, disjunct from California. Collections from MN, SM, SJ.

J. sphaerocarpa. A common subalpine species. AA, BL, CF, GA, LR, PA, RT.

J. subulata. Common on streamsides in spruce-fir forests. GA, GN, LR.

Nardia [for S. Nardi]

Nardia is characterized by its entire (in our species), succubous orbicular-reniform leaves; underleaves distinct, lanceolate to triangular; gemmae none; female bracts similar to the leaves; perianth short, within bracts, with crenulate mouth and oval-globose capsule with bistratose walls. Rare in Colorado.

1a.   Plants large, 1.0–2.5 mm wide, 1.0–5.0 cm long, pale green; cells large, 25–30 x 30–40 μm in mid-leaf; oil bodies glistening and homogeneous; dioicous. **N. scalaris**

1b.   Plants small, 0.7–1.4 mm wide, 0.4–2.5 mm long, brownish; cells small, 20–25 x 25–30 μm in mid-leaf; oil bodies granular and opaque. **N. geoscyphus**

N. geoscyphus [earth cup]. BL: Green Lakes Valley, Indian Peaks, *Gradstein 2263 (B-45621)* (!Vana, !Hong); *Kiener 3377* (YU) (!Hong). All of our species have entire, emarginate leaves.

N. scalaris [ladder-like]. One report, unverified, from "Colorado", *Röll (BM)*.

LEPICOLEACEAE (LPC)

Gymnocolea [naked sheath (perianth)]

G. inflata. Probably the most common, even dominant liverwort in subalpine fens, especially in shallows or around the edges of rock pools. It is a tiny dark brownish or blackish plant producing simple, unbranched stems. The colonies are usually dense. The leaves are incubous but hardly overlapping, almost round, shallowly two-lobed above the middle and the leaf lobes curve in, cup-like. There are no underleaves. The perianth is like an inflated football with several short broad teeth around the narrow mouth. The leaf cells are uniformly thick-walled, without trigones.

LEPIDOZIACEAE (LPD)

The Lepidoziaceae is a family notable for having genera with variously terminally toothed or multi-cleft leaves, some having the leaves so deeply and narrowly divided that they look like cottony ropes! Many are tropical.

Lepidozia [scaly twig (imbricate female bracts)]

L. reptans. Schuster says: "The incubous leaves, with the apices decurved and 3–4-lobed, giving them the appearance of a cupped hand (when seen ventrally), at once identifies this species. This, together with the frequent presence of terminal flagella-like branches and the regularly pinnate branching, serve to characterize this species in the field." This is an uncommon plant, but possibly it is overlooked because it is so small and tends to grow on rotting logs. BL: Sandbeach Lake trail, 2 mi W of Copeland Lake, Rocky Mountain National Park, *Hermann B-27389*; LR: On rotting wood in the stream, Hidden Valley, *Weber B-36502, B-56609*.

LOPHOZIACEAE (LPH)

1a.   Leaves all or mostly 3–4-lobed, at least on well-developed shoots, underleaves sometimes present; perianths plicate . . . . (2)
1b.   Leaves all or nearly all bilobed on sterile shoots; underleaves lacking . . . . (3)

2a.   Leaves all or mostly 4-lobed, nearly flat; plants growing horizontally, with the leaves at about a 15–30° angle with stem, nearly horizontally spreading. **Lophozia** subgenus *Barbilophozia*
2b.   Leaves 3- or 2-lobed (a least in part 3-lobed), cochleariform; plants erect or nearly so with the leaves inserted at a 40–60° angle with the stem; leaf lobes usually incurved. **Tritomaria**

3a.   Perianth cylindric, with the apex suddenly and abruptly truncated into a short beak. **Lophozia** subgenus *Leiocolea*

3b.   Perianth distinctly plicate, gradually narrowed, never beaked. **Lophozia** subgenus *Lophozia*

Lophozia subgenus *Lophozia* [lobed twig]

Lophozia is our largest and most difficult genus. It is also one of the most commonly collected groups. A few of the species are easy to recognize, and Schuster (1953) has given us a great deal of information from his unique knowledge the genus in the field. Authorities disagree as to whether *Lophozia* is one genus or many (including *Barbilophozia, Dilophozia, Leiocolea, Massula,* and a few others).

1a.   Leaves opaque, densely chlorophyllose and a peculiar pale bluish-green, 2–3-lobed, the lobes acute to acuminate, often provided with supplementary teeth; cells thin-walled, with about 50 minute oil bodies; apical leaves imbricate and forming a crispate head; plants less than 1 cm long with thick, fleshy stems; gemmae green. **L. incisa**

1b.   Not as above . . . . (2)

2a.   Leaf cells thin-walled, never with trigones. **L. obtusa**

2b.   Leaf cells thickened (with trigones) at the corners, but these sometimes weakly defined . . . . (3)

3a.   Paroicous species, fruiting abundantly, with cochleariform, dentate male bracts below the perianth; small, less than 1cm long; leaves as wide or wider than long, suborbicular, with a broad, shallow sinus; gemmae reddish-brown or purple; cells lacking obvious trigones. **L. excisa**

3b.   Dioicous species; male and female plants often in separate patches; margins of male bracts not dentate (except for an occasional basal tooth) . . . . (4)

4a.   Shoots ascending or sub-erect, with the leaves nearly transversely inserted and mostly erect-spreading, the leaves much longer than wide, narrowly ovate; perianth mouth ciliate-dentate . . . . (5)

4b.   Shoots creeping or decumbent, only the tips slightly ascending; leaves more or less spreading, never erect, suborbicular to broadly ovate-rectangular, as wide or nearly as wide as long; perianth with teeth 1–4 cells long (Section Ventricosae; *these are the toughest ones to distinguish!*) . . . . (6)

5a.   Gemmae greenish-yellow, 10–14 x 16–22 µm; plants usually 3–10 mm high; oil bodies nearly smooth; plants pale yellow-green; on decaying logs. **L. ascendens**

5b.   Gemmae reddish-brown or reddish-yellow, 14–18 x 20–28 µm; shoots 0.8–2.5 mm high; oil bodies appearing coarsely papillose; plants deep or pure green. **L. longidens**

6a.   Leaves with more or less incurved lobes, more or less strongly cupped, cochleariform or canaliculate-convex; perianth mouth with teeth 1 (or at most 2) cells long; leaves 0.9–1.2 as wide as long. **L. wenzelii**

6b.   Leaves with more or less spreading lobes, not strongly canaliculate-cochleariform; leaf sinuses descending 1/4–1/3 the leaf height, generally obtusely to acutely angular . . . . (7)

7a.   Leaves 0.75–0.95 times as wide as long, clearly, even if slightly, elongate; perianth mouth with teeth 2–4 cells long; trigones strongly bulging; oil bodies 6–10 per cell. **L. longiflora**

7b.   Leaves 0.95–1.10 times as wide as long (at least on mature robust shoots); perianth mouth with teeth 1–2 cells long; trigones moderately to slightly bulging; oil bodies 10–15 per cell. **L. ventricosa**

L. ascendens. A small, pale yellowish-green plant 0.8–1.3 mm wide, 0.3–1.5 cm long, with strongly ascending growth form, narrowly triangular leaf-lobes with scarcely bulging trigones, abundant 1–2-celled yellowish-green gemmae, the cylindric perianth with a laciniate-lobulate mouth with 10–15 lobes and teeth 5–10 cells long and 3–5 cells wide at the base. LR: Rocky Mountain National Park; on decayed wood, Timber Lake trail, *Hermann 79-934*, cited by Hong (1980); GA, *Hermann 25473 (B-57202)*, BL: Niwot Ridge, *Hermann 27425*.

L. excisa. Hong characterizes this as being abundantly fertile, green or tinged above with red, large and thin-walled leaf cells (28–30 x 30–40 µm, with minute trigones, 1–2-celled vinaceous to purplish gemmae and strongly plicate perianth with a crenulate mouth. Hong separates this from the other species by the fact that the medulla of the stem is differentiated dorsiventrally and contains mycorrhizal fungi. We have a specimen so named, from CR: South Colony Creek basin, 11700 ft, *Kiener 10238*.

L. incisa. A common species in the subalpine coniferous forests. It is easily recognized with just a hand lens by the pale green color and the shoots that are shaped like loose lettuce heads; the leaves are variously incised with sharp pointed lobes. Under the microscope the leaf cells are filled with masses of chloroplasts and the tiny oil bodies (about 1 µm diameter may be found among them. (*L. opacifolia* Culmann & Meylan is recognized by Hong but has been made a subspecies of *L. incisa* by Schuster & Damsholt).

L. longidens. A strongly ascending species with ovate-rectangular, longer than wide, leaves with straight, elongate, hornlike lobes, masses of red-brown 1–2-celled gemmae at the lobe apices, and a long-exserted perianth with a ciliate or ciliate-dentate mouth. We have one collection: LR: Rocky Mountain National Park; infrequent on rotten logs, moist banks and rocks in the subalpine, Hidden Valley, *Hermann 27611* (!Hong). Frye & Clark (1945) reported this from Independence Pass, *Rakestraw* in 1938.

L. longiflora. Evans (1915) cited the following specimens: AA: Near Pagosa Peak, *C. F. Baker*; GL: Rollins Pass, *Bethel*. We have a recent collection, BL:

Lake Eldora, *Weber B-22082*. Hong accepted the species for Colorado but cited no specimens.

L. obtusa. This is characterized, according to Hong, by the frequent presence of underleaves, uniformly bilobed (to 1/4 their length) leaves with rounded or obtuse lobes, almost isodiametric thin-walled cells (30–35 µm in mid-leaf) and slenderly cylindric perianth with a dentate mouth. Hong (1980) reported this from LR: Rocky Mountain National Park, Poudre Lake, *Hong 78-848*.

L. ventricosa. Probably the most common species in the forested area. A prostrate or creeping plant 08–2.5 mm wide, 1.0–2.5 cm long, with ovate, bilobed leaves with bluntly acute lobes, lack of coarse trigones, abundant greenish, angular, 2-celled gemmae and long emergent perianth with pluriplicate mouth and 1–2-celled teeth. The stems are short and erect and tightly packed; the leaves, seen from the top, are distinctly cupped, the lobes not turning outward. BL, GN, LR, ST:

L. wenzelii [for Wenzel von Kroyanke]. Hong describes this as follows: "This large species (1.2–1.6 mm wide, 1.0–8.0 cm long) is characterized by its prostrate growth form, sub-vertically inserted broadly orbicular leaves with strongly arched margins, and broad crescent-shaped sinus, homogeneous ellipsoidal oil-bodies (4–10 per cell), angular, 2-celled, pale green gemmae and deeply plicate ob-pyriform one-celled perianth teeth." Hong cites no collections and presents no map, but lists this from Colorado. It is possible that what we have been calling *L. wenzelii* is actually *L. sudetica*.

Lophozia subgenus *Barbilophozia* [bearded, for ciliate-based leaves]

The most easily recognized group of species in the genus. The plants are fairly large, that is, they are elongate creeping plants with leaves that are not densely crowded into heads. Their leaf arrangement is succubous, and the leaves are large enough and spreading enough to tell with a hand lens that they are distinctly 3- or 4-lobed. *L. hatcheri* is the species most frequently encountered from the foothills up into the montane. Its leaves have a few long cilia at the base. Unless you are myopic, the leaves need to be examined with a microscope (see the key). With experience some of the species can be recognized on sight.

1a.   Leaves cochleariform, obliquely inserted . . . . (2)
1b.   Leaves nearly flat, horizontally inserted . . . . (4)

2a.   Leaves mostly 2–3-lobed, the sinus descending 1/4–1/3 of their length; plants with attenuated erect shoots whose leaves are smaller and appressed. **L. attenuata**
2b.   Leaves mostly 2 (rarely 3–4)-lobed, the sinus descending half to 3/5 of their length . . . . (3)

3a.   Leaves divided into two (occasionally 3–4) lobes; sinus descending about half their length; lobes blunt to sub-acute; dorsal leaf base usually with a short cilium. **L. kunzeana**

3b.   Leaves divided into three lobes, the sinus descending less than 1/3 their length; lobes sharply acute. **L. floerkei**

4a.   Ventral leaf bases lacking cilia; underleaves rare or none and little or not ciliate; leaves never mucronate or apiculate; gemmae absent; leaves symmetrical, with more or less triangular lobes. **L. barbata**

4b.   Ventral leaf bases with a few distinct cilia, formed by several cells 3–10 times as long as wide; underleaves present, ciliate; at least some leaves mucronate . . . . (5)

5a.   Leafy shoots normally 1.5–2.7 mm wide; brown gemmae common on the proximal leaves; leaves flat, the lobes ovate-triangular, usually mucronate. **L. hatcheri**

5b.   Leafy shoots normally 4–5 mm wide and 4–8 cm long; gemmae rarely present; leaves undulate-crispate. **L. lycopodioides**

L. attenuata. Damsholt (2002) says this species is "usually distinguishable by (1) having numerous erect, stiff and filiform gemmiparous branches from the apex of the main shoots, and (2) having main shoots with 3-lobed leaves. In the few cases where such gemmiparous branches are lacking, *Lophozia attenuata* may be confused with *Lophozia hatcheri* and *L. floerkei*. *L. attenuata* differs from *L. hatcheri* and *L. floerkei* in lacking underleaves, having sub-quadrately lobed leaves, and lacking thread-like appendages of the ventral leaf bases. CC: Summit Lake, Mt. Evans, in a moist pot-hole, *Wikel & Colson B-114669*. Frye & Clark (1945) cite specimens (as *Orthocaulis gracilis*) from "Silverton, 1931, *Frye.*" and "Longs Peak, *Kiener.*" Hong does not mention the species. Our collection lacks the elongate shoots, but in all other respects if fits *B. attenuata*.

L. barbata [ciliate]. Frequent in the subalpine and upper montane forests. The least common of the three larger species.

L. floerkei [for H. G. Floerke]. Distinguished by having almost consistently 3-lobed leaves, short-celled appendages of the ventral leaf margins, and distinct underleaves. It may be confused with small *B. hatcheri*, but that species always has apiculate lobes and at least some leaves have long, contorted thread-like appendages on the ventral leaf margin. CC: Summit Lake, Mt. Evans, 13500 ft, *Weber, Porsild. & Holmen B-4347* (!Hong).

L. hatcheri [for J. B. Hatcher]. By far the most common species in the foothills and montane zone. In the old growth subalpine spruce forests it is largely replaced by *L. lycopodioides*, a handsome and conspicuous species in comparison. It also occurs on South Georgia Island and in Antarctica, and is the only species of the genus that is bipolar in distribution (Bednarek-Ochyra et al. 2000).

L. kunzeana [for Gustav Kunze]. Evans (1915), reported a collection: EP: Minnehaha, 1913, *G. E. Nichols*. We have collections from the San Juan Mountains: OR: Ironton Park, *Wikel B-113571*. There it is closely associated with *Tritomaria quinquedentata*.

L. lycopodioides. A beautiful, large species characteristic of forest floors in mature, relatively undisturbed moist subalpine spruce forests. It is like a large edition of *L. hatcheri* but differs in its undulate leaves and lack of gemmae.

Lophozia subgenus *Leiocolea*

Fertile plants of this genus are easy to recognize in the field. It is the only bilobed group of liverworts having smooth, cylindrical perianths whose apices are suddenly constricted into a distinct, short beak. A similar beak occurs in *Jungermannia*, but in that genus the leaves are entire.

1a.   Plants small (less than 1.0 mm wide and 12 mm long); underleaves absent. **L. badensis**

1b.   Plants large (over 1.0–5.0 mm wide and 1.0–8.0 cm long); underleaves present . . . . (2)

2a.   Gemmae present. **L. heterocolpos**

2b.   Gemmae absent . . . . (3)

3a.   Leaves bilobed up to 25%; cells large, 35–40 x 40–55 μm in mid-leaf. **L. bantriensis**

3b.   Leaves bilobed more deeply; cells small, ca. 25–30 x 25–35 μm in mid-leaf. **L. alpestris**

L. alpestris. Hong reported this from LR: Poudre Lake and Timber Creek Trail, *Hermann 78-843 and 79-934*. "This green to brown species is characterized by the lack of a mycorhizal band in the stem, broadly ovate to rotund-quadrate leaves, small cells, bulging trigones, grayish, finely granular-papillose oil bodies (2–6 per cell), and distinct lanceolate to subulate underleaves. It occurs on humus in woods, in litter over limestone, and on moist boulders beside streams."

L. badensis [from Baden, Germany]. This tiny species is characterized by its translucent, slender stems, leaves uniformly and shallowly bilobed, almost circular, with lobes many cells wide at their bases, large leaf cells (20–30 μm) in proportion to plant size, short-beaked perianth and general absence of underleaves (which rarely occur as slime papillae). The species occurs among mosses on damp rotten wood in spruce forests. Even sterile plants can be recognized by their small size, pale translucent, shallowly and broadly bilobed leaves. At first one thinks of *Cephalozia*, but the lobes are broad and not connivent; then perhaps a minute *Lophozia*. One collection was cited by Hong: CF: Buena Vista, *Conard, 40-870* (G). Our collection from SJ: Weminuche Wilderness, NE flank of Mt. Silex, just above confluence of Vallecito and Trinity Creeks, 10600 ft, *Jamieson 11490*. It is known otherwise in western North America from northern Montana to Alaska.

L. bantriensis [from Bantry, Ireland]. "This green to reddish brown, large species is characterized by its scarce rhizoids, shallowly bilobed leaves with large cells, distinct linear-subulate to lanceolate underleaves, large, pale brownish oil-bodies (2–8 per cell) and emergent cylindrical perianth with a short beak. This species occurs on moist sandy soil and rocks near streams." Hong reports this, without citations, from what appear on his dot map to be LR and MZ. Our specimen is from LR: Trail ridge, 10800 ft, *Kiener 7120 (B-21846, B-50498)*.

L. heterocolpos [variable womb]. Hong reports this, without citations, from four localities in western Colorado. It appears to be the most common and widespread species in western North America. "This green to golden-brown to reddish-brown species is characterized by its abundantly occurring gemmiparous shoots with brownish 2-celled oblong to ovoid gemmae, leaves slightly longer than wide, strongly collenchymatous small cells with bulging and often contiguous trigones, grayish ovoid to ellipsoid oil bodies (2–5 per cell) and distinct ovate-lanceolate to lanceolate underleaves." CF: Osburne Creek, San Isabel Nat. Forest, *Hong 79-1012*. GN: Gothic Natural Area, 10000 ft, *Weber B-9127*; SJ: Trail from South Mineral campground to Ice Lake Basin, 11000–11800 ft, *Weber B-9541a* (!Persson); ST: Monte Cristo Creek, 11000 ft, *N. G. Miller B-115618*.

Tritomaria [thrice-divided]

1a.   Leaves symmetrical, equally trilobed with blunt or rounded apices; perianth mouth entire or subentire. **T. polita**

1b.   Leaves asymmetrical, unequally 2–3-lobed with acute or acuminate apices; perianth mouth strongly dentate . . . . (2)

2a.   Leaves plane, wider than long; gemmae rare, yellowish to yellowish-brown. **T. quinquedentata**

2b.   Leaves canaliculate-complanate, longer than wide; gemmae abundant, reddish-brown to rust-red . . . . . (3)

3a.   Gemmae smooth, ovoid-ellipsoidal; leaf cells 10–15 x 15–25 μm in mid-leaf; cell walls thick, trigones not bulging. **T. exsecta**

3b.   Gemmae angular, polygonal; leaf cells 15–20 x 20–40 μm in mid-leaf; cell walls thin, trigones bulging. **T. exsectiformis**

T. exsecta [cut out]. We have a single specimen. CC: Wet tundra, Summit Lake, Mt. Evans, 3900 m, *Weber & Anderson B-10878*.

T. exsectiformis. Evidently frequent from the outer foothills to subalpine. BL: Vertical sandstone cliff, White Rocks, 5500 ft, 8 km NE of Boulder, *Weber B-5588*; LR: Rocky Mountain National Park; vertical soil bank of Alpine Brook, 3020 m, *Hermann 25995 (B-57244)*; Fall River Road, *Weber & Grove B-36540*;

decayed wood, Chiquita Creek, Roaring River, *Hermann* (!Hong).

T. polita [smooth]. Infrequent in wet places along rills in the subalpine ecotone. CC, GA, PA, PT, SJ.

T. quinquedentata. BL: '*Haplomitrium* Hill', Green Lake Valley, Indian Peaks, 3400–3700 m, *Weber B-44150;* Pawnee Lake cirque, 3300–3500 m, *Komarkova B-41952.* BL, GA, OR, ST.

PLAGIOCHILACEAE (PLC)

Plagiochila [sloping lip (perianth mouth)]

A fairly common genus occurring on moist forest floors and streamsides from the foothills to the subalpine. It is easily recognized because the leaves are attached obliquely to the stem, alternate, and strongly *succubous*, with the antical (upside) base of the leaf strongly decurrent, and the lamina recurved and overlapping the leaf downside of it.

P. asplenoides subsp. **porelloides.** Common in the montane and subalpine zones. In the wet mossy alpine tundra a minute race occurs, which differs in having strongly laterally compressed shoots, resembling *Nardia*, and larger cells with numerous oil-bodies.

P. asplenoides subsp. **arctica.** CC: Wet tundra, N side of Summit Lake, on vertical soil banks near the water line, *N. G. Miller B-115059.* The subspecies (perhaps a distinct species, *P. arctica),* represents a great disjunction from the circum-arctic region.

PORELLACEAE (PRL)

This is our largest leafy liverwort. It has long branches (several inches) and forms mats of creeping stems 2–4 mm wide and up to 10 cm long, on moist cliff sides, at fairly low altitudes, especially in the outer foothills of the Front Range. The color is a dark dull green. The leaves are incubous, closely overlapping an forming a rather smooth surface. The underside is a different matter altogether. There are large underleaves, which tend to be ruffled on the lower margins, which also curve outward. The leaf is two-lobed. Instead of the smaller lobe lying flat upon the much larger lobe as it would be in *Scapania,* it is tucked in underneath, narrow, and is oriented parallel to the stem, so altogether the underside of a *Porella* looks like it is filled with odds and ends of leaf-like things. The genus does not produce gemmae. Short side shoots with crowded overlapping leaves produce androecia. The perianth is near the shoot apex and is inflated, narrowed to a small mouth.

Porella [little pore]

1a. Ventral lobes long-decurrent; margins strongly reflexed; underleaves twice as long as the ventral lobes. **P. cordaeana**

1b. Ventral lobes not or only short-decurrent; margins plane or weakly reflexed; underleaves 1–2 times as wide as the ventral lobes. **P. platyphylla**

P. cordaeana [for A.J.K. Corda]. Common on cliffs in the Front Range canyons. BL: Bluebell canyon, *Weber B-9021;* LR: Little Thompson Canyon, *Hermann 27055;* PA: Estabrook (report by Evans, 1915).

P. platyphylla. Hong stated that "*P. platyphylla* is easily

distinguished from *P. cordaeana* by its regular 2–3-pinnate branching, underleaves that are slightly wider than the ventral lobes, and a ciliate perianth mouth." This species occurs in the same areas as *P. cordaeana,* but perhaps it is more of a calciphile. BL: Bluebell canyon, *Weber B-9021;* CC: Clear Creek canyon, on calcareous schist, *Weber B-40723*; FN: Phantom Canyon, *Weber B-37547*; LR: Drake, *Hermann 25760*; Little Thompson Canyon, *Hermann 27055;* JF: Platte Canyon, *Holzinger s.n.*

RADULACEAE (RDL)

This family has only one genus, *Radula*, which is distinguished by having a rectangular lower 'lobule', which is not seen from above because it is hidden by the larger ovate or orbicular dorsal lobe. There are no underleaves. Otherwise *Radula* looks much like a small *Porella*. It is light green as opposed to the dull, dark green of *Porella*, which also differs by having underleaves. Fertile specimens have a narrow, flattened perianth with a wide, truncate, flat mouth. There are no rhizoids whatever on the stem.

Radula [a scraper, the perianth edge]

R. complanata [flattened]. A plant of cliffs, often growing over imperfect lichens. One conspicuous feature is that each leaf cell has one large football-shaped brownish oil body. Usually oil bodies disappear soon after the plants die, but in one of our collections, they are still obvious in specimens a year old. BL, LR, GA, GN. The genus *Radula* is mostly tropical.

SCAPANIACEAE (SCP)

Scapania [a spade or hoe, shape of the flattened perianth]

1a.   Minute species, 2–6 mm long, 0.5–2.25 mm wide; gemmae reddish-brown, often at the ends of erect, slender shoots with more or less reduced leaves; leaf margins usually entire, both lobes ovate-pointed, less than 0.7 as wide as long and non-decurrent .... (2)
1b.   Larger species, 8–80 mm long; gemmae always on normal, unmodified, large-leaved shoots with spreading leaves .... (3)

2a.   Leaf margins with equally thickened 2–4 cell rows forming a differentiated border; oil bodies 3–9 per cell. **S. glaucocephala**
2b.   Leaf margins with thin-walled cells, not forming a differentiated border; oil body solitary. **S. gymnostomophila**

3a.   Ventral leaf lobe distinctly (often conspicuously) decurrent, the decurrent strip several cells broad, gradually tapered (this is difficult to see in stems in which the leaves are closely overlapping, but diligence usually gets results) .... (4)
3b.   Leaves not decurrent ventrally or rarely with a 1-seriate line of cells below the keel insertion .... (8)

4a.   Mature leaves not sharply keeled, the keel rounded; dorsal lobe squarrose; minute plants 1.0–2.2 mm wide x 5–12 mm long; ventral lobes one half to two-thirds as wide as long; gemmae deep reddish brown, 2-celled. **S. cuspiduligera**
4b.   Mature leaves with a sharply folded keel; dorsal lobe flattened against the ventral or variously spreading, not strongly squarrose; plants usually robust, 2–5 mm wide x 20 mm long; ventral lobes broadly ovate to rotund .... (5)

5a.   Dorsal lobe with an arcuate insertion, distinctly (usually long-) decurrent .... (6)
5b.   Dorsal lobe transversely inserted, indistinctly decurrent .... (7)

6a.   Leaf margins almost entire; dorsal lobes almost reniform; gemmae rare, 1-celled, greenish to reddish. **S. uliginosa**

6b.   Leaf margins slightly dentate; dorsal lobes broadly cordate; gemmae never produced. **S. paludosa**

7a.   Dorsal lobe not more than 2/3 the ventral lobe in size, the keel less than half the length of the ventral lobe; marginal leaf cells (in ours) not incrassate. **S. undulata**

7b.   Dorsal lobe 0.7–0.8 times the ventral in size; keel 0.5–0.65 the length of the ventral lobe; marginal rows of leaf cells incrassate. **S. subalpina**

8a.   Ventral lobe broadly ovate to suborbicular, width mostly 0.85–1.25 times the length; plants robust, 2.1–5.0 mm wide x 2–10 cm long . . . . (9)

8b.   Ventral lobe narrow, 0.45–0.8 times the length; plants small, 0.5–2.2 mm wide x 2–18 mm long . . . . (11)

9a.   Gemmae greenish at maturity, 2-celled; leaf lobes mostly acute to subacute; submerged in subalpine slow streams and pools; color dark green or black. **S. irrigua**

9b.   Gemmae reddish to brown at maturity; leaf lobes mostly rounded; in fens and moist tundra; color usually brownish-green to brown . . . . (10)

10a.   Leaves with highly arched, often semicircular keel; dorsal and ventral lobes obtusely pointed, more or less denticulate; gemmae brown, narrowly elliptical, one-celled; oil bodies large, persistent; plants mostly 2.4–3.5 mm wide. **S. brevicaulis**

10b.   Leaves with keel more or less concave or almost straight; dorsal and ventral lobes obtuse to broadly rounded, always entire; gemmae reddish-brown, 2-celled, broad; oil bodies mostly small, disappearing soon after death; plants usually 1.9–2.8 mm wide. **S. hyperborea**

11a.   Leaves with the marginal cell rows of non-gemmiparous lobes uniformly thick-walled (usually strongly so), forming a distinct border; subalpine, forested areas. **S. curta**

11b.   Leaves with marginal cells with trigones as large as those in the inner cells, the lobes thus not distinctly bordered; alpine tundra. **S. mucronata** (see also *S. fulfordiae*)

S. brevicaulis [short-stemmed]. A rare subalpine species. "Easily distinguished by its cordate-reniform dorsal lobes, yellow-brown 1-celled gemmae and large oil bodies" (Hong 1980). Collections cited by Hong: BL: Left Hand peat fen, 10600 ft, *Weber & Dahl B-6982*; Isabelle Glacier, *Komarkova B-42040*. Compare with *S. hyperborea.*

S. curta [short]. A small species fairly common in forested areas. The best distinguishing feature is the one or two rows of smaller, thick-walled cells bordering the leaf. The dorsal lobe is broader than high, distinctly pointed. In the field this is a light green plant more or less erect with the lobes not strongly flattened against each other. The dorsal lobe may have some small teeth near the apex. We have not seen gemmae in our material. BL, CF, CC, LR, ST.

S. cuspiduligera. Uncommon along streamlets, from upper foothills to subalpine. The small size and spreading, almost reflexed, dorsal lobes are distinctive field characters. BL, GL, LR, ST.

S. fulfordiae [for Margaret Fulford]. We have not been able to examine the type. This is said to be similar to *S. mucronata,* but "distinguished from that species by its rather broadly rounded ventral lobes, large dorsal lobes (ca. 0.65–0.8 the ventral in size in contrast to ca. 0.5–0.6 the ventral in *S. mucronata* [sic.]), small marginal cells which average 10–14 µm, fewer oil bodies (2–4 per cell), exclusively green 1-celled, small gemmae and strongly developed keel (0.5–0.75 the length of the ventral lobe)." It is a small plant, about the size of *S. curta.* Known from one collection from near Centennial, Wyoming, and two Colorado collections: GA: Lulu City Trail, Rocky Mountain National Park, *Hong 79-895, 79-903.*

S. glaucocephala. One collection: SJ: Coal Bank Hill, *Hong 79-1061 (B-71209).*

S. gymnostomophila [related to *S. gymnostoma*]. This species is so minute, a few mm tall, with only two or three leaves, and not known to produce sporophytes, is hardly to be seen or recognized except by an experienced specialist. The dorsal leaf lobe is much smaller than the ventral one. The leaves are said to bear reddish-brown masses of two-celled gemmae. Our single collection is fragmentary, from ST: On small bare spots on calcareous seepage area, 11000 ft, Blue Lake, *N. G. Miller B-115103*.

S. hyperborea. "Characterized by the consistently rounded apices of the dorsal lobes, entire margins of the ventral and dorsal lobes, and the 2-celled, broadly ovoid, brownish-red gemmae" (Hong 1980). Probably the most common species on the tundra along rocky snow-melt rills. CC: Gray's Peak, *Nelson & Weber B-41681*; slopes around south shore of Summit Lake, Mt. Evans, 12800 ft, *Weber B-8909* (!Schuster).

S. irrigua. Common in subalpine forests where water spreads from snow-melt streamlets. A dark species with rather remotely spaced leaves. BL, CC, GA, LR

S. mucronata. CF, CC. Schuster (vol. 3, p. 435) cites a specimen collected by Holzinger, Pikes Peak, 7000–10000 ft, 1896 (NY).

S. paludosa. One collection from GA: West St. Louis Creek, 9000 ft, ca. 6 mi W of Fraser, *Weber et al. B-11092* (!Hong).

S. subalpina. Common in subalpine iron fens: BL, CC, SJ, ST.

S. uliginosa. Our only record is one cited by Evans (1915): LK: Twin Lakes, 1873, *Wolf & Rothrock*.

S. undulata. Probably the most common species of montane and subalpine forest floors.

TRICHOCOLEACEAE (TRC)

Blepharostoma [ciliate, + mouth (of perianth)]

This genus is distinguished by the leaves, which are normally four-lobed to the base, with each lobe one cell wide; the underleaves are similarly three-lobed.

B. trichophyllum. *Trichophyllum* is an apt epithet, because the leaves are divided into four one-cell wide 'fingers'. The slender, yellowish-green plants are small, delicate, and easily overlooked among other liverwort clumps, but sometimes they occur in large, pure stands, as we found them under dwarf birches at Blue Lake.

The underleaves are similar but only trifid. Some other liverworts have leaves that are variously dissected, but In Colorado there is nothing like *Blepharostoma*. It is generally restricted to the subalpine zone and lower alpine ecotone, where it is common.

PEOPLE ASSOCIATED WITH BRYOPHYTE NAMES

Rich sources of information on the lives of these scientists will be found in Wittrock (1903–1905), Rydberg (1907), Barnhart (1965), and Frahm & Eggers (2001). For a history of North American bryology, see Rodgers, Andrew Denny, *Noble Fellow: William Starling Sullivant.*

Agassiz, Louis (1807–1873), Swiss-American, Harvard Professor, first to propose there were glacial periods.

Amann, Jules (1859–1939), Swiss bryologist, a pharmacist in Davos, then chemist and bacteriologist in Lausanne. *Flore des mousses de la Suisse* and *Bryogeographie de la Suisse.*

Anderson, Lewis Edward (1912–2007), contemporary bryologist of Duke University, co-author with Howard Crum of *Mosses of Eastern North America.*

Andreae, Johannes Gerhard Reinhard (1724–1793), a Hannover pharmacist, was a lover of fine literature in many languages and the English poets, and a fine pianist. He became interested in chemistry and mineralogy and did field work in Switzerland, describing 300 types of soils. *Andreaea* was named for him by Ehrhart.

Andrews, Albert Leroy (1878–1961), bryologist of New England, specialist in *Bryum*. Obit: Steere in *Bryologist* 65:25–37. 1962.

Ångström, Johan (1813–1879), Swedish physician and bryologist, collected in Swedish and Finnish Lappland.

Arnell, Hampus William (1848–1932), Swedish hepaticologist.

Arnott, George Walker (1799–1868), Scottish botanist (colleague of William Jackson Hooker).

Austin, Coe Finch (1831–1880), American bryologist.

Bachelot de la Pylaie, Auguste Jean Marie (1786–1856), French botanist.

Balbis, Giovanni Battista (1765–1831), Italian botanist

Balsamo-Crivelli, Giuseppe Gabriel (1800–1874), Italian botanist.

Barnes, Charles Reid (1858–1910), American bryologist, first biological science professor at University of Chicago, author of *Genera of North American Mosses*. Obit: Howe in *Bryologist* 13:66–67. 1910.

Bartram, Edwin Bunting (1878–1964), Philadelphia bryologist, did relatively little field work but published prodigiously on mosses of the tropics (Guatemala, Hawaii, New Guinea, Philippines. Obit: Crum in *Bryologist* 69:124–134. 1966.

Bartram, John (1699–1777), American pioneer botanist.

Beauvois (see Palisot de Beauvois)

Berkeley, Miles Joseph (1803–1889), English bryologist

Bescherelle, Émile (1828–1903), French bryologist. Obit: Holzinger in *Bryologist* 6:46. 1903.

Best, George Newton (1846–1926), New Jersey bryologist, published monographs of *Thuidium, Pseudoleskea*. Obit: Beals in *Bryologist* 30:20–22. 1927.

Beyrich, Heinrich Karl (1796–1834), German-American botanist, collected in southeastern U.S., joined a military expedition west of the Mississippi, died of cholera.

Bischoff, Gottlieb Wilhelm (1797–1834), German, son of a pharmacist, an artist and hepaticologist, coined the terms archegonium and antheridium.

Blandow, Otto Christian (1778–1810), German, pharmacist and bryophyte collector (*Helodium blandowii*.)

Blazius-Biagi (?–?), a Benedictine monk of Florence.

Blind, J. J., Pastor in Münster, Germany, 1834–1848. *Bryum blindii, Blindia acuta.*

Blom, Hans Haavardsholm (1955–), contemporary Norwegian bryologist, specializing in *Schistidium*.

Blytt, Axel Gudbrand (1843–1898), Norwegian bryologist.

Blytt, Matthias Numsen (1789–1862), father of Axel Blytt

Bolander, Henry Nicholas (1831–1897), born in Germany, a teen-aged disciple of Lesquereux in Columbus, Ohio, who taught him botany. State Botanist of California 1864–1871.

Braithwaite, Robert (1824–1917), English bryologist.

Breidler, Johann (1828–1913), Austrian, collector, wrote books on the mosses of the Austrian Tyrol: *Breidleria*.

Bridel-Brideri, Samuel Elysée (1761–1828), Swiss, private teacher of royal children in Gotha, poet, divided the mosses into classes, orders and families. Wrote a classic two volume work, *Bryologia universa seu systematica ad novum methodum disposito, historia et descriptio omnium muscorum frondosorum hucusque cognitum.*

Britton, Elizabeth Gertrude (1858–1934), American tropical bryologist, wife of Nathaniel Lord Britton, director of New York Botanical Garden. She founded the Sullivant moss Society, now the American Bryological and Lichenological Society.

Brotherus, Viktor Ferdinand (1849–1929), Finnish bryologist, author of the two volumes on mosses of Engler's *Natürliche Pflanzenfamilien*, (1924–1925). See Koponen & Piippo (2002).

Brown, Robert (1773–1858), Scottish botanist and explorer.

Bruch, Philipp (1781–1847), pharmacist son and grandson of pharmacists, collaborated with Schimper in the great *Bryologia Europaea. Bruchia*.

Bryhn, Niels (1854–1916), Norwegian botanist. *Bryhnia*.

Buch, Hans Robert Viktor (1883–1964), Finnish bryological geographer.

Buxbaum, Johann Christian (1693–1730), German, collected mosses in Persia and the Caucasus, Russia. *Buxbaumia*.

Cain, Roy Franklin (1906–?), Canadian mycologist and bryologist. He received the American Mycological Society's Distinguished Mycologist Award in 1983.

Carestia, Antonio (1825–1908), Italian botanist.

Chen, Pan-Chieh (1907– 1970), Chinese bryologist.

Churchill, Steven Paul (1948–), contemporary American tropical bryologist, monographed *Jaffueliobryum*.

Clark, Lois (1884–1967), botanist of Washington State, collaborated with T. C. Frye in *Hepaticae of North America*. Obit: Howard in *Bryologist* 71:140–141 (portrait). 1968.

Coppey, Amédée (1874–1913), French bryologist and lichenologist.

Corda, August Karl Josef (1809–1849), Bohemia, Custodian of the National Bohemian Museum in Prague. Wrote important treatments on liverworts in books of Opiz, Sturm, and author of many genera. Died in shipwreck en route from Texas to Bremen.

Cosson, Ernest St. Charles (1819–1889), French botanist.

Crum, Howard Alvin (1922–2002), American bryologist, Univ. of Michigan, specialist in mosses of Great Lake forests, *Sphagnum* mosses, co-author with Lewis Anderson of *Mosses of Eastern North America*.

Culmann, Paul (1860–1936), Swiss, trained in physics, optics, collector of mosses in Switzerland, published on various moss groups.

Delgadillo Moya, Claudio (1945–), contemporary Mexican botanist who monographed *Aloina* and *Crossidium*.

De Notaris, Giuseppe (1805–1877), Italian bryologist.

Dickson, James (1738–1822), Scottish bryologist.

Dixon, Henry Neville (1861–1944), English bryologist. Obit: Bartram in *Bryologist* 47:135 (portrait)–146.1944.

Don, George (1789–1856), Scottish botanist

Douin, Charles Isidore (1858–1944), French hepaticologist.

Drummond, Thomas (1780–1835), Scottish botanist, collector in American Arctic, Texas.

Dumortier, Barthélemy Charles Joseph (1797–1878), French, a 'father' of Hepaticology.

Dupret, Rev. Francois Hippolyte (1853–1932), Canadian botanist.

Evans, Alexander William (1868–1959), Connecticut bryologist and lichenologist, great mentor of subsequent students, published first comprehensive list of Colorado liverworts. Obit: Schuster in *Bryologist* 63:73–81; Hale, l.c., pp. 80–83; Anderson, l.c. pp. 84–88.

Ehrhart, Friedrich (1742–1795), Swiss student of Linnaeus, known for his pre-Hedwigian moss exsiccatae and naming of many new species and genera (*Georgia*, after King George III, and *Catharinaea*, after Catherine II of Russia).

Fabbroni (?–?), once director of the mint in Florence and friend of Raddi. This is possibly the same person as L. Pelli-Fabbroni (see below). *Fabronia*.

Fendler, August (1813–1883), East Prussian. Came to Missouri in 1836 collected for Asa Gray in New Mexico, Panama, Venezuela, and Trinidad. His collections were used by Sullivant.

Fleischer, Max (1861–1930), Upper Silesia, famous artist, went to Java to paint for the Buitenzorg Botanical Garden, became interested in bryophytes and became the greatest bryologist of south-east Asia.

Floerke, Heinrich Gustav (1764–1835), Germany, Professor of Zoology, Botany, and Natural History in Rostock. *Barbilophozia floerkei*.

Flotow, Julius (1788–1856), German botanist.

Flowers, William Seville (1900–1968), renowned teacher and botanist of Salt Lake City, Utah, author of Mosses: *Utah and the West*. Obit: *Bryologist*, portrait, death notice only (1975).

Forster, Johan Georg (George) Adam (1754–1794), Danzig, known for studies of Australian bryophytes.

Frölich, Friedrich (1769–1845), German pastor.

Frost, Charles Christopher (1805–1880), intermediary between the collector, Sereno Watson, and the describer, C. F. Austin, of a species of *Riccia*.

Fulford, Margaret (1904–1999), Ohio hepaticologist.

Funck, Heinrich Christian (1771–1839), Austrian botanist, specialist in flora of the Tirol.

Fürnrohr, August Emanuel (1804–1861), Regensburg, Bavaria, pharmacist and naturalist, author of various textbooks and especially on the flora of Regensburg.

Frye, Theodore Christian (1869–1962), Washington State hepaticologist, author, *Hepatics of North America*. Obit: Howard in *Bryologist* 66:124–136. 1963.

Gärtner, Gottfried (Philipp) (1754–1825), co-author with Meyer & Scherbius of *Flora der Wetterau*

Giesecke, Charles Lewis (1761–1833), formerly Karl Ludwig Metzler von Giesecke, German hepaticologist. Originally a member of a dramatic group, co-author of libretto of Mozart's *Magic Flute*; later Professor of Mineralogy, Dublin, with Metternich in Constantinople, seven years in Greenland. A person worth investigating!

Girgensohn, Gustav Carl (1786–1872), Esthonian pastor, teacher. *Sphagnum girgensohnii*.

Gottsche, Carl Moritz (1808–1892), Director of Botanical Gardens in Hamburg, great student of liverworts.

Gray, Samuel Frederick (1766–1828), English botanist.

Green, Thomas (ca. 1820). Bruce Allen suggests this is the person for whom *Pohlia greenii* is named. He collected in Shropshire, Wales. Bridel (1819) acknowledged the collection as follows: *"in monte Snowdon Angliae generosissimus Green detexit et communicavit."*.

Greville, Robert Kaye (1794–1866), Scottish bryologist, collaborated with Sir William Hooker. *Grevillea*.

Grimm, J.F.K. (1737–1821, a friend of Bridel.

Grolle, Riclef (1934–2004), world-famous hepaticologist, Jena. Permanently crippled with polio in his teens, he was the greatest modern specialist in tropical liverworts.

Grout, Abel Joel (1867–1947), American bryologist, high school teacher on Staten Island, wrote the classic 3-volume *Moss Flora of North America*. Obit: Steere in *Bryologist* 51:201–212. 1948. (1948).

Guépin, J.B.P. (1778–1858), French botanist and correspondent of Darwin.

Gümbel, Theodor (1812–1858), wrote and illustrated the *Moss Flora of the Rheinfpfalz* and illustrated many plates in the *Bryologia Europaea*. He is the G in B.S.G,

Hagen, Ingebrigt Severin (1852–1917), Norwegian botanist of Trondheim.

Hall, Elihu (1814–1890), Illinois plant collector, with C. C. Parry, in Colorado, 1862.

Haller, Albrecht von (1708–1777), Swiss physician, botanist and poet. *Campylophyllum halleri*.

Hampe, Ernst Georg Ludwig (1795–1880), pharmacist, Univ. Of Göttingen; prolific writer on mosses and liverworts of the world.

Handel-Mazzetti, Heinrich, Freiherr von (1882–1940), Austrian botanist, specializing in Asiatic floras. Author of *Symbolae Sinicae*.

Hartman, Carl Johan (1790–1849), Swedish floristic botanist.

Hatcher, John Bell (1861–1904), American palaeontologist who found *Lophozia hatcheri* in Tierra del Fuego. Author of *Bone Hunters in Patagonia*, discoverer of *Taurosaurus*.

Hattori, Sinske (1915–1992), founder of the Hattori Botanical Laboratory and mentor of numerous important Japanese bryologists. Obit: Sharp in J. Hattori Bot. Lab. 74: (portrait) +1.1993.

Hedenäs, Lars Gustaf (1957–), contemporary Swedish bryologist, Natural History Museum, Stockholm, specializing in wetland mosses.

Hedwig, Johannes (1730–1799), physician, Professor, Leipzig, the 'Linnaeus of mosses', published *Species Muscorum* and other basic bryological books.

Heim, Ernst Ludwig (1747–1834), German physician of the Humboldt family.

Hennedy, Roger (1809–1877, Robert Brown's mentor.

Heinemann, F. (?–?), pharmacist, discovered *Andreaea heinemannii* on the Grimsel in the Alps.

Hervey, Eliphalet Williams (1834–1925), Massachusetts bryologist.

Herzog, Theodor K. (1880–1961), German bryogeographer. *Geographie der Moose.* Obit: *Bryologist* 54:287. 1961 [death notice].

Hessler, Karl (1799–?), German botanist, wrote doctoral dissertation on *Timmia* at Göttingen.

Hill, John (1716–1775), English botanist, Prefect of Royal Botanic Gardens, Kew.

Hoffmann, Georg Franz (1760–1826), Bavarian, Professor of botany in Göttingen and Moscow, author of *Deutschlands Flora* (1795–1796).

Höhnel, Franz Xavier Rudolf Ritter von (1852–1920), Hungarian, collected throughout Africa, Asia Minor, the Americas, Ceylon and Java, described *Webera* (= *Pohlia*) *andalusica*.

Holzinger, John Michael (1853–1929), German born, Professor of natural history and botany, Winona, Minnesota. Collected and prepared large exsiccati, including *Musci Acrocarpi Boreali-Americani* and *Mosses of Colorado* (collected by C.F. Baker). His herbarium is at the Univ. of Minnesota.

Hong, Won Shic (1919–), contemporary hepaticologist at Great Falls College, Montana, specializing in liverworts of western North America.

Hooker, Sir Joseph Dalton (1817–1911), son of W. J. Hooker, also Director of the Royal Botanical Garden, 1865–1885. The name is cited as Hooker *f.* [*fils*]. J. D. Hooker spent five days in Colorado in 1877 with Asa Gray (see Weber 2003b).

Hooker, Sir William Jackson (1785–1865), British botanist, Director of Royal Bot. Gardens, Kew. The name is cited as Hooker or Hook.

Hoppe, David Heinrich (1760–1846), German botanist, specialist in the flora of the Tirol.

Horikawa, Yoshiwo, (1902–1976), Japanese bryologist.

Hornschuch, Christian Friedrich Benjamin (1793–1850), German pharmacist, professor in Greifswald, co-author of *Bryologia Germanica* (1831).

Hornemann, Jens Wilken (1770–1841), Danish botanist.

Hudson, William (1730–1793), English botanist.

Hübener, Johann Wilhelm Peter (1807–1847), Hamburg, great collector, published books and exsiccati on German mosses and liverworts.

Iwatsuki, Zennoske (1929–), Japanese bryologist, now director of the Hattori Botanical Laboratory.

Jaffuel, Felix (1857–1931), cleric, plant collector in Chile and Bolivia.

Jäger, August (1842–1877), Freiburg, pharmacy teacher, published several books on bryophytes, his greatest work being *Genera et species muscorum systematicae disposita seu Adumbratio florae muscorum totius orbis terrarum* (1870–1880) completed by his friend F. Sauerbeck

Jamieson, David (1943–), Colorado bryologist, specializing on the southwestern San Juan Mountains. monographed *Hygrohypnum.*

Jennings, Otto Emery (1877–1964), Ohio-born bryologist of Pennsylvania. Obit: Netting in *Bryologist* 68:353–359. 1965. His bryophyte herbarium is at MO, his liverwort collection at COLO.

Jensen, Christian Erasmus Otterström (1859–1941), Danish hepaticologist.

Jensen, Thomas (1824–1877), Danish bryologist.

Jörgensen, Eugen Honoratus (1862–1938), Norwegian hepaticologist, University of Bergen.

Jungermann, Ludwig (1572–1653), Professor of Botany, Leipzig, collected in Greece and Palestine. *Jungermannia* Linnaeus.

Juratzka, Jakob (1821–1878), Austrian bryologist. *Anthelia juratzkana.*

Kaalaas, Baard Bastian Larsen, (1851–1918), Norwegian hepaticologist.

Kanda, Hiroshi, (1946–), contemporary bryologist, especially Amblystegiaceae.

Kaulfuss, Georg Friedrich (1786–1830), German bryologist, professor of botany in Halle.

Kiaer, Frantz Caspar (1835–1893), Norwegian botanist

Kindberg, Nils Conrad (1832–1910), Swedish bryologist, published on North American mosses.

Klinggräff, Hugo E. M. von (1820–1902), West Prussian bryologist, wrote *Liverworts and Mosses of West and East Prussia. Gemmabryum klinggraeffii.*

Knowlton, Frank Hall (1860–1926), American paleobotanist, published on Yellowstone National Park, Florissant Fossil Beds and Dinosaur National Monuments.

Koch, Leo F. (1916–), North Dakota bryologist, Univ. of Illinois. Left bryology too early. See his paper, Academic freedom at the University of Illinois, U.S.A. *J. Hattori Bot. Lab.* 23:1–2. 1961.

Koch, Wilhelm Daniel Joseph (1771–1849), German bryologist.

Koponen, Timo, (1939–), contemporary Finnish bryologist, specialist in Mniaceae.

Kunze, Gustav (1793–1851), director of the Leipzig Botanic Garden, specialized in the bryophytes of Cuba and Middle America.

Kuwahara, Yukinobu (1927–), contemporary Japanese bryologist.

La Pylaie (see Bachelot de la Pylaie)

Laurer, Johann Friedrich (1798–1873), Bayreuth, wrote a paper on the cryptogamic flora of the island Rügen (hence *Mnium rugicum = Plagiomnium ellipticum*). *Tortula laureri.*

Lawton, Elva (1896–1993), bryologist of the Pacific Northwest. Obit: Denton in *Bryologist* 96:641–644.1993.

Leiberg, John (1853–1913), Swedish-American botanist and forester, particularly active in the Pacific Northwest, particularly Oregon and Idaho as field agent for USDA and USFS.

Leske, Nathanael Gottfried (1751–1786), Moscow, Professor at Leipzig and Marburg. *Leskeaceae.* Not to be confused with Loeske.

Lesquereux, Louis (1806–1899), Swiss-born Ohio bryologist, collaborator with Sullivant. *Lescuraea.*

Li, Xing-Jiang, (1932–), contemporary Chinese bryologist.

Lightfoot, John (1735–1788), English cleric and botanist.

Limpricht, Karl Gustav (1834–1902), Schleswig, Moss flora of Schleswig, He and his son published the Mosses of Germany, Austria, and Switzerland.

Lindberg, Sextus Otto (1835–1889), Director of the Botanical Garden in Helsingfors, Finland, one of the most prominent bryologists of his time. See Rydberg (1907, p. 26) and Koponen & Isoviita (2005).

Lindenberg, Johann Bernhard Wilhelm (1781–1851), Lübeck, wrote the classic *Hepaticarum Europaeorum* and was joint author with Gottsche and Nees of *Synopsis Hepaticarum. Asterella lindenbergii.*

Lisa, Domenico (1801–1867), Italian bryologist.

Loeske, Leopold (1865–1935), Harz, Germany, a watchmaker, published a great many important papers and monographs on European mosses.

Lorentz, Paul Gunther (1835–1881), Austrian, Uruguayan bryologist.

Lyall, David (1817–1895), surgeon and botanist on U. S.–Canada Boundary Survey, 1858–1860, made collections on Vancouver Island and western Cascades. Obit. J. D. Hooker in J. Bot. 33: 209–211.1895.

Lyell, Charles (1817–1895), Royal Navy Surgeon, served with J. D. Hooker in Antarctic expedition on HMS *Erebus* and *Terror.*

Macoun, John (1831–1920), Irish-Canadian bryologist. Obit: Andrews in *Bryologist* 24:39–41. 1921.

Mann, Wenzel Blasius (1799–1839), practicing physician in Bohemia, colleague of Floerke. *Mannia.*

Marchant, Nicholas (circa 1678).

Mårtensson, Olof (1915–),Swedish plant ecologist, did exceptional studies of the vegetation of the region on Lake Torne Träsk, Swedish Lappland.

Martius, Carl Friedrich Philip von (1794–1868), Professor of Botany at Munich, particularly noted for his *Flora Brasiliensis,* for which Nees wrote the Hepaticae.

Massalongo, Caro Benigno (1852–1928), Italian hepaticalogist, son of the lichenologist Abramo Bartolommeo Massalongo.

Mathieu, Charles Marie Joseph (1791–1873), Belgian bryologist and mycologist.

Meese, David (1723–1770), Dutch gardener.

Menzies, Archibald (1754–1842), Scottish botanist and explorer who accompanied Capt. Vancouver on his round-the-world voyage, 1791–1795, and made extensive plant collections on the Australian coast. Pronounced "Minjes".

Merrill, Gary L. See Smith, Gary Lane

Meyer, Bernard (1767–1836), German botanist, co-author with Gärtner and Scherbius of *Flora der Wetterau.*

Meylan, Charles (1868–1941), Swiss hepaticologist.

Micheli, Pietro Antonio (1679–1737), Italian botanist.

Mielichhofer, Mathias (1772–1847), Austrian, went on collecting trips with Hornschuch, Schrader, and Schwägrichen in the Alps and Germany. *Mielichhoferia.*

Milde, (Karl August) Julius (1824–1871), Breslau. He early contracted tuberculosis and had to endure a number of tragedies in his family. His main work was the *Flora of Silesia* (Schleswig).

Mitten, William (1819–1906), English bryologist. His extensive and important collections from South America are at the New York Botanical Garden. See Nicholson, W. E. William Mitten; A sketch and bibliography. 10:1-5.1907.

Molendo, Ludwig (1834–1902), Bavarian botanist, *Flora of Tirol.*

Mönkemeyer, (Adolph August) Wilhelm (1862–1938), Stadtoldendorf, Germany, collected bryophytes in west Africa as a young man, held position in botanical gardens of Göttingen and Leipzig, produced Vol. 10 of *Die Laubmoose Europas.*

Mohr, Daniel Matthias Heinrich (1780–1808), Holstein, professor at Kiel, collected in Sweden, co-author with F. Weber, the *Index musci plantarum cryptogamarum.*

Molendo, Ludwig (1833–1902), Bayreuth, assisted Martius but broke up with him, collected cryptogams in the Alps for Rabenhorst and Schimper, became a strong supporter of Darwin. Molendoa.

Montagne, Jean Pierre François Camille (1784–1866), French botanist.

Moore, David (1808–1879), Scottish bryologist.

Mougeot, J. B. (1776–1858), French botanist.

Mühlenbeck, Heinrich Gustav (1798–1845), Alsatian pharmacist, accompanied Bruch and Schimper on a collecting trip to the Alps. *Dicranum muehlenbeckii.*

Mühlenberg, Gotthilf Heinrich Ernst (1753–1815), Pennsylvania, sent to Germany to study for the priesthood, returned to become the earliest qualified plant collector in the U.S., a correspondent of Schreber, Willdenow, Hoffmann, and visited by Humboldt & Bonpland. *Funaria muehlenbergii.*

Müller, Carl (Karl) Johann August (1818–1899). There are at least 15 German Müllers, two of them Carl or Karl, both famous bryologists. This one is now properly referred to as **Müller Halle**, for Halle, where he lived. Trained as a pharmacist, his great work was the posthumous *Synopsis Muscorum Frondosorum.* His great herbarium of 12000 species in 70000 specimens, was destroyed in the bombing of Berlin; some of his type duplicates exist in a few other herbaria.

Müller, Karl (1881–1955), German hepaticologist, specialist on European liverworts, also writer of a wine-industry lexicon. His proper identification is **Müller Fribourg**.

Myrin, Claës Gustav (1803–1835), Swedish botanist.

Nardi, S. (?–?), an Italian abbot. *Nardia.*

Necker, Noel Josephe de (1730–1793), Lille, France. Botanist in Mannheim and author of four early bryological treatises. *Neckera* described by Hedwig.

Nees von Esenbeck, Christian Gottfried Daniel (1776–1858), Reichenberg Castle, Odenwald. Acquaintance of Goethe and Marx, neither a philosopher nor a communist, but had several lively extramarital female affairs and was active politically. Nevertheless, he described many liverwort genera and moss species, most importantly in the Javanese flora. He died penniless after having to sell his herbarium. His mosses were destroyed in the Berlin air raids. He deserves a novel. A brother, T.F.L. Nees, was a minor bryological figure.

Nelson, Elias E. (1876–1949), one of the first students of Aven Nelson at University of Wyoming. Collected with Nelson and Leslie M. Goodding in Yellowstone Park in 1899.

Norris, Daniel Howard (1933–), contemporary California bryologist, *Bryoflora of California, Madroño* 51:1–269. 2004. Also contributed to papers on New Guinea bryophytes.

Nyholm, Elsa (1911–2002), Swedish bryologist, self-taught author of *Illustrated Moss Flora of Scandinavia.* Obit, Hedenäs in *Lindbergia* 50–51. 2003.

Ochyra, Ryszard (1949–) and **Halina Bednarek-Ochyra**, (1949–) contemporary Polish bryologist couple, specializing in Polish and Antarctic bryophytes and *Racomitrium, sensu lato.*

Oeder, Georg Christian von (1728–1791), Ansbach, Germany, physician, Prof. at Univ. of Copenhagen. His *Flora Danica* (2170) included cryptogams. *Plagiopus oederianus.*

Palisot de Beauvois, Ambroise Marie François Joseph (1755–1820), French cryptogamic botanist, collected in West Africa and North America. See Bednarek-Ochyra et al. (2001), Ochyra et al. (2004).

Paris, Jean Édouard Gabriel Narcisse (1827–1911), French bryologist.

Pelli-Fabbroni, Leopoldo (?–?), a Florentine lawyer and good friend of Raddi. *Pellia.*

Persson, Natan Peter Herman (1893–1978), Swedish bryologist, Natural History Museum, Stockholm. Collected in the Queen Charlotte Islands, B. C.

Pfeiffer, Carl Georg Ludwig (1805–1877), German hepaticologist, collected in Cuba, 1838.

Piper, Charles Vancouver (1867–1926), American botanist, author of *Flora of the State of Washington*, taught botany and founded the herbarium at Washington State College, Pullman. Published a monograph on the Soybean.

Podpera, Josef (1878–1954), Bohemian bryologist.

Poelt, Josef (1924–1995), Bryologist-lichenologist, University of Graz, Austria. Obit, Hertel & Oberwinkler in *Lichenologist* 28:182–187. 1996.

Pohl, Johann Emanuel (1782–1834), Bohemian physician-botanist, traveled in the interior of Brasil in 1817–1821, but died before the second volume of his travels was published.

Preiss, Ludwig (1811–1883), German collector *par excellence*. On his Australian journey (1838–1841) he collected about 200,000 specimens of plants. He was much honored by eponymy.

Proskauer, Johannes Max (1923–1970), Göttingen. Jewish, escaped with his parents to England in 1943, father died in Auschwitz. Studied in London and Wales, Professor Univ. of California, Berkeley, specialist in hornworts. Contributed the complete treatment of Anthocerotae for K. Müller's *Lebermoosen Europas*. Obit: Fulford, McBride, & Embree, *Bryologist* 75:168–173. 1972.

Raddi, Giuseppe (1770–1829), Italian botanist, collected in Brasil.

Ramsay, Helen Patricia, (1928–), contemporary Australian bryologist.

Rau, Eugene Abraham (1848–1932), Pennsylvania bryologist.

Reboul, Eugene de (1781–1851), Florentine botanist.

Renauld, Ferdinand (1837–1910), French bryologist.

Ricci, P. F. (?–?), 18th Century Italian botanist.

Richardson, Sir John (1787–1865), Scottish Arctic explorer, physician and zoologist with the Franklin expeditions to find a Northwest Passage

Robinson, Harold (1932–), contemporary American bryologist, specialist in Brachytheciaceae, also extraordinary taxonomist of vascular plants (Asteraceae) and insects.

Röll, Julius (1846–1928), Ostheim von der Rhön, Professor at Aue in the Erz Mountains, specialist in peat moss taxonomy and ecology. *Roellia.*

Rubers, Wim Vincentius, (1944–), contemporary Dutch bryologist, Rijksherbarium, Leiden.

Russow, Edmund (August Friedrich) (1841–1897), Dorpat, Esthonia. Studied botany at the University of Dorpat; his dissertation was on the peat mosses. His first new species was dedicated to his friend as *Sphagnum girgensohnii*, and his specialties were *Sphagnum* and *Splachnum*.

Saelán, Anders T. (1834–1921), Finnish botanist.

Saito, Minoru (1919–), contemporary Japanese bryologist.

Sanio, Carl Gustav (1832–1891), East Prussian, career in medicine and anatomy, with special interest in development of spores of *Equisetum,* and analyses of bryophytes. *Sanionia,* and an anatomical feature, the "bars of Sanio" in coniferous wood tracheids.

Sapehin, Lev Andreevich (1906–1931), Russian geneticist.

Sauerbeck, Friedrich Wilhelm (1820–1882), German specialist in diatoms, joint author with Jäger of *Genera et species muscorum systematicae disposita seu Adumbrartio florae muscorum totius orbis terrarum.*

Sauter, Anton Eleutherius (1800–1881), Austrian, practical physician; botanically he published works on floristics, bryology and mycology. It is said that he never needed a microscope and published his descriptions from study with naked eye or hand lens. *Paraleucobryum sauteri.*

Schiffner, Victor Felix (1862–1944), Czech, Professor of systematic botany at Vienna, one of the most productive bryologists; contributed the Hepaticae section of Engler, *Natürichen Pflanzenfamilien*, collected in Brazil, published on mosses and especially liverworts world–wide.

Schimper, Wilhelm Philipp (1808–1880), Alsatia, trained for the ministry but held a great variety of posts including professor of Paleontology in Paris and Strasburg. His great work was the 20-volume *Bryologia Europaea*. Although Bruch was the first author, he died early on; Gümbel, the third author, was responsible mostly for some illustrations; the major work was that of Schimper. Nevertheless, all three authors are cited as Bruch, Schimper & Gümbel or, later, merely Bruch & Schimper. Our Schimper is not to be confused with his son, Andreas (1856–1901) who wrote the great classic botanical work entitled *Plant Geography from a Physiological Basis*.

Schleicher, Johann Christoph (1768–1864), German-Swiss botanist, director of one of the first botanical gardens in Switzerland, published on liverworts and general flora of Switzerland.

Schljakov, Roman Nikolaevich (1912–1999), Russian bryologist. Obit: Constantinova in *Arctoa* 8:89–100.1999.

Schmidel, Casimir Christoph (1718–1792), a court physician, professor, and pre–Linnean bryologist who first described *Tetraphis pellucida* and *Marchantia polymorpha*, and whose students published dissertations on liverworts. He gave a microscope and a selection of his papers to Hedwig, the future 'father of bryology'.

Schofield, Wilfred Borden (1927–), contemporary Canadian bryologist and bryogeographer, British Columbia and Alaska.

Schrader, Heinrich Adolf (1767–1836), German physician and Professor of Botany at Göttingen, published *Spicilegium Florae Germanicae* in 1794, and a very important cryptogamic exsiccatum.

Schrank, Franz von Paula von (1747–1835), Austrian and Bavarian bryologist, described, among others, *Metzgeria pubescens*.

Schreber, Johann Christian Daniel (1739–1810), German from Thuringia, studied under Linnaeus at Uppsala and earned his doctorate under him, published the first bryological monograph, on *Phascum: De Phasco observationes quibus hoc genus muscorum vindicatur atque illustratur*. Memorialized in *Pleurozium schreberi*.

Schultz, Karl Friedrich (1766–1837), German pharmacist and physician. Published a revision of the genera *Barbula* and *Syntrichia*.

Schuster, Rudolf Mathias (1921–), preeminent Bavarian-born American bryologist, published six-volume *Hepaticae and Anthocerotae of North America east of the Hundredth Meridian*.

Schwägrichen, Christian Friedrich (1775–1853), German physician, later professor of natural history in Leipzig. On Hedwig's death, Schwägrichen completed the *Species muscorum frondosorum*, produced the *Historiae muscorum hepaticarum prodromus* and re-worked the mosses for the fourth edition of Linnaeus' *Species Plantarum*. His name is firmly linked to that of Hedwig, 'father of bryology'.

Schweinitz, Ludwig David von (1780–1834), Bethlehem, Pennsylvania bryologist. His herbarium is at the Philadelphia Academy of Science.

Scopoli, Johann Anton (1723–1788), Austrian physician, professor of many branches of natural history. Published the *Flora Carniolica*, treating liverworts as algae. First used Linnaeus' system in describing birds.

Scouler, John (1804–1971), Scottish naturalist.

Sebille, Abbé René Léon (1851-1938), French botanist, published with Cardot.

Seliger, I. (1752–1812), Silesian pastor.

Sendtner, Otto (1813–1859), Bavarian bryologist.

Shaw, Arthur Jonathan, contemporary American bryologist, Duke University, specialist on *Pohlia*.

Smith, Gary Lane (1939–), contemporary botanist, Field Museum of Natural History, specialist in the Polytrichaceae.

Smith, Sir James Edward (1759–1828), founder and lifetime president of the Linnean Society of London, author of the *Flora Britannica*, who purchased privately, for England, Linnaeus' library, botanical and zoological specimens, and manuscripts.

Sommerfelt, Sören Christian (1794–1838), Norwegian pastor and botanist, wrote *Flora of Lappland*.

Spence, John R. (1956–), Contemporary American bryologist, specializing in *Bryaceae*.

Sprengel, Kurt Polycarp Joachim von (1766–1833), Pomeranian physician, Professor at Halle, Saxony. Author of *Coscinodon* and *Bryum weigelii*.

Spruce, Richard (1817–1893), English bryologist, made outstanding collections and published memoirs of his work in South America, 1849–1864.

Starke, Johann Christian (1744–1808), Silesian pastor, honorary member of the Botanical Society of Regensburg. *Kiaeria starkei*.

Steere, William Campbell (1907–1998), bryologist, specialist in mosses of Ecuador and Alaska; Director of New York Botanical Garden.

Stephani, Franz (1842–1927), Berlin, a buyer in a business firm from 1884 to his retirement in 1907. In his youth he collected some mosses in the Erzgebirge, but all of his scientific work was with liverworts that were sent to him by others, He was undoubtedly the most important hepaticologist of all time, especially involved with liverworts of the world's tropical areas. He made thousands of line drawings, which his wife and daughter copied and are now available on microfiche. He contracted Alzheimer's disease but continued describing liverworts long after he could no longer recognize his family. One of his later type specimens, it is reported, turned out to be a bit of his pipe tobacco.

Sullivant, William Starling (1803–1873), Cincinnati, Ohio, pioneer bryologist of North America, published classic work with Louis Lesquereux, and *Icones Muscorum*, illustrations of American mosses.

Swartz, Olaf Peter (1760–1818), Swedish botanist, pupil of Linnaeus and the 'father' of fern studies, botanized extensively in the American Antilles; his publications covered bryophytes, flowering plants, lichens, and his monumental *Synopsis Filicum*.

Syed, Hadiuzzaman, contemporary Pakistani bryologist.

Taylor, Thomas (1775–1848), Irish botanist, collaborated with Sir Joseph Dalton Hooker on mosses of Australia.

Thedenius, Knut Fredrik (1814–1894), Swedish botanist (Uppland)

Thériot, Marie Hypolité Irénée (1859–1947), French bryologist.

Thomson, Thomas (1817–1878), traveler, botanist, published (1852) *Western Himalaya and Tibet: A narrative of a journey through the mountains of Northern India during the years 1847–1848.*

Timm, Joachim Christian (1734–1805), Pomeranian pharmacist. He was a zealous plant collector and wrote a *Flora of Mecklenburg* in 1788. Hedwig honored him with the moss genus *Timmia*. The specific epithet, *"megapolitana"* stands for "of Mecklenburg."

Treviranus, Ludolf Christian (1779–1864), Bremen physician. Published a work on the cryptogamic plants of the Rhine Province.

Trevisan de Saint Leon, Vittore Benedetto Victorio (Conte) (1818–1897), Italian botanist.

Tuomikoski, Risto Kalevi (1911–1989), Finnish bryologist and entomologist, specialist in bog and fen mosses.

Turner, Dawson (1775–1858), banker, botanist, and antiquary, father-in-law of Sir William Hooker.

Underwood, Lucien Marcus (1853–1907), bryologist and pteridologist of New England.

Vánă, Jiri, (1940–), contemporary Czechoslovakian hepaticologist.

Vaucher, Jean Pierre Étienne (1763–1841), Swiss algologist.

Venturi, Gustav (1830–1898), Italian bryologist.

Viviani, Domenico (1772–1840), Italian botanist.

Voit, Johann Gottlieb Wilhelm (1786–1813), physician in Frankfurt-am-Main, wrote the first moss flora of North Bavaria. *Voitia.*

Wahlenberg, Göran (1780–1851), Swedish botanist, collected in Lappland, Switzerland and the Carpathians. Author of *Flora Lapponica*, 1812, and *Flora Suecica*, 1824–26.

Wallroth, Carl Friedrich Wilhelm (1792–1857), German bryologist and lichenologist, published *Flora Cryptogamia Germanica* in 1831.

Warnstorf, Carl Friedrich (1837–1921), was a German bryologist who taught in a boys' school for 32 years and wrote most of his works during this period. He was a worldwide authority on *Sphagnum* and other peat mosses: *Cryptogamic Flora of Brandenburg* and *Sphagnologia Universalis*. His 30000 specimens of *Sphagnum* were destroyed in the World War II bombings of Berlin and Budapest.

Weber, Georg Heinrich (1752–1828), Göttingen, Professor of Medicine in Kiel, contributed the moss section in Wigger's *Flora Holsatica*. Hedwig honored him with the genus *Webera* (now *Pohlia*).

Weber, William Alfred (1918–), eclectic Colorado botanist, (floristics, plant geography, phanerogamy, bryology, lichenology, etc., etc., etc., . . .)

Weigel, Johann Adam Valentin (1740–1806), a pastor in Würzburg. *Bryum weigelii*, which he collected in the Riesengebirge.

Weiss, Friedrich Wilhelm (1744–1826), Göttingen, a physician, wrote one of the first moss floras, *Plantae Cryptogamicae Florae Gottingensis. Weissia.*

Wenzel von Krojanke (?–?), an apothecary friend of Nees von Esenbeck on his botanical trips.

Willdenow, Carl Ludwig (1765–1812), Berlin, Pharmacist and professor in Berlin. His herbarium contained over 30000 specimens, including 3000 type collections of Humboldt and Bonpland from South America. By some miracle it survived the destruction of the rest of the Berlin herbarium in World War II.

Williams, Robert Statham (1859–1945), New York Botanical Garden, botanical work in Montana and South American Andes.

Wilson, Paul S. (1965–), contemporary American bryologist, New York State University (*Pseudoleskeella*).

Wilson, William M. (1799–1871), English bryologist, author of *Bryologia Britannica*, 1855, and co-author with J. D. Hooker, of Musci in *Flora Antarctica, 1867.*

Withering, William (1741–1799), English physician-botanist.

Wright, Charles (1811–1885), American botanical explorer, Texas (Asa Gray, *Plantae Wrightianae*), U. S. and Mexican Boundary Survey, North Pacific coasts of U.S. and Asia.

Zetterstedt, Johan Emanuel (1828–1880), Swedish bryologist.

Zier, John (?–1796). Dedication by Hedwig of *Bryum zierii*, but other details not found.

GLOSSARY

Abaxial: Facing away from the stem (as the 'underside' of a moss leaf).

Acrocarpous: Producing archegonia and sporophytes terminally on the main stem. Acrocarpous plants grow in erect tufts and are simple or sparsely branched. See also pleurocarpous.

Acumen: A slenderly tapered point; thus, pointed.

Adaxial: Facing the stem (as the 'upper side' of a moss leaf) (erroneously called ventral).

Aeruginose: Bluish–green, the color of weathered copper.

Aff.: Abbreviation for Latin *affinis*, meaning related to.

Air chambers: Cavities within the thallus, usually opening to the dorsal surface by a pore.

Alar cells: Cells at the basal angles of the leaf, often differentiated in shape, size, wall thickness, or color.

Amphigastria: The underleaves of leafy liverworts.

Androecium: 'The male house', including the antheridia, paraphyses, and surrounding leaves.

Annulus: A ring of enlarged, specialized elastic cells at the mouth of the capsule which turns inside out, forcing the operculum from the capsule urn.

Antheridiophore: In *Marchantia* and relatives, a stalk bearing the antheridia.

Antheridium: The bryophyte male reproductive organ. Multicellular and stalked, it resembles an ear of corn and contains many sperm cells.

Antical: In liverworts, alluding to the upper surface or margin (as of a leaf as opposed to postical).

Apiculus: Of a leaf or capsule, an apex which is short and rather suddenly narrowed (stouter variations of this are mucro and cusp).

Apophysis: Same as hypophysis.

Appendiculate: Particularly in Bryaceous species, having short transverse projections on the cilia of the endostome.

Appressed: Flattened or appressed to the stem.

Arcuate: Curved or bent in an arc.

Archegoniophore: In *Marchantia* and relatives, a stalk bearing the archegonia.

Archegonium: The female reproductive organ. It is flask-shaped with a narrow neck and a swollen base (venter). Fertilization of the egg occurs in the venter, and the embryonic sporophyte anchors itself and proceeds to grow the seta and immature capsule upwards, carrying the remains of the neck which tears away, grows for a time and, as a calyptra, rests unattached atop the capsule.

Arcuate: Curved like a bow.

Areolation: The pattern of the cells in the leaf (dense, open, etc.).

Areoles: In liverworts, the compartments of the thallus indicated by lines of division seen on the thallus surface.

Arista: A hard, straight awn or bristle, usually a projection of the costa.

Articulate: Having thickened, transverse lines or joints, as in a peristome.

Astomous: Lacking an operculum (**cleistocarpous**).

Attenuate: Narrowly tapered.

Auricle: An ear-like lobe at the basal margin of the leaf.

Autoicous: See Sexuality.

Awn: A bristle at the leaf tip, usually consisting of an excurrent costa.

Axis: The main stem.

Basal membrane: A delicate undivided base of the inner peristome (endostome).

Beak: A slender projection of the operculum.

Begleiter: Thin -walled cells accompanying the median guide cells in the costa, especially in Pottiaceae; same as hydroids.

Bi: A prefix meaning two or twice; ex. bistratose, with cells in two layers.

Biseriate: In two rows.

Bog: A wetland in which moisture is provided entirely by precipitation. They are ombrotrophic (cloud-fed) and acidic, with little plant nutrients. Bogs do not exist in Colorado. So-called peat bogs here are fens, which see.

Bracteole: In liverworts, a modified underleaf.

Bracts: In liverworts, specialized leaves surrounding the sex organs.

Branch primordia: A cell or cells that give rise to a branch. Where the primordium develops determines the growth form of the moss. Koponen (1982a) discusses the various types. Plants with few primordia are little branched. The location of branch primordia determines the symmetry of the branching, whether radial, one-sided, or distichous.

Branching pattern: (In pleurocarpous fen mosses, after Hedenäs, 1993). Distichously branched (branched in one plane); radially branched (when viewed from the top of the shoot, with branches pointing outwards from the stem in all directions); one-sided (with all branches pointing in the same direction); sparsely branched (too little branched to determine the pattern).

Brood bodies: Any vegetative structures that function as vegetative propagula.

Bulbiform: Shaped like a bulb, as some minute stems.

Bulbils: Large, often colored brood bodies either in the leaf axils or on the rhizoids.

Caespitose: Growing in erect cushions or sods (often called tufted, as opposed to matted).

Calciphile: A plant associated with a calcium-containing substrate and presumed to be so restricted.

Calyptra: An unattached covering formed by the remains of the archegonium, which partly or entirely fits over the capsule. It may be cucullate (split along one side), mitrate (covering the whole capsule like a tea-cozy) or mitrulate, with a long tubular apex and a basal portion split into several lobes.

Canaliculate: Channeled, as the distal costal region of some moss leaves.

Cancellinae: Large groups or 'windows' of hyaline cells characteristic of the leaf bases of some genera (notably *Syntrichia*).

Capillary: Very slender, as the stems of some mosses.

Capitulum: The mass of crowded short branchlets at the apex of the stem of *Sphagnum*.

Capsule: The spore case, often differentiated into a lid or operculum, and an urn; the urn commonly has an upper spore-bearing part and a basal neck of sterile tissue. In mosses, it is the capsule, a sporophyte structure, which has a few stomates.

Carpocephalum: The elevated female 'receptacle' of certain Marchantiaceae, consisting of a disk and stalk issuing from the thallus.

Carr: A wetland consisting of aquatic vegetation consisting of shrubs, particularly willows, inhabiting slow-flowing water.

Catenulate: With the leaves incurved toward the stem and closely arranged, resembling a chain (as in *Pseudoleskeella*).

Caulidium: An alternative and more technically accurate term for the bryophyte stem.

Central strand: A differentiated cylinder of tissue visible in cross–section in the center of the stems of some bryophytes.

Cf.: Abbreviation of Latin, *confer*, compare with, usually meaning possibly, perhaps or related to.

C. fr.: Latin, *con fructus*. meaning fruiting; in bryophytes, with sporophytes.

Chlorocysts: The small, elongate green cells forming a network around the large, empty hyaline cells in *Sphagnum*.

Cilia: Slender, often uniseriate filaments alternating with the segments of the endostome; filamentous strands of cells on the margins of some moss leaves.

Circinate: Curved into a circle or near-circle (as in *Scorpidium revolvens*).

Cleistocarpous: Lacking an operculum and hence irregularly dehiscent; opposite of **stegocarpous**.

Cochleariform: Rounded and deeply concave, like the bowl of a spoon.

Columella: A central column of sterile tissue forming the central axis of a sporangium. Sometimes this is permanently attached to the operculum, as in *Tortula systylia*. Rarely noticed unless it remains attached to the operculum upon dehiscence. However, in Polytrichaceae, the columella spreads out at the top and effectively closes the capsule except when dry.

Comose: Bushy or tufted, like a horse's mane, having enlarged leaves crowded in an apical tuft or rosette.

Complanate: with the leaf arrangement altered to appear flattened, as the stems of *Plagiothecium*.

Complicate-bilobed: In *Scapania*, two-lobed, with one lobe folded against the other.

Compressed: Flattened lengthwise, squeezed together, not condensed.

Con-: A prefix meaning together; ex. convolute, rolled together, as the margins of leaves.

Confertate: crowded together, or pressed together closely (ex. *Bryum argenteum*, in which the stems are extremely confertate)

Concave: Dished. A term that is ambiguous when speaking structures like leaves, which have two surfaces. The same leaf can be convex and concave at the same time. When looking at the abaxial surface of leaf it is usually flat or convex; looking at the adaxial side it is flat or concave; however, looking at the surface of a liverwort cell the trigones of the cell walls may be convex of they swell toward the interior of the cell, and concave if they recede from the cell center.

Convex: See concave.

Cordate: Stylized heart-shaped.

Cortex: The outer 'skin' of a moss stem, often composed of differentiated thin- or thick-walled cells (as opposed to the interior tissue or **medulla.**

Corticolous: Growing on bark.

Cortex: The outer row or rows of cells of the bryophyte stem, sometimes clearly differentiated from those abaxial to them.

Costa: The mid-vein of the moss leaf.

Costate: Having a costa.

Crenate: With rounded marginal teeth.

Cribrose: Perforated, as peristome teeth.

Crisped: Strongly contorted, curled, or twisted.

Cryptopore: Of stomates in *Orthotrichum* that are partly hidden by epidermal cells overlapping the guard cells.

Cucullate: (Of calyptra) hood-like, split along one side; (of leaf apex) cupped.

Cupulate: Cup-shaped, often referring to swollen perigonia (antheridial branchlets).

Cushion: A dense hemispherical or rounded tuft of crowded erect stems (see **mat**).

Cuspidate: Ending abruptly in a stout, rigid point (there are so many terms for gradations of similar things!).

Cuticle: A non-cellular, waxy coating on the surface of the leaf (sometimes faintly striate as in *Dicranoweisia*).

Cygneous: Curved like a swan's neck (as setae of *Grimmia pulvinata*).

Cymbiform: Cochleariform and broadly boat-shaped (mostly used for leaves of some species of *Sphagnum*).

De-: Prefix meaning down, off, from.

Decurrent: Said of the proximal margins of moss leaves that extend attached along the stem.

Dendroid: Having an erect main stem and terminal branching, tree-like (as in *Climacium*).

Dentate: With outward rather than forward-facing teeth that may be multicellular.

Denticulate: With fine teeth, often just cell tips that project from the leaf margin.

Det.: For Latin, *determinavit*, identified by. See also **exclamation point.**

Di-: Prefix meaning two or denoting separation.

Dioicous: See **Sexuality.**

Diplolepidous: Having a double peristome (an exostome and endostome).

Discoid: Flattened or platter-like, as the male inflorescence of *Philonotis.*

Disjunct: A distribution pattern in which a species is found in widely separated areas.

Distal: Farthest from the base; apical.

Distant: Spaced wide apart (same as *remote*).

Distichous: In two ranks, as the leaves of *Distichium* and *Fissidens.*

Dorsal: In mosses, a confusing term better called adaxial. However, it is correct to use for the upper surface in thalloid liverworts.

Dwarf male: In mosses, a miniature male gametophyte borne on a full-sized female plant.

e-: Prefix meaning without (ex: eciliate, ecostate).

Ecostate: Lacking a costa.

Elater: A long, spirally thickened cell that changes shape markedly with humidity (associated with spores, in liverworts.

Elimbate: Lacking a differentiated margin.

Emarginate: Broadly notched at the apex (as the leaves of some *Syntrichia*).

Emergent: Partly exposed, said of moss capsules that protrude only partly beyond the tips of the perichaetial leaves.

Endemic: Restricted to a single floristic area.

Endostome: The inner peristome, usually consisting of a basal membrane with teeth (segments) alternating with slender cilia.

Entire: Smooth-margined, without teeth.

Ephemeral: Short-lived.

Eutrophic: (Of a habitat) rich in mineral nutrients, particularly nitrate and phosphate.

ex-: Prefix meaning from, out of, outer.

Exclamation point: Used to indicate verification by someone.

Exannulate: Lacking an annulus.

Excentric: Off center, as the attachment of the seta to the capsule (English, eccentric). See Stearn (2004).

Excurrent: Of costae which exceed the length of the leaf.

Exostome: The outer row of a double peristome.

Exserted: Protruding beyond the tips of surrounding structures.

Exsiccata (pl. exsiccatae): A set of specimens distributed to herbaria as standards for research.

Exsiccatum (pl. exsiccati): A single herbarium specimen.

Falcate: Sickle-shaped.

Falcate-secund: A leaf form that is strongly curved and turned only to one side of the stem (as *Drepanocladus*).

Fascicle: A cluster of branches arising from the same point on the stem.

Fastigiate: With branches nearly parallel and of more or less equal length.

Fen: A quiet-water wetland that is fed by adjacent mineral material rather than rain-water alone. Depending on their mineral supplies, Colorado has many calcareous fens, and relatively fewer iron fens. Like bogs, they produce peat.

Fenestrate: Having a differentiated area of clear cells resembling windows (especially in the leaves of *Syntrichia*).

Fertile: Having sporophytes (an unusual use of the term, which should apply to the female sex organs or the egg).

Fibrils: Fine fiber-like wall thickenings in the leucocysts (hyaline cells) of *Sphagnum*.

Filamentous gemmae: Brood-bodies that are uniseriate, multicellular, and packed with chloroplasts (as in the leaf axils of *Rosulabryum laevifilum*).

Filiform: Thread-like.

Fimbriate: Fringed.

Flabelliform: Fan-shaped.

Foot: The base of the sporophyte that is embedded in the gametophyte and extracts water, mineral nutrients, and photosynthate from it.

Fusiform: Spindle-shaped, narrowed and tapered at each end.

Gemmae: Vegetative propagula of various shapes and sizes.

Guide Cells: Large, empty, firm-walled cells stretching across the costa from side to side as seen in cross-section.

Gymnostomous: Lacking a peristome.

Gynoecium: 'The female house', including the perichaetium and the archegonia.

Habitat: The ecological situation of the plant.

Habitus: The aspect of the plant.

Hair-point: A hair-like and usually hyaline leaf tip, strictly speaking not containing the excurrent costa, but the term often loosely applied to any slender hyaline extension of the leaf tip.

Haploid: With a single set of chromosomes, the condition in the vegetative moss plant.

Haplolepidous: Having a single peristome.

Heterophyllous: Producing clearly different sizes or shapes of leaves on stems versus branches.

Homomallous: All pointing in the same direction, as in *Homomallium mexicanum*, where the creeping stem has straight leaves that face upward and forward).

Hyaline: Transparent, lacking chloroplasts or pigments.

Hyalocyst: The empty, transparent porose cells of *Sphagnum*.

Hyaloderm: A cortex composed of enlarged, empty, usually thin-walled and fragile, colorless cells.

Hydroids: In the costa (especially important in Pottiaceae), scattered large thin-walled cells between the abaxial stereid layer and the large firm-walled guide cells that stretch across the costa in cross-section. Unlike the guides, the hydroids are so thin-walled that they often appear as collapsed cells.

Hygroscopic: Readily absorbing moisture.

Hypnoid peristome: A perfect double peristome with 16 lanceolate teeth, cross-striolate at base and trabeculate at middle, and a well-developed endostome consisting of a high basal membrane and 16 lanceolate, keeled segments alternating with 1 or more narrow cilia; in its typical development, associated with inclined and asymmetric capsules.

Hypophysis: The sterile basal neck of a capsule where it joins the seta, often swollen and with a few stomata.

Imbricate: Crowded and tightly overlapping, like shingles.

Immersed: Hidden or submerged, said of capsules that are completely overtopped by the perichaetial leaves.

In-: Prefix meaning inwardly, sometimes meaning not.

Incrassate: With thick cell walls.

Incubous: A leaf arrangement in liverworts. I use the analogy of Venetian blinds. If the slats overlap going up the window, this is analogous to incubous leaf arrangement. If the slats overlap downwards the condition is succubous. You can't slide down a roof if the tiles are succubously arranged. This has to do with the attachment of the leaf bases. Leaves without slanted attachments are transverse. If you were a bug, you would easily slide forward along an incubous leaf arrangement; you would get stuck if you try it with a succubous one. Unless leaves are inserted on the stem transversely, they will be either incubous or succubous.

Inflexed: Abruptly bent upwards or toward the axis.

Inflorescence: A group of sex organs surrounded by bracts (an unfortunate term for mosses borrowed from vascular plant terminology but still used frequently).

Initials: Special cells often of different shape and size on stems or leaves, which give rise to rhizoids. Also called nematogones (Koponen 1982a). They can be made visible by coloring them with basic stains, e.g. toluidin blue. Rhizoidal initials are important on the distal parts of leaves of some Amblystegiaceae.

Inner peristome: See endostome.

Innovations: New growth.

Insertion: The manner in which leaves are attached to the stem. See **incubous**.

Involucre: A tube of thallus tissue protecting the archegonia of thalloid liverworts.

Involute: Inrolled lengthwise; a margin or an entire leaf may be involute.

Iso-: Prefix meaning the same or equal.

Isodiametric: Said of cells that are about as broad as long. See also **quadrate**.

Isophyllous: Having leaves of similar shapes and sizes on stems and branches.

Julaceous: The usual definition is 'smoothly cylindric, terete-foliate, like a worm or catkin, referring to stems or branches with crowded or imbricate leaves'. This, it seems, will be confusing to many novices. We would expand this to require that the leaves of julaceous stems are deeply cochleariform as well as imbricate, resulting in a swollen appearance. Examples: *Myurella julacea, Bryum argenteum, B. calobryoides,* and *Entodon concinnus.*

Juxtacostal: Next to the costa; such cell features have special importance in *Grimmia*.

Keeled: Folded lengthwise or with a prominent costa suggesting this.

Lamella: A wall-like rib or flap, several cells high, running lengthwise perpendicularly down the adaxial side of some mosses (as in *Pterygoneurum, Polytrichum*).

Lamina: The flat blade of the leaf.

Lanceolate: Narrow and tapered from base to apex.

Leaf: An incorrect but entrenched word for the moss **phyllidium**, which see.

Leg.: Abbreviation of Latin, *legit,* collected by.

Leucocyst: The large, empty hyaline cell in the leaves of *Sphagnum.*

Lid: Operculum.

Lignicolous: Growing on wood.

Ligulate: Narrow and elongate, with parallel sides; strap-shaped or ribbon-like.

Limbidium: A differentiated margin, usually of narrow cells, as in some Bryaceae).
Linear: Very long and narrow, with parallel sides (narrower than ligulate).
Lingulate: Tongue-shaped, oblong with a broadened apex.
Lumen (pl., **lumina):** The cell cavity.

Macronemata: Rhizoids that are large, freely branched, and restricted to the leaf axils and branch insertions (particularly in Mniaceae).
Mammilla: low, rounded projection on the operculum; a bulging surface of a cell.
Marsupium: swollen and elongate pouch of stem tissue, arising from the ventral surface of the stem, that encloses and protects the sporophyte in some leafy liverworts.
Mat: A growth form in which stems are flattened on the substrate and densely interwoven.
Medulla: The inner tissue of a stem, surrounded by the outer cortical layer.
Micron: One thousandth of a mm (abbreviated μm). Moss leaf cells and spores average less than 50 μm.
Micronemata: Short, slender, more or less simple rhizoids on stems or leaf-bases, generally distributed as opposed to macronemata.
Microphyllous: With minute leaves, referring especially to modified ones of brood branches.
Mitrate: Shaped like a bishop's cap or candle-snuffer (said of the calyptra when it is not split down one side).
Monoicous: Having antheridia and archegonia on the same plant.
Monotypic: Represented by only one taxon (as a monotypic family, genus or species).
Mucilage cells: In liverworts, cells of the leaf margins, and sometimes within a thallus, which produce mucilage (also called slime papillae.
Mucro: A short, abrupt point (another term for a short point).

Multi-: Prefix meaning many.
Muticous: Lacking a hair-point, mucro, or awn.

Neck: The sterile base of a capsule, sometimes distinctly narrower or wider than the urn.
Nematodontous: Of peristomes consisting of whole cells rather than the residual walls of two cell layers: ex, *Polytrichum, Tetraphis.*
Nerve: The midrib (costa) of the bryophyte leaf.

Ob-: Prefix meaning the reverse: ex, **oblate,** broader than long; **obovoid,** broadest at the top.
Obsolete: Scarcely evident, near lacking (implying loss).
Obtuse: Broadly pointed or rounded at the apex.
Oil bodies: Oil-containing inclusions in the cells of liverworts. They tend to be ephemeral so are best observed in fresh specimens.
Oligotrophic: (Of a habitat) poor in mineral nutrients.
Ombrotrophic: Said of vegetation that gets its water only from rainfall and so is mineral-poor. Bogs have ombrotrophic vegetation. Colorado does not have true bogs.
Operculum: The lid of the capsule; capsules from which the operculum has fallen away are *deoperculate*; capsules that lack detachable lids are *cleistocarpous*.
Oval: Shortly elliptic.
Ovate: Egg-shaped in outline (ovoid, egg-shaped, is used for a three-dimensional object).

Papilla (-ae): Decorative outgrowths of the cell walls of many mosses, consisting of sharp or blunt points, sometimes branched or C-shaped. These are probably important in scattering light since they are common in arid, hot areas.
Paraphyllia: Tiny filaments, scales, or leaf-like structures scattered on the stems and branches of some leafy liverworts and pleurocarpous mosses (compare with **pseudoparaphyllia**).
Paraphysis (plural, -ses): uniseriate septate filament associated with antheridia and archegonia.
Paroicous: See Sexuality.
Patent: (Of leaves), making a 20–45 degree angle with the stem.
Pendent: Hanging down (do not use pendant as an adjective; it is a noun!).

Percurrent: (Of a costa) extending to but not beyond the leaf apex.

Peri-: Prefix meaning surrounding.

Perichaetium: A series of bracts (perichaetial bracts) subtending the archegonia.

Perigonium: The bracts surrounding the antheridia.

Peristome: The sieve mechanism of the moss capsule, consisting of hygroscopic teeth composed of residual cell walls, usually in multiples of four (four, sixteen, thirty-two and 64). Species with erect capsules tend to have a single ring of teeth (**haplolepidous**), while nodding capsule commonly have two rows, an inner and an outer (**diplolepidous**).

Phaneropore: Of stomates in *Orthotrichum* that are flush with the epidermis and not covered by overlapping epidermal cells.

Phyllidium (phyllidia): The moss 'leaf'. The term now used in current Spanish literature, but evidently not likely to replace in America the use of 'leaf', which, strictly speaking, is a diploid structure of the vascular plants.

Pleurocarpous: Producing the sex organs and capsules laterally along the stem.

Plicate: Longitudinally pleated.

Pluri-: Prefix meaning several; for example, **pluripapillose**.

Poly-: Prefix meaning many; for example, **polymorphic**, of many forms.

Porose: Having a pit or hole in the walls of adjacent cells, best seen by focusing up and down; the porosity is seen as thinning of the wall rather than showing an actual break. In *Sphagnum* the pores are large round openings in the walls of the cells seen on the surface rather than between cells.

Primordia: Very early stages in organ development.

Propagulum (-a): Any part of the moss plant that is capable of breaking away to form a new plant: gemmae, leaf tips, buds, microphyllous branches.

Prorula: The protruding end of a cell (often inaccurately termed an *end papilla*). Prorulae are best observed by slowly raising and lowering the microscope objective so that different planes emerge. Cells may be prorulate or prorulose. See also *scindula*.

Proximal: Nearer to the base (the opposite of *distal*).

Pseudoparaphyllia: Structures that superficially resemble paraphyllia but are found only in pleurocarpous mosses and are clustered around the bases of branches or branch buds. See Ireland (1971).

Quadrate: Squared, as some isodiametric cells with only four visible sides.

Radial: Arranged uniformly around a central axis (as opposed to bilateral or flattened).

Radicle: Same as **rhizoid.**

Rank: Row, as in three-ranked.

Remote: Spaced well apart.

Resorption: The lysing or erosion of parts of the cell walls in the tips of *Sphagnum* leaves.

Reticulate: Forming a network.

Revoluble: Rolling away and turning inside out, as the ring formed by an annulus that falls away from the capsule mouth.

Rhizoids: Unicellular, simple or branched strands of usually brown elongate cells arising from various parts of the bryophyte plant and not containing chloroplasts. Tuomikoski (1982a) distinguishes five conditions:
  1. Rhizoids originating on leaves. Rhizoids (or rhizoidal initials) may be situated at the leaf apex (e.g. in *Calliergon*), on the costa (*Tomentypnum*), or diffusely among the ordinary leaf cells (*Plagiomnium*).
  2. Rhizoids originating along the stem without having any special location (micronemata). This type of rhizoid topography is found in *Plagiomnium, Calliergon,* and some species of *Bryum*.
  3. Rhizoids originating at the leaf axils from the cells surrounding the branch primordia or 'dormant buds' (macronemata). This type occurs in *Mnium* and *Calliergonella*.
  4. Rhizoids concentrating in dense bunches on the ventral side of the stem just at or below the insertion of the costa. Examples of genera having this arrangement are *Acrocladium, Callicladium,* and *Drepanocladus*.
  5. Rhizoids absent. there are several genera (*Scorpidium, Limprichtia*), in which neither rhizoids nor rhizoidal initials have been detected so far.

Rhombic: Diamond-shaped

Rhomboidal: Longer than rhombic, with slanted cell wall.

Rostellate: Having a very short beak.

Rostrate: Having a long beak.

Rosulate: Forming a spreading rosette (as the apices of male stems of Polytrichaceae), or the terminal cluster of leaves of some Bryaceae (as *Rosulabryum*).

Rugose: Strongly wavy, wrinkled, or undulate (as in *Rhytidium rugosum*).

Saxicolous: Growing on rock.

Scindula (-e): The protruding end of one cell overlapping the base of another; same as *prorula*.

Secund: Strongly turned to one side.

Segments: The teeth of the endostome.

Septate: Partitioned with dividing walls.

Seriate: Arranged in distinct rows (*uniseriate*, on one row; *biseriate*, in two parallel rows).

Serrate: Regularly toothed like a saw blade, the teeth of one or more cells and pointing toward the leaf apex.

Serrulate: Minutely serrate with the teeth composed of only part of a single cell.

Seta: The stalk of the capsule.

Setaceous: Bristle-like.

Sexuality

Autoicous: the plant is monoicous (monoecious), with the archegonia and antheridia in separate 'inflorescences' on the same stem.

Cladautoicous: Having the antheridia on a separate branch.

Goniautoicous: Having the male inflorescence bud-like and axillary on the same stem or branch as the female one.

Pseudo-autoicous: Having dwarf male plants growing on stems of the female plant (especially some species of *Dicranum*).

Rhizautoicous: Having the male inflorescence on a very short branch attached to the female stem by rhizoids and appearing to be a single bud-like plant.

Dioicous: Having the sex organs on separate plants (tufts).

Monoicous: an 'umbrella term' for mosses having both sexes on the same plant (the term has several variations).

Paroicous: Monoicous, with the archegonia and antheridia in the same inflorescence but not mixed, the antheridia in the axils of the perichaetial leaves or less commonly in groups outside a cluster of archegonia.

Polyoicous: Having combinations of autoicous conditions.

Synoicous: Monoicous, having the antheridia and archegonia mingled in the same inflorescence

Shoulder: An abruptly narrowed transition between the base and the rest of the leaf (as in *Timmia*).

Sigmoid: S-shaped.

Simple: Not forked or branched.

Sinuose: Having a wavy or fluted cell wall.

Sinus: The gap between adjacent lobes of leaves of some liverworts.

Slime papilla: In some liverworts, a cell that secretes mucilage.

Solifluction: Slow downward flow, especially on tundra, of saturated soil and vegetation.

Spatulate: Narrow below and gradually broadening above (compare with **lingulate**).

Spear: In mosses, a very young sporophyte before the capsule has differentiated.

Splash-cup: A rosette of leaves subtending a mass of gemmae or antheridia. The splashing of raindrops disperses the gemmae or sperm.

Squarrose: (Of leaves) spreading widely and with the tips recurved downward.

Stereids: Thick-walled cells with very little lumen, occurring *en masse* as seen in the cross-section of the costa, most common on the abaxial side, especially in Pottiaceae.

Sterile: Lacking sporophytes.

Stolon: A horizontal stem or runner that is above ground but often attached to the substrate at the tip; ex. *Plagiomnium*.

Strangulate: Of capsules, especially those of *Orthotrichum*, constricted below the rim.

Stratose: Layered (uni-, bi-, multi-stratose).

Strict: Rigidly straight.

Strumose: With a goiter-like swelling on one side of the base of the capsule (in some Dicranaceae).

Sub-: Prefix meaning under, nearly, or somewhat, depending on the context.

Succubous: In liverworts, the leaf arrangement in which the proximal part of a leaf overlaps the distal part of the leaf proximal to it (think of Venetian blinds where one slat overlaps the slat below). See **incubous**.

Sulcate: With deep longitudinal folds, ex. the capsule of *Ceratodon*.

Synoicous: See Sexuality.

Systylious: Having the operculum attached to the top of the columella after dehiscence.

Terete: Cylindric.

Terricolous: Growing on soil.

Tomentum: A dense covering of rhizoids.

Trabeculate: With cross-bars (as many peristome teeth).

Transverse: In liverworts, the condition of leaf attachment in which the leaf base is not slanted.

Trigone: A triangular wall thickening in the corners of many cells of liverworts.

Tuber: In mosses, a gemma-like propagulum, borne underground on rhizoids, particularly in Bryaceae

Tumid: Extremely julaceous, bulging, swollen.

Underleaves: In liverworts, the reduced (often vestigial) lower row of leaves, often hidden amongst rhizoids.

Undulate: Transversely wavy, less so than rugose.

Urceolate: Urn-shaped, said of capsules constricted below the mouth and with a short neck.

Urn: The spore-bearing portion of the capsule, usually broader than the next.

Vaginant lamina: In *Fissidens*, the folded basal part (evidently a single very large lamella) of the leaf, that clasps the stem.

Valve: The splitting segments of the capsules of liverworts; also, the split parts of capsules of **Andreaea**.

Ventral: Referring to the underside of the thallus in liverworts; in mosses a confusing term best replaced by abaxial.

Vermiform or **vermicular:** Long, narrow, and wavy (worm-like) cells, especially in the Amblystegiaceae).

Verrucose: Roughened, covered with warty projections, as the setae of some *Brachythecium*.

BIBLIOGRAPHY

Amann, J. 1925. *Mnium arizonicum* sp. n. *Révue Bryologique* 52:23.

Anderson, L. E.1960. Personal reflections on Alexander W. Evans. *Bryologist* 64:84–88.

Anderson, L. E.1976. Antero Vaarama, 1912–1976. *Bryologist* 79:377–379.

Andrews, A. L. 1949. Taxonomic notes: *Splachnobryum kieneri*. *Bryologist* 52:78–83.

Arnell, S. 1956. *Illustrated Moss Flora of Fennoscandia. I. Hepaticae.* 308 pp, index, map. Gleerup, Lund.

Austin, C. F. 1874. *Hepaticae,* [in] Rothrock, J. T. Preliminary report on the botany of central Colorado. p. 62. Geogr. & Geol. Exploration & Survey west of the Hundredth Meridian, George M. Wheeler in charge [known as the Wheeler Expedition].

Austin, C. F. 1875a. New Hepaticae [*Riccia frostii* n. sp., Nevada . . . and Colorado, *Wolf & Rothrock*, associated with R. crystallina]. *Bull. Torrey Bot. Club* 6:17–21.

Austin, C. F. 1875b. New mosses from Colorado. *Bull. Torrey Bot. Club* 6:45–47.

Bai, Z. L. 2002. *Crossidium aberrans* Holzinger & Bartr., a new record from Asia. *Hikobia* 13: 637–640.

Barnes, C. R., & F. D. Heald. 1896. *Analytical keys to the genera and species of North American mosses.* Bull. Univ. Wisconsin, Madison.

Barnhart, J. H. 1965. *Biographical Notes Upon Botanists.* 3 volumes, reprint of a card index. New York Botanical Garden.

Bednarek-Ochyra, H. 2006. A Taxonomic monograph of the moss genus *Codriophorus* P. Beauv. (Grimmiaceae). W. Szafer Inst. Bot., Polish Acad. Sci., Krakow.

Bednarek-Ochyra, H., D. Lamy, & R. Ochyra. 2001. A note on the moss genus *Codriophorus* P. Beauv. *Cryptogamie, Bryologique.* 22:105–111.

Bednarek-Ochyra, H., & R. Ochyra. 2006. Typification of *Codriophorus fascicularis* (Grimmiaceae). *Taxon* 54:1065–1074.

Bednarek-Ochyra, H. J. Váňa, R. Ochyra, & R. I. Lewis-Smith. 2000. *The Liverwort Flora of Antarctica.* Polish Academy of Sciences.

Behle, W. H. 1984. In Memoriam: Seville Flowers (1900–1968). *Great Basin Naturalist* 44:210–222.

BFNA. Preliminary treatments of bryophyte families intended for the *Flora of North America*, vols. 27, 28, 29. These manuscripts are available to the public on the Internet at:
http://www.mobot.org/plantscience/BFNA/bfnamenu.htm
To date, only volume 27 has been published. See Flora of North America.

Blom, H. H. 1996. A revision of the *Schistidium apocarpum* complex in Norway and Sweden. *Bryophytorum Bibliotheca* 49:1–333.

Bowers, M. C. 1966. New bryophytes for Colorado. *Bryologist* 69:368–369.

Bowers, M. C. 1969. A cytotaxonomic study of the genus *Mnium* in Colorado. *Rev. Bryologique Lichenologique.* 36:167–202.

Brandegee, T. S. 1896. Flora of southwestern Colorado. *Bull. U. S. Geol. Geogr. Surv. Terr.* (Hayden) 2:227–246.

Bridel-Brideri, Samuel Elysée. 1819. *Mantissa generum specierumque Muscorum frondosorum universa.*

Britton, E. G. 1894. Contributions to American bryology. VII. A revision of the genus *Physcomitrium*, with descriptions of five new species. *Bryologist* 21: 189–208. Plates 197–203.

Buck, W. R., & H. Crum. 1978. A re-interpretation of the Fabroniaceae with notes on selected genera. *J. Hattori Bot. Lab.* 44:347–369.

Buck, W. R., & D. H. Norris. 1996. *Hedwigia stellata* and *H. detonsa* (Hedwigiaceae) in North America. *Nova Hedwigia* 62:361–370.

Churchill, S. 1987. Systematics and biogeography of *Jaffueliobryum* (Grimmiaceae). *Mem. N. Y. Bot. Garden* 45:691–708.

Clark, L., & T. C. Frye. 1937. Extension of ranges among northwestern Hepaticae. *Bryologist* 40:13–16.

Conard, H. S. 1956. *How to Know the Mosses and Liverworts.* Revised by Paul Redfearn, 1979. Wm. C. Brown Co., Dubuque, IA.

Cooper, D. J. 1991a. The habitats of three boreal fen mosses new to the Southern Rocky Mountains of Colorado [*Sphagnum contortum* report]. *Bryologist* 94:49–59.

Cooper, D. J. 1991b. Additions to the peat-land flora of the Southern Rocky Mountains: Habitat descriptions and water chemistry. *Madroño* 38: 139–143.

Cooper, D. J. 1995. Water and soil chemistry, floristics, and phytosociology of the extreme rich High Creek Fen, in South Park, Colorado, U.S.A. *Can. J. Bot.* 74:1801–1811.

Cooper, D. J. 1997. A montane *Kobresia myosuroides* fen community type in the Southern Rocky Mountains of Colorado, U.S.A. *Arctic and Alpine Research* 29: 300–303.

Cooper, D. J., R. Andrus, & C. D. Arp. 2002. *Sphagnum balticum* in a Southern Rocky Mountain fen. *Madroño* 49:186–188.

Cox, C. J., & T.A.J. Hedderson. 1999. Phylogenetic relationships among the ciliate arthrodontous mosses: evidence from chloroplast and nuclear DNA sequences. *Plant Systematics and Evolution* 215:119-139.

Craft, J. H., & I. H Craft. 1952. Bryophytes of south-central Colorado. Part I. *J. Iowa Acad. Sci.* 59:82–84.

Cronberg, N., R. Natcheva, & K. Hedlund. 2006. Microarthropods mediate sperm transfer in mosses. *Science* 313:1255.

Crosby, M. R., R. E. Magill, B. Allen, & S. He. 2000. *Checklist of the Mosses.* 320 pp. Missouri Botanical Garden, St. Louis.

Crum, H. 1966. A tribute to Edwin B. Bartram. [with bibliography]. *Bryologist* 69:124–134.

Crum, H. 1973a. Mosses of the Great Lakes Forest. *Contrib. Univ. Michigan Herb.* 10:1–404.

Crum, H. 1973b. (editor) Mosses of Utah new to science [described by Seville Flowers]. *Bryologist* 76:286–292.

Crum, H. 1985. Lewis E. Anderson, [in] *Contributions to systematic bryology, dedicated to Lewis E. Anderson.* Monographs in Systematic Botany. 11, Missouri Bot. Garden.

Crum, H. 1991. *Liverworts and Hornworts of southern Michigan.* Univ. of Michigan.

Crum, H. 1997. A seasoned view of North American *Sphagna. J. Hattori Bot. Lab.* 82:77–98.

Crum, H., & L. E. Anderson. 1955. Taxonomic studies in the Funariaceae. *Bryologist* 58: 1–15.

Crum, H., & L. E. Anderson. 1981. *Mosses of Eastern North America.* 2 volumes. 1328 pp., 637 plates. Columbia University Press.

Crum, H., & D. Hall. 1995. An evaluation of *Didymodon ferrugineus*, more commonly known as *Barbula reflexa. Contrib. Univ. Michigan Herbarium* 10:141–146.

Crundwell, A. C., & E. Nyholm. 1964. The European species of the *Bryum erythrocarpum* complex. *Trans. Brit. Bryol. Soc.* 4:597–636.

Crundwell, A. C., & E. Nyholm. 1974. *Funaria muhlenbergii* and related European species. *Lindbergia* 2:222–229.

Damsholt, K., & J. Váňa. 1977. The genus *Jungermannia* L. emend. Dumortier in Greenland. *Lindbergia* 4:1–26.

Delgadillo M., C. 1975. Taxonomic revision of *Aloina, Aloinella,* and *Crossidium* (Musci). *Bryologist* 78:245–303.

Delgadillo M., C. 2007. *Crossidium,* in *Flora of North America North of Mexico.* Volume 27, *Bryophyta, Part 1.* Pp. 611–614. Oxford University Press.

Derda, G. S., & R. Wyatt. 1999. Levels of genetic variation and its partitioning in the wide–ranging moss *Polytrichum commune. Syst. Bot.* 24:512–528.

Derda, G. S., & R. Wyatt. 2003. Genetic variation and population structure in *Polytrichum juniperinum* and *P. strictum* (Polytrichaceae). *Lindbergia* 28:22–40.

Dobzhansky, T. 1951. *Genetics and the Origin of Species.* Columbia Univ. Press.

Eckel, P. M. 1998. Re-evaluation of *Tortella* (Musci, Pottiaceae) in conterminous U.S.A. and Canada, with a treatment of the European species *Tortella nitida. Bull. Buffalo Soc. Nat. Sci.* 36:117–191.

Engel, J. J., & R. M. Schuster. 1984. An overview and evaluation of the genera of Geocalycaceae, subfamily Lophocoleoideae (Hepaticae). *Nova Hedwigia* 39:385–463.

Evans, A. W. 1914. Notes on New England Hepaticae, XI. *Rhodora* 16:62–76. [Mention of *Clevea hyalina* in Colorado, no specimens or localities cited].

Evans, A. W. 1915a. Preliminary list of Colorado Hepaticae. *Bryologist* 18:44–47.

Evans, A. W. 1915b. The genus *Plagiochasma. Bull. Torrey Bot. Club* 42:259–308.

Evans, A. W. 1940. A list of hepatics found in U. S., Canada, and Arctic America. *Bryologist* 43:133–138.

Ewan, J., & N. Ewan. 1981. *Biographical Dictionary of Rocky Mountain Naturalists—1682–1932.* 253 pp. W. Junk, The Hague.

Fife, A. J. 1970. Taxonomic observations on three species of North American *Funaria. Bryologist* 82:204–214.

Flock, J. W. 1978. Lichen-bryophyte distribution along a snow cover/soil moisture gradient, Niwot Ridge, Colorado. *J. Arctic & Alpine Res.* 10:31–37.

Flora of North America Editorial Committee 2007. *Flora of North America North of Mexico.* Volume 27, *Bryophyta, Part 1.* 713 pp. Oxford University Press.

Flowers, W. S. 1942. Mary Parry Haines. *Proc. Indiana Acad. Sci.* 51:78–82.

Flowers, W. S. 1961. The Hepaticae of Utah. *Univ. Utah Biol. Ser.* 12:1–108. 18 plates.

Flowers, W. S. 1973. *Mosses: Utah and the West.* 567 pp., 149 plates. Brigham Young Univ., Provo, UT.

Foster, M. 1994. *Strange Genius: The Life of Ferdinand Vandeveer Hayden.* 443 pp. Roberts Rinehart, Niwot, Colorado.

Frahm, J.-P., & J. Eggers. 2001. *Lexikon deutschsprachiger Bryologen.* 672 pp., over 500 portraits. Books on Demand (BoD) Norderstedt.

Fransén, S. 2004. A taxonomic revision of extra-neotropical *Bartramia*, section *Vaginella* C. Müll. *Lindbergia* 29:73–107.

Fransén, S. 2005. Taxonomic revision of the moss genus *Bartramia* Hedw., sections *Bartramia* and *Vaginella* C. Müll., Ph. D. thesis, Göteborg University, Sweden, 122 pages.

Frye, T. C., & L. Clark. 1937, 1945. Hepaticae of North America. *Univ. of Washington Publ. Biol.* 6:1–1022.

Gallego, M. T. 2006. *Syntrichia*, pp. 120–143 in *Flora Briofítica Ibérica*, vol.3. Sociedad Española de Briología (SEB), Murcia.

Gardiner, A., M. Ignatov, S. Huttonen, & A. Troitsky. 2005. The resurrection of the families Pseudoleskeaceae and Pylaisiaceae Schimp. (Musci, Hypnales). *Taxon* 54:651–663.

Geiser, S. W. 1937. *Naturalists of the Frontier.* 341 pp. Southern Methodist University.

Gier, L. J. 1953. N.L.T. Nelson, 1862–1932, bryologist. *Bryologist* 56:147–148, portrait.

Greven, H. C. 1999. A synopsis of *Grimmia* in Mexico, including *Grimmia mexicana, sp. nov. Bryologist* 102:426–436.

Greven, H. C. 2003. *Grimmias of the World.* 247 pp., 93 Figs., CD–ROM (color illustrations). Backhuys Publ., Leiden.

Griffin, D., III. 1998. Axillary hairs in *Bartramia*, section Strictidium. *Evansia* 15:81–83.

Griffin, D., III, & W. R. Buck. 1989. Taxonomic and phylogenetic studies on the Bartramiaceae. *Bryologist* 92:368–380.

Grout, A. J. 1916. Mosses of Colorado from Tolland and vicinity. *Bryologist* 19:1–8.

Grout, A. J. 1928–1939. *Moss Flora of North America North of Mexico.* 3 volumes. Published by the author.

Hartman, E. 1969. The ecology of the "copper moss" *Mielichhoferia mielichhoferi* in Colorado. *Bryologist* 72:56–59. [Hartman's collections actually contain both *M. elongata* and *M. mielichhoferiana*, Ed.]

Hastings, R. I. 1999. Taxonomy and biogeography of the genus *Coscinodon* (Bryopsida, Grimmiaceae) in North America. *Bryologist* 102: 265–286.

Hastings, R. I. 2002. Biogeography of *Grimmia teretinervis* (Bryopsida, Grimmiaceae) in North America. *Bryologist* 105:262–266.

Hastings, R. I., & H. C. Greven. 2007. *Grimmia*, in *Flora of North America North of Mexico*. Volume 27, *Bryophyta, Part 1*. Pp. 225–258. Oxford University Press.

Hayden, F. V. 1869. Third Annual Report of the United States Geological Survey of the Territories, embracing Colorado and New Mexico, conducted under the authority of the Secretary of the Interior.

Hedenäs, L. 1970. A partial revision of *Campylium* (Musci). *Bryologist* 100:65–88.

Hedenäs, L. 1987. The taxonomic position of *Tomentypnum*. *J. Bryol.* 14:729–736.

Hedenäs, L. 1989a. The genera *Scorpidium* and *Hamatocaulis*, gen. nov., in northern Europe. *Lindbergia* 15:8–36.

Hedenäs, L. 1989b. The taxonomic position of *Conardia* H. Robinson. *J. Bryol.* 15:779–783.

Hedenäs, L. 1989c. The genus *Sanionia* (Musci) in northwestern Europe; a taxonomic revision. *Ann. Bot. Fenn.* 26:399–419.

Hedenäs, L. 1990a [1992]. The genus *Pseudocalliergon* in northern Europe. *Lindbergia* 16:80–99.

Hedenäs, L. 1990b [1992]. Taxonomic and nomenclatural notes on the genera *Calliergonella* and *Breidleria*. *Lindbergia* 16:161–168.

Hedenäs, L. 1993a. Field and microscope keys to the Fennoscandian species of the *Calliergon-Scorpidium-Drepanocladus* complex, including some related or similar species. 79 pp. Privately published.

Hedenäs, L. 1993b. A generic revision of the *Warnstorfia-Calliergon* group. *J. Bryol.* 17:447–479. [*Straminergon* proposed here]

Hedenäs, L. 1996. Taxonomic notes on *Brachythecium erythrorrhizon* B. S. & G., based mainly on studies of Swedish material. *Lindbergia* 21:21–25.

Hedenäs, L. 1997. A partial generic revision of *Campylium* (Musci). *Bryologist* 100:65–88.

Hedenäs, L. 1998. An overview of the *Drepanocladus sendtneri* complex. *J. Bryol.* 20:83–102.

Hedenäs, L. 2002. Korvgulmossa *Pseudocalliergon turgescens*, en spännande mossa I våra kalkrikaste trakter [*Pseudocalliergon turgescens*, a fascinating moss in our most strongly calcareous areas]. In Swedish, with English summary. *Sv. Bot. Tidskr.* 96:29–40.

Hegewald, E. 1972. *Dicranum groenlandicum* in Finland. *Memoranda Soc. Fauna Flora Fennica* 48:85–87.

Hermann, F. J. 1976. Additions to the bryophyte flora of Colorado. *Bryologist* 79:215–221.

Hermann, F. J. 1980. Distinctions between *Anacolia menziesii* (new to Colorado) and *Bartramia stricta*. *Bryologist* 83:253–254.

Hermann, F. J. 1987. The bryophytes of Rocky Mountain National Park, Colorado. *Mem. N. Y. Bot. Garden* 45:219–231.

Hofmann, H. 1998. A monograph of the genus *Homalothecium*. *Lindbergia* 23:119–159.

Holmgren, P. K., & N. H. Holmgren. 1998 onwards (continuously updated). *Index Herbariorum*. New York Botanical Garden. http://sciweb.nybg.org/science2/IndexHerbariorum.asp

Holyoak, D. T., & N. Pedersen. 2007. Conflicting molecular and morphological evidence of evolution within the Bryaceae (Bryopsida) and its implications for generic taxonomy. J. Bryology 29:111-124.

Hong, W. S. 1979. The genus *Scapania* in western North America, I. Historical background. *Bryologist* 82:181–188.

Hong, W. S. 1980a. The genus *Scapania* in western North America, II. Taxonomic treatment. *Bryologist* 83:40–59.

Hong, W. S. 1980b. Hepaticae of Rocky Mountain National Park. *Bryologist* 83:351–354.

Hong, W. S. 1983a. The genus *Gymnomitrium* in North America west of the Hundredth Meridian. *Lindbergia* 9:169–177.

Hong, W. S. 1983b. The genus *Porella* in North America west of the Hundredth Meridian. *Bryologist* 86:145–155.

Hong, W. S. 1986. The family Cephaloziellaceae in North America west of the Hundredth Meridian. *Bryologist* 89:145–155.

Hong, W. S. 1988a. The family Lepidoziaceae in North America west of the Hundredth Meridian. *Bryologist* 91:326–333.

Hong, W. S. 1988b. The family Cephaloziaceae in North America west of the Hundredth Meridian. *Lindbergia* 14:79–88.

Hong, W. S. 1989. The genus *Frullania* in North America west of the Hundredth Meridian. *Bryologist* 92:362–367.

Hong, W. S. 1990. The family Calypogeiaceae in North America west of the Hundredth Meridian. *Bryologist* 93:313–318.

Hong, W. S. 1993. The family Geocalycaceae (Hepaticae) in North America west of the Hundredth Meridian. *Bryologist* 96:592–597.

Hong, W. S. 1994. *Tritomaria* in Western North America. *Bryologist* 97:166–170.

Hong, W. S. 2002a. The distribution of *Lophozia* in North America west of the Hundredth Meridian. *Lindbergia* 17:49–62.

Hong, W. S. 2002b. *Leiocolea* in Western North America. *Lindbergia* 27:97–103.

Hong, W. S. 2002c. A key to the Hepaticae of Montana. *Northwest Science* 76:271–285.

Hong, W. S., & J. Váňa. 2000. The distribution of *Nardia* in western North America. *Lindbergia* 25:90–14.

Horton, D. G. 1983. A revision of the Encalyptaceae (Musci), with particular reference to the North American taxa. Part II. *J. Hattori Bot. Lab.* 54:353–532.

Howe, M. A. 1917. Notes on North American species of *Riccia*. *Bryologist* 20:33–36.

Ignatov, M. A., & S. Huttonen. 2002. Brachytheciaceae (Bryophyta)—a family of sibling genera. *Arctoa* 11:245-296.

Ignatov, M. S., & E. A. Ignatova. 2003. *Flora mkhov srednei chasti evropeiskoi Rossii. Moss flora of Middle European Russia. Vol.1. Sphagnaceae–Hedwigiaceae.*

Ignatov, M. S., & E. A. Ignatova. 2004. *Flora mkhov srednei chasti evropeiskoi Rossii. Moss flora of Middle European Russia. Vol. 2. Fontinalaceae–Amblystegiaceae.*

Ireland, R. R. 1971. Moss pseudoparaphyllia. *Bryologist* 74:312–330.

Ireland, R. R. 2004. [Review] Henk C. Greven. Grimmias of the World. *Bryologist* 207:274–276.

Iwatsuki, Z. 1970. Notes on *Isopterygium* Mitt. (Plagiotheciaceae). *J. Hattori Bot. Lab.* 63:445–451.

James, T. P. 1871. Mosses [of Utah and Nevada]. Report of the Geological Expedition of the Fortieth Parallel, by Clarence King: Botany, by Sereno Watson, etc. Volume 5. 398 pp. Government Printing Office, Washington.

Jamieson, D. 1986a. *A monograph of the genus* Hygrohypnum *Lindberg.* 125 pp., 80 figs., maps. (Unpublished Ph.D. Thesis, Univ. of British Columbia.)

Jamieson, D. 1986b. Notes on new and rare mosses in southwestern Colorado. *Evansia* 3:8–11.

Jørgensen, P. M. 1999. The shaping of early lichenology: Linnaeus and his pupils. *Symbolae Botanicae Upsaliensis* 32(3): 11–21.

Khanna, K. R. 1967. A cytological investigation of the mosses of the Rocky Mountains. *Univ. of Colorado Studies, Series in Biology* 26:1–39.

Komarkova, V. 1979. *Alpine vegetation of the Indian Peaks area, Front Range, Colorado. Flora et Vegetatio Mundi* 7. 591 pp. Cramer, Vaduz. 2 vols.

Koponen, T. 1977. Antero Vaarama, 1912–1975. *J. Bryology* 9:575–578

Koponen, T. 1980. A synopsis of Mniaceae (Bryophyta). IV. Taxa in Europe, Macaronesia, NW Africa and the Near East *Ann. Bot. Fenn.* 17:125–162.

Koponen, T. 1982a. Rhizoid topography and branching patterns in moss taxonomy. *Nova Hedwigia,* Beiheft 71:95–99.

Koponen, T. 1982b. On the taxonomic value of habitat ecology in mosses. *Nova Hedwigia,* Beiheft 71:101–109.

Koponen, T. 1988. The phylogeny and classification of Mniaceae and Rhizogoniaceae (Musci). J. Hattori Bot. Lab. 64:37–46.

Koponen, T., & P. Isoviita. 2005. Sextus Otto Lindberg and his collection of letters. *Bryobrotherella* 8:1–38.

Koponen, T., & S. Piippo. 2002. Viktor Ferdinand Brotherus and his collection of letters. *Bryobrotherella* 5:1–46.

Kramer, W. 1980. *Tortula* Hedwig sect. Rurales De Not. (Pottiaceae, Musci) in der östlichen Holarktis. *Bryophytorum Bibl.* 21:165 pp., 29 plates.

Kruckeberg, A. 2002. *Geology and Plant Life: The Effects of Land Forms and Rock Types on Plants.* 362 pp. Univ. of Washington Press, Seattle.

Kucera, J., & H. Köckinger. 2000. The identity of *Grimmia andreaeoides* Limpricht and *Didymodon subandreaeoides* (Kindberg) R. Zander. *J. Bryol.* 22:49–54.

Lawton, E. 1957. A revision of the genus *Lescuraea* in Europe and North America. *Bull. Torrey Bot. Club* 84:281–307.

Lawton, E. 1971. *Moss Flora of the Pacific Northwest.* 362 pp., 195 plates. Hattori Botanical Laboratory.

Lesquereux, L. 1874. *Mosses* [in] Porter, Thomas C., & John M. Coulter. *Synopsis of the Flora of Colorado.* U.S.G.S. (F. V. Hayden). Misc. Publ. No. 4. 180 pages.

Lesquereux, L., & T. P. James. 1884. *Manual of the Mosses of North America.* v+447 pp. plates 1–6. S. E. Cassino, Boston.

Lewinsky, J. 1974. The genera *Leskeella* and *Pseudoleskeella* in Greenland. *Bryologist* 77:601–611.

Limpricht, K. G. 1890-1904. *Die Laubmoose Deutschlands, Österreichs und Der Schweiz.* 3 volumes, Leipzig.

Long, G. 2000. *Catoscopium nigritum.* [in] New national and regional bryophyte records. *J. Bryol.* 22:69–70.

Löve, Á. & D. Löve. 1953. Studies on *Bryoxiphium. Bryologist* 56:73–94; 183–203.

Malcolm, Bill & Nancy. 2006. *Mosses and other bryophytes: An illustrated glossary*, 2nd ed. 336 pp., 1400 color illustrations. Micro-Optics Press, Nelson, N.Z.

Margadant, W. D. 1968. *Early Bryological Literature.* xi + 277 pp. Hunt Botanical Library, Pittsburgh.

Mårtensson, O. 1955–1956. Bryophytes of the Torne Träsk area, northern Swedish Lappland. *Kgl. Sv. Vetensk.-Akad. avhandl.,* Naturskyddsärenden Nr. 14. I. Hepaticae. 1–107. (1955); II. Musci. 1–321. (1956); III. General Part, 1–94 + maps. (1956).

McIntosh, T. T. 2007. *Schistidium,* in *Flora of North America North of Mexico.* Volume 27, *Bryophyta, Part 1.* Pp. 207–225. Oxford University Press.

McQueen, C. B. 1990. *Field Guide to the Peat Mosses of Boreal North America.* 138 pp., 30 plates. Univ. Press of New England.

McQueen, C. B. 1999. Macroscopic key to the Sphagnaceae of North America, *Evansia* 16:1–10.

Miller, N. G., A. M. Fryday, & J. W. Hinds. 2005. Bryophytes and lichens of a calcium-rich spring seep isolated on the granitic terrain of Mt. Katahdin, Maine, U.S.A. *Rhodora* 107:339–358.

Mishler, B. D. 1985. Biosystematic studies of the *Tortula ruralis* complex, I. Variation of taxonomic characters in culture. *J. Hattori Bot. Lab.* 58:225–253.

Mishler, B. D. 2000. The biology of bryophytes—bryophytes aren't just small tracheophytes. *Am. J. Bot.* 88:2129–2134.

Mishler, B. D. 2007. *Syntrichia,* in *Flora of North America North of Mexico.* Volume 27, *Bryophyta, Part 1.* Pp. 618–627. Oxford University Press.

Moulton, G. E. (ed.). 1999. *The herbarium of the Lewis and Clark Expedition.* 358 pp. Vol. 12, *The Journals of the Lewis and Clark Expedition.* Univ. Nebraska Press.

Muñoz, J. 1998a. Materials toward a revision of *Grimmia*: nomenclature and taxonomy of *Grimmia longirostris. Ann. Mo. Bot. Gard.* 85:352–363.

Muñoz, J. 1998b. A taxonomic revision of *Grimmia* subgenus Orthogrimmia. *Ann. Mo. Bot. Gard.* 85:367–403.

Muñoz, J. 1998c. Amended description of *Coscinodon calyptratus* (Grimmiaceae) and major range extensions into Eurasia and New Zealand. Bryologist 101:89–92.

Muñoz, J. 1999. A revision of *Grimmia* in the Americas. 1. Latin America. *Ann. Mo. Bot. Gard.* 86:118–191.

Muñoz, J., & F. Pando. 1983. A world synopsis of the genus *Grimmia. Ann. Mo. Bot. Gard. 83:1–133.*

Murray, B. M. 1987. *Systematics of the Moss Class Andreaeopsida.* Unpublished Ph.D. Thesis, Univ. of Alberta.

Nelson, P. P. 1973. *Leptodon smithii* (Musci, Neckeraceae), a genus new to North America. *Bryologist* 76:434–437.

Norris, D. H., & T. Koponen. 1999. Bryophyte flora of the Huon Peninsula, Papua New Guinea. LXVII. *Amphidium* (Rhabdoweisiaceae, Musci). *Ann. Bot. Fennici* 36:265–269.

Norris, D. H, & J. R. Shevock. 2004a. Contributions toward a bryoflora of California. I. Specimen-based catalogue of mosses. *Madroño* 51:1–131.

Norris, D. H., & J. R. Shevock. 2004b. Contributions toward a bryoflora of California. II. A key to the mosses. *Madroño* 51:133–269.

Nyholm, E. 1954–1969. *Moss Flora of Fennoscandia.* 799 pp. Gleerup, Lund.

Nyholm, E. 1971. Studies in the genus *Atrichum*: A short survey of the genus and its species. *Lindbergia* 1:1–33.

Nyholm, E. 1993. *Illustrated Flora of Nordic Mosses*, Fasc. 3. Bryaceae, Rhodobryaceae, Mniaceae, Cinclidiaceae, Plagiomniaceae. Nordic Bryological Society.

Ochi, H. 1981. A revision of the neotropical Bryoidae, Musci (first part). J. F. Educ. Tottori Univ. Nat. Sci. 29:49–154.

Ochyra, R. 1989. Animadversions on the moss genus *Cratoneuron* (Sullivant) Spruce. *J. Hattori Bot. Lab.* 67:203–242.

Ochyra, R. 1998. *The Moss Flora of King George Island, Antarctica.* 278 pp. 163 plates and maps. W. Szafer Inst. of Bot., Polish Acad. Sci., Cracow.

Ochyra, R. 2004. *Tortula hoppeana*, the correct name for *Desmatodon latifolius* in *Tortula* (Bryopsida, Pottiaceae). *Bryologist* 107:497–500.

Ochyra, R., J. Zarnowiec & H. Bednarek-Ochyra. 2003. *Census Catalogue of Polish Mosses.* Polish Acad. Sci., Krakow.

Ochyra, R., & D. Lamy. 2004. New names for mosses proposed by Palisot de Beauvois in his "Muscologie" of 1822. *Cryptogamie, Bryol.* 25:179–196.

Ochyra, R., & G. Zijlstra. 2005. The basionym of *Eucladium verticillatum* (Pottiaceae). *Taxon* 54:808–810.

Pederson, N., C. J. Cox, & L. Hedenäs. 2003. Phylogeny of the moss family Bryaceae inferred from chloroplast DNA sequences and morphology. *Syst. Bot.* 28:471–482.

Pedrotti, C. C. 2001. *Flora dei Muschi d'Italia.* Vol. 1. 817 pp. Antonio Delfine Editore,

Perry, A. R., & D. G. Long. 2001. Alan Cyril Crundwell, B. Sc. (1923–2000). *J. Bryol.* 23:267–272.

Persson, H. 1956. Studies in copper mosses. *J. Hattori Bot. Lab.* 17:1–18.

Peterson, W. 1979. A revision of the genera *Dicranum* and *Orthodicranum* (Musci) in North America north of Mexico. 453 pp. Unpublished Ph.D. thesis, Univ. of Alberta, Edmonton.

Pfister, D. 1993. Geneva Sayre (1911–1992). *Bryologist* 96:465–478.

Porter, T. C., & J. M. Coulter. 1874. *Synopsis of the Flora of Colorado.* Musci by Leo Lesquereux. *U. S. Geol. Geogr. Surv. Terr. Misc. Publ.* 4:154–161.

Proctor, M. F. 2000. Physiological ecology, Chapter 8 in A. J. Shaw, & B. Goffinet, *Bryophyte Biology.* Cambridge.

Pursell, R. A. 1997. A comparison of *Fissidens obtusifolius* and *F. sublimbatus. J. Hattori Bot. Lab.* 82:203–212.

Rau, E. A., C. F. Austin, & T. P. James. 1876. List of Colorado Musci and Hepaticae collected by T. S. Brandegee in 1873–75. *Bull. Torrey Bot. Club* 6: 89, 90).

Reese, W. D. 1975. [Review] Seville Flowers. *Mosses: Utah and the West. Quarterly Review of Biology* 50:206.

Robinson, H. 1962. Generic revisions of North American Brachytheciaceae. *Bryologist* 65:73–146.

Robinson, H. 1987. Notes on generic concepts in the Brachytheciaceae and the new genus *Steerecleus. Mem. N. Y. Bot. Garden* 45:678–681.

Rodgers, A. D., III. 1940 (reprinted 1968). *"Noble Fellow": William Starling Sullivant. 1803–1873.* 361 pp. Hafner.

Rothrock, J. T. 1878. Catalogue of plants. Musci by T. P. James. *U. S. Geogr. Survey, west of the 100th Meridian (Wheeler Exped.)* 64:341–349.

Rydberg, P. A. 1907. Scandinavians who have contributed to the knowledge of the flora of North America. *Contrib. New York Bot. Garden* No. 100. 49 pp.

Sayre, G. 1938. *The Moss Flora of Colorado.* 153 pp., 5 figs. Unpublished Ph.D. Thesis, University of Colorado. Abstract. *Univ. of Colorado Studies* 26:123–126.

Sayre, G. 1944. Colorado species of *Grimmia. Bryologist* 47:118–122.

Sayre, G. 1945. The distribution of *Fontinalis* in a series of moraine ponds. *Bryologist* 48:34–36.

Sayre, G. 1952. Key to the species of *Grimmia* in North America. *Bryologist* 55:251–259.

Sayre, G. 1959a. *Dates of Publications Describing Musci, 1801-1821.* 102 pp. Russell Sage College, Troy, N.Y.

Sayre, G. 1959b. *Cryptogamae Exsiccati: An annotated bibliography of published exsiccati of algae, lichens, hepatics, and Musci.* Part 1, General cryptogams, algae, and lichens. *Mem. N. Y. Bot. Garden* 19:1–124.

Sayre, G. 1964. The authorities for the epithets of mosses, hepatics, and lichens. *Bryologist* 67:113–135. (with C.E.B. Bonner & L. Culberson).

Sayre, G. 1971. Cryptogamae Exsiccatae: Bryophytes. *Mem. N.Y. Bot. Garden* 19:175–276.

Sayre, G. 1975. Cryptogamae Exsiccatae: Collectors. *Mem. N. Y. Bot. Garden* 19:277–423.

Schenk, G. 1997. *Moss Gardening—including lichens, liverworts, and other miniatures.* Timber Press, Portland, OR.

Schofield, W. B. 1966. *Crumia,* a new genus of the Pottiaceae endemic to western North America. *Can. J. Bot.* 44:609–614.

Schofield, W. B. 1969. Phytogeography of northwestern North American bryophytes and vascular plants. *Madroño* 20:155–207.

Schofield, W. B. 1980. Phytogeography of the mosses of North America (north of Mexico). Pp. 131–170 in *The Mosses of North America.* 59th Annual meeting, Pacific Division AAAS, Seattle, 1978.

Schofield, W. B. 1985. *Introduction to Bryology.* 431 pp. MacMillan.

Schuster, R. M. 1953. *Boreal Hepaticae: A Manual of the Liverworts of Minnesota and Adjacent Regions. Amer. Midl. Nat.*49: 257–683.

Schuster, R. M. 1966–1992. *The Hepaticae and Anthocerotae of North America east of the hundredth Meridian.* 6 volumes. Columbia (Vol. 1–5), Field Museum of Natural History, Chicago (Vol. 5–6).

Schuster, R. M. 1983. [editor] *New Manual of Bryology.* Hattori Botanical Laboratory, and the Cryptogamic Laboratory, Hadley, Mass.

Shacklette, H. 1967. Copper mosses as indicators of metal concentrations. *Bull. USGS* 1198-G.

Sharp, A. J., H. Crum, & P. Eckel, eds. 1994. *The Moss Flora of Mexico.* 2 volumes. 1113 pp., Index, i–xvi. New York Botanical Garden.

Shaw, A. J. 1982. *Pohlia* Hedwig (Musci), in North and Central America and the West Indies. *Contr. Univ. Michigan Herb.* 15:219–295.

Shaw, A. J., & B. Goffinet [eds.]. 2000. *Bryophyte Biology.* x + 476pp. Cambridge Univ. Press.

Shaw, A. J., N. J. Guidula, & T. M. Wilson. 1992. Reproductive biology of the rare copper moss, *Mielichhoferia mielichhoferiana. Contrib. Univ. Michigan Herb.* 18: 131–140.

Shaw, A. J., & P. E. Rooks. 1994. Systematics of *Mielichhoferia* (Bryaceae: Musci) I. Morphological and genetic analyses of *M. elongata* and *M. mielichhoferiana. Bryologist* 97: 1–12.

Shaw, A. J. [with S. Rams, R. S. Ros, & O. Werner]. 2004. *Pohlia bolanderi* from Sierra Nevada, Spain, new to the European bryophyte flora. *Bryologist* 107:312–315.

Smith, A.J.B. 2006. *The Moss Flora of Great Britain and Ireland.* Second edition with corrections. 1012 pages, 315 figures. Cambridge Univ. Press.

Smith, D. K. 1994. Funariaceae, in Sharp, A. J., et al. *Moss Flora of Mexico,* vol. 1:427–442. N. Y. Botanical Garden.

Smith, G. L. (later Merrill). 1971. A Conspectus of the Genera of the Polytrichaceae. *Mem. N. Y. Bot. Gard.* 31:1–83.

Spence, J. R. 1986. *Bryum calobryoides,* a new species from Western North America. *Bryologist* 89:215–218.

Spence, J. R. 1996. *Rosulabryum,* genus novum (Bryaceae). *Bryologist* 99:221–225.

Spence, J. R. 2004. A preliminary treatment of the Bryaceae of the Bryophyte Flora of North America Region. *Evansia* 21:1–16.

Spence, J. R. 2005. New genera and combinations in Bryaceae (Bryales, Musci) for North America. *Phytologia* 87:15–28.

Spence, J. R. 2007. Nomenclatural changes in the Bryaceae (Bryopsida) for North America II. *Phytologia* 89:110–114.

Spence, J. R. 2007b. Four species of the Bryaceae new to the U.S.A. *Evansia* 24:29–30.

Spence, J. R., & J. W. Flock. 1988. *Alpine mosses of the Indian Peaks Region, Front Range, Colorado.* Mountain Res. Sta., Univ. of Colorado. Nat. Hist. Paper 2. 19 pages.

Stearn, W. T. 2004. *Botanical Latin*, 4th ed., 546 pp. Timber Press, Portland, OR.

Stebbins, G. L., Jr. 1952. Aridity as a stimulus to plant evolution. *American Naturalist* 86:33–44.

Stech, M. 1999. A molecular systematic contribution to the position of *Amphidium* Schimp. (Rhabdoweisiaceae, Bryopsida). *Nova Hedwigia* 68:291–300.

Steere, W. C. 1948. Abel Joel Grout (1867–1947), *Bryologist* 51:201–210.

Steere, W. C. 1969. Asiatic elements in the bryophyte flora of western North America. *Bryologist* 72:507–512.

Steere, W. C. 1975a. Occurrence of *Entodon cladorrhizans* in Alaska, with notes on geographical distribution of *E. concinnus. Bryologist* 78:334–342.

Steere, W. C. 1975b. Kjeld Axel Holmen (1921–1974). *Bryologist* 78:240–242.

Steere, W. C. 1975c. The occurrence of *Pseudoleskeella papillosa* in Canada and Arctic Alaska. *Lindbergia* 3:91–92.

Steere, W. C. 1978. *The Mosses of Arctic Alaska. Bryophytorum Bibliotheca* 14: 1–508 + 23 figs., 48 maps.

Stoneburner, A. 1985 [May 1986]. Variation and taxonomy of *Weissia* in the southwestern United States. *Bryologist* 88:293–314.

Sullivant, W. S. 1864. *Icones Muscorum,* or figures and descriptions of most of those mosses peculiar to eastern North America which have not been heretofore figured. 216 pp., 128 plates. Facsimile edition, 1969.

Sullivant, W. S. 1874. *Icones Muscorum,* Supplement (posthumous).109 pp., 81 plates. Facsimile edition, 1969.

Szweykowski, J., K. Buczkowska, & I. S. Odrzykoski. 2005. *Conocephalum salebrosum* (Marchantiopsida, Conocephalaceae), as new Holarctic liverwort species. *Pl. Syst. Evolution* (on-line), pp. 1–26.

Tan, B., J.-C. Zhao, & R.-L. Hu. 1995. An updated checklist of mosses of Xinjiang, China. *Arctoa* 4:1–14.

Thompson, D. W. 1942. *On Growth and Form,* 2nd edition. Cambridge.

Váňa, J., & W. S. Hong. 1999. The genus *Jungermannia* in western North America. *Lindbergia* 24:133–144.

Vanderpoorten, A. 2004. A simple taxonomic treatment for a complicated evolutionary story: the genus *Hygroamblystegium* (Hypnales, Amblystegiaceae). [in] Goffinet, B., V. Hollowell, & R. Magill (eds.) *Molecular Systematics of Bryophytes. Monogr. Syst. Bot. Missouri Bot. Gard.* 98:xvi + 448 pp.

Vanderpoorten, A., M. S. Ignatov, S. Huttonen, & B. Goffinet. 2005. A molecular and morphological recircumscription of *Brachytheciastrum* (Brachytheciaceae, Bryopsida). *Taxon* 54:369–376.

Vanderpoorten, A., & C. E. Zartman. 2002. The *Bryum bicolor* complex in North America. *Bryologist* 105:128–139.

Van der Wijk, R., W. D. Margadant, & P. A. Florschütz (eds.). 1959–1969. *Index Muscorum.* 5 volumes. Utrecht.

Vitt, D. H. 1973. A revision of the genus *Orthotrichum* in North America, north of Mexico. *Bryophytorum Bibliotheca* 1: 1-208, 60 plates.

Vitt, D. H. 1976. The genus *Seligeria* in North America. *Lindbergia* 3:241–175.

Vitt, D. H. 1993. The distribution of North American Bryophytes: *Roellia roellii. Evansia* 10:112–113.

Vitt, D. H., & W. R. Buck. 1992. Key to the moss genera of North America north of Mexico. *Contr. Univ. Mich. Herb.* 18:43–71.

Voss, E. G., & A. A. Reznicek. 1988. Frederick J. Hermann (1906–1987): The evolution of a botanical career. *Michigan Botanist* 27:59–73.

Warncke, E. 1968. *Marchantia alpestris* in Denmark. *Bot. Tidsskr.* 63:358–368.

Warncke, E. 1975. Kjeld Axel Holmen, 1921–1974. *Lindbergia* 3:115–117.

Weber, W. A. 1952. Moss flora of Colorado. Additions in *Aulacomnium. Bryologist* 55:297.

Weber, W. A. 1959. Some features of the distribution of arctic relicts at their austral limits. Proc. IX International Botanical Congress 2:141–145.

Weber, W. A. 1961a. Studies of Colorado bryophytes. *Univ. Colorado Studies, Ser. Biol.* 7:27–52.

Weber, W. A. 1961b. A second American record for *Oreas martiana,* from Colorado. *Bryologist* 63:241–144.

Weber, W. A. 1963. Additions to the bryophyte flora of Colorado. *Bryologist* 66:192–200.

Weber, W. A. 1965a. Mount Evans, concentration point for Pleistocene relict plants. Guidebook for one-day field conferences, Boulder Area, Colorado, pp. 10–12. International Association for Quaternary Research, VIIth Congress.

Weber, W. A. 1965b. *Plant Geography in the southern Rocky Mountains,* in *The Quaternary of the United States* (eds. H. E. Wright, Jr. & D. G, Frey), pp. 453–468. Princeton Univ. Press.

Weber, W. A. 1973. Guide to the Mosses of Colorado: Keys and ecological notes based on field and herbarium studies. *Inst. Alpine & Arctic Res., Occ. Paper* 6:1–48.

Weber, W. A. 1976. New combinations in the Rocky Mountain Flora. *Phytologia* 33:105–106. [*Schistidium atrichum* and *S. dupretii*].

Weber, W. A. 1991. The Alpine Flora of Summit Lake, Mount Evans, Colorado. *Aquilegia* 15(4):3–10.

Weber, W. A. 1997. *King of Colorado Botany: Charles Christopher Parry, 1823–1890.* 183 pp., 5 plates. Univ. Press of Colorado, Boulder.

Weber, W. A. 2000. *Leptopterigynandrum*: Field observations and relevance to family designation. *Evansia* 17:81–83.

Weber, W. A. 2002a. Colorado bryological hot spots. 1. Boulder Mountain Park. *Evansia* 18:(4):143–147. [published January 2002].

Weber, W. A. 2002b. Colorado bryological hot spots, 2. Mount Evans. *Evansia* 19:71–73.

Weber, W. A. 2003a. Seville Flowers' *Mosses, Utah and the West.* Nomenclatural clarifications and updates. *Evansia* 20:1–9.

Weber, W. A. 2003b. The Middle Asian Element in the Southern Rocky Mountain Flora of the western United States: A critical review. *J. of Biogeography* 30:649–685.

Weber, W. A., & F. J. Hermann. 1985. Occurrence of *Cynodontium gracilescens* in North America (Colorado). Bryologist 88:26.

Weber, W. A., & P. Nelson. 1974. *Random Access Key to the Genera of Colorado Mosses.* University of Colorado Museum.

Weber, W. A., & R. C. Wittmann. 1992. *Catalog of the Colorado Flora; A Biodiversity Baseline.* 215 pp. University Press of Colorado.

Weber, W. A., & R. C. Wittmann. 2003. Colorado bryological hot spots. 3. High Creek Fen. Evansia 20:51–52.

Weber, W. A., R. C. Wittmann, & R. Worthington. 2003. *Grimmia bernoullii* in the United States. *Evansia* 24: 104–106,

Williams, R. 1984. *Aven Nelson of Wyoming.* 407 pp. Colorado Assoc. Univ. Press, Boulder.

Wilson, P., & D. H. Norris. 1989. *Pseudoleskeella* in North America and Europe. *Bryologist* 92:387–396.

Wittrock, V. B. 1903. *Catalogus Illustratus Iconothecae Botanici Horti Bergiani Stockholmiensis* [Illustrated catalog of the Bergius Botanical Garden's collection of portraits of botanists, with biographical information]. *Acta Horti Bergiani* 3(2,3). 1–198+37 plates; 1–198i+150 plates.

Zander, R. H. 1978a. New combinations in *Didymodon*, and a key to the taxa in North America north of Mexico. *Phytologia* 41:11–32.

Zander, R. H. 1979. Notes on *Barbula* and *Pseudocrossidium* in North America, and an annotated key to the taxa. *Phytologia* 44:177–214.

Zander, R. H. 1981. *Didymodon* (Pottiaceae) in Mexico and California: taxonomy and distribution of continuous and non-discontinuous taxa. *Cryptog., Bryol. Lichénol.* 2:379–422.

Zander, R. H. 1989. Seven new genera in Pottiaceae (Musci) and a lectotype for *Syntrichia*. *Phytologia* 65:424–436.

Zander, R. H. 1993. *Genera of the Pottiaceae: Mosses of Harsh Environments.* Bull. Buffalo Soc. Nat. Sci. 32:i–vi, 1–378. 113 plates.

Zander, R. H. 1999. A new species of *Didymodon* (Bryopsida) from Western North America and a regional key to the taxa. *Bryologist* 102:112–115.

Zander, R. H., & R. R. Ireland, Jr. 1979. Propaguliferous *Ceratodon purpureus* in riparian environments. *Bryologist* 82: 474–478.

Zander, R. H., & R. Ochyra. 2001. *Didymodon tectorum* and *D. brachyphyllus* in North America. *Bryologist* 104:372–377.

Zander, R. H., & W. A. Weber. 1997. *Didymodon anserinocapitatus* (Musci, Pottiaceae) new to the New World. *Bryologist* 100:237–238.

Zander, R. H., & W. A. Weber. 2005. *Anoectangium handelii* (Bryopsida, Pottiaceae) in the New World. *Bryologist* 108:47–49.

INDEX BY SPECIFIC EPITHETS:
MOSSES

longifolium, Paraleucobryum
longifolius, Drepanocladus
longirostris, Grimmia
longiseta, Meesia
longisetum, Polytrichastrum
luridum, Hygrohypnum
lyallii, Meiotrichum

macounii, Bucklandiella
macrocarpum, Haplodontium
magnifolium, Rhizomnium
marchica, Philonotis
marginatum, Mnium
martiana, Oreas
medium, Plagiomnium
megapolitana, Timmia
menziesii, Anacolia
mexicanum, Homomallium
microstoma, Funaria
mielichhoferiana, Mielichhoferia
mildeanum, Imbribryum
mollis, Hydrogrimmia
montana, Grimmia
montanum, Dicranum
mougeotii, Amphidium
mucronifolia, Tortula
muehlenbeckii, Dicranum, Imbribryum
muhlenbergii, Funaria
muralis, Tortula

nelsonii, Brachythecium
neodamense, Ptychostomum
neomexicana, Fontinalis
nervosa, Leskeella
nevadense, Didymodon
nevadensis, Tortula
nigritum, Catoscopium
nitens, Tomentypnum
nivalis, Hedwigia, Voitia
norvegica, Syntrichia, Timmia
norvegicum, Bryoxiphium
nutans, Pohlia

obtusifolia, Pohlia, Tortula
obtusifolium, Orthotrichum
obtusifolius, Fissidens
occidentale, Schistidium
ochraceum, Hygrohypnum
oederianus, Plagiopus
oedipodium, Brachythecium
oligocarpa, Neckera
osmundoides, Fissidens
ovalis, Grimmia

ovatum, Pterygoneurum

pagorum, Syntrichia
pallens, Orthotrichum, Ptychostomum
pallescens, Hypnum, Ptychostomum
palustre, Aulacomnium
papillosa, Syntrichia
papillosissima, Syntrichia
patens, Lescuraea
pellucida, Tetraphis
pellucidum, Dichodontium, Orthotrichum
pendulum, Ptychostomum
pilifera, Grimmia,
piliferum, Polytrichum
piperi, Buxbaumia
plagiopodia, Grimmia
planifolium, Trichostomum
platyphyllum, Sphagnum
plinthobia, Tortula
plumosum, Brachythecium
polyantha, Pylaisiella
polycarpa, Leskea
polygamus, Drepanocladus
polysetum, Dicranum
potosica, Bartramia
pratensis, Breidleria
procera, Encalypta
proligera, Pohlia
protensum, Campylium
protobryoides, Tortula
pseudopunctatum, Rhizomnium
pseudotriquetrum, Ptychostomum
pulchella, Isopterygiopsis
pulchellum, Eurhynchium
pulchrum, Schistidium
pulvinata, Grimmia
pumilum, Orthotrichum
purpureus, Ceratodon
pusilla, Fabronia
pyrenaicum, Hylocomiastrum
pyriforme, Leptobryum, Physcomitrium

radicosa, Lescuraea
rauii, Jaffueliobryum
recurvirostrum, Bryoerythrophyllum, Hymenostylium
repens, Platygyrium
replicatum, Pseudocrossidium
revolutum, Hypnum
revolvens, Scorpidium
rhabdocarpum, Dicranum
rhaptocarpa, Encalypta
richardsonii, Calliergon
rigida, Aloina

INDEX BY SPECIFIC EPITHETS:
LIVERWORTS AND HORNWORTS

CATALOG: MOSSES

This list attempts to account for all names used in publications on the Colorado bryophyte flora, and their present dispositions. Names in bold face are accepted in the book. Names in italic face represent synonyms or erroneous reports (Err. rep.).

ABIETINELLA Müller Halle, 1896 (THD)
  abietina (Hedwig) Fleischer
ALOINA Kindberg, 1882 (PTT)
  bifrons (De Notaris) Delgadillo
  hamulus (Müller Halle) Brotherus
  rigida (Hedwig) Limpricht
AMBLYODON P. Beauvois, 1805 (MEE)
  dealbatus (Hedwig) Bruch & Schimper
AMBLYSTEGIELLA (AMB)
  adnata = Homomallium
  confervoides = Platydictya. Err. rep.
  sprucei = Platydictya jungermannioides
  subtilis = Platydictya. Err. rep.
AMBLYSTEGIUM Bruch & Schimper, 1853 (AMB)
  compactum = Conardia
  fluviatile = A. varium
  humile = A. varium
  juratzkanum = A. serpens subsp.
  kochii = A. varium?
  noterophilum = A. varium
  radicale. Err. rep.
  riparium (Hedwig) Bruch & Schimper
  serpens (Hedwig) Bruch & Schimper subsp.
    juratzkanum (Schimper) Renauld & Cardot
  sprucei = Platydictya jungermannnioides
  tenax = A. varium
  trichopodium = A. varium
  varium (Hedwig) Lindberg
AMPHIDIUM Schimper, 1856 (ORT)
  lapponicum (Hedwig) Schimper
  mougeotii (Bruch & Schimper) Schimper
ANACALYPTA (PTT)
  latifolia = Stegonia
ANACOLIA Schimper, 1876 (BRT)
  laevisphaera (Taylor) Flowers
  menziesii (D. Turner) Paris
ANDREAEA Hedwig, 1801 (AND)
  heinemannii Hampe & Müller Halle
  rupestris Hedwig
ANISOTHECIUM (DCR)
  crispum = Dicranella
  schreberianum = Dicranella
  vaginale = Dicranella crispa
  varium = Dicranella
ANOECTANGIUM Schwägrichen, 1811 (PTT)
  aestivum (Hedwig) Mitten

  handelii Schiffner
ANOMOBRYUM (BRY)
  filiforme = Bryum julaceum
ANOMODON Hooker & Taylor, 1818 (ANM)
  attenuatus (Hedwig) Hübener
  rostratus (Hedwig) Schimper
ANTITRICHIA (CRY)
  californica. Err. rep.
APHANORHEGMA (FNR)
  serratum. Err. rep.
ATRICHUM P. Beauvois, 1804 (PLT)
  selwynii Austin
  undulatum var. *selwynii = A selwynii*
AULACOMNIUM Schwägrichen, 1827 (AUL)
  androgynum (Hedwig) Schwägrichen
  palustre (Hedwig) Schwägrichen var. **palustre**
  palustre var. *dimorphum =* var. *palustre*
  palustre var. **imbricatum** Bruch & Schimper
  palustre var. *lingulatum =* var. *palustre*
  papillosum = A. palustre
  turgidum. Err. rep.
BARBULA Hedwig, 1801 (PTT)
  acuta = Didymodon rigidulus
  andreaeoides = Didymodon subandreaeoides
  brachyphylla = Didymodon
  convoluta Hedwig
  fallax. Err. rep.
  icmadophila = Didymodon rigidulus var.
  latifolia = Tortula hoppeana
  manniae = Crossidium. Err. rep.
  membranifolia = Crossidium squamiferum
  mucronifolia = Tortula
  recurvifolia = Barbula reflexa
  recurvirostris = Bryoerythrophyllum
  reflexa. Err. rep.
  rigida = Aloina
  rigidula = Didymodon
  ruralis = Syntrichia
  spiralis = Pseudocrossidium replicatum. Err. rep.
  tophacea = Didymodon
  unguiculata Hedwig
  vinealis = Didymodon
BARTRAMIA Hedwig, 1801 (BRT)
  breviseta. Err. rep.
  fontana = Philonotis
  ithyphylla Bridel
  ithyphylla var. *breviseta* Kindberg *= subulata* subsp.

americana
oederi = *Plagiopus oederianus*
pomiformis. Err. rep.
potosica Montagne
potosina. Orthographic error
stricta. Err. rep.
subulata Bruch & Schimper subsp. **americana**
   Fransén
viridissima = *B. potosica*
BLINDIA Bruch & Schimper, 1846 (SLG)
   acuta (Hedwig) Bruch & Schimper
BRACHYTHECLASTRUM (BRC)
   collinum = *Brachythecium*
   fendleri = *Brachythecium*
   leibergii = *Brachythecium*
   velutinum = *Brachythecium*
BRACHYTHECIUM Schimper, 1853 (BRC)
   acutum = *B. salebrosum*
   albicans. Err. rep.
   asperrimum. Err. rep
   barnesii. Err. Rep.
   brandegei = *Cirriphyllum cirrosum*
   calcareum. Err. rep.
   campestre. Err. rep.
   curtum = *B. oedipodium*
   collinum (C. Müller Halle) Bruch & Schimper
   digastrum. Err. rep.
   erythrorrhizon Bruch & Schimper
   fendleri (Sullivant) Jäger
   glareosum. Err. rep.
   hylotapetum B. & N. Higinbotham
   latifolium = *B. nelsonii*
   leibergii Grout
   nelsonii Grout
   oedipodium (Mitten) Jäger
   oxycladon. Err. rep.
   plumosum (Hedwig) Bruch & Schimper
   rivulare Bruch & Schimper
   rutabulum. Err. rep.
   salebrosum. Err. rep.
   starkei = *B. oedipodium*
   stereopoma (Spruce ex Mitten) Jäger
   suberythrorrhizon = *B. collinum*
   turgidum (C. J. Hartman) Kindberg
   utahense = *B. fendleri*
   velutinum (Hedwig) Bruch & Schimper
BREIDLERIA Loeske, 1910 (HPN)
   pratensis (Koch *ex* Spruce) Loeske
BRYHNIA (BRC)
   novae-angliae. Err. rep.
BRYOERYTHROPHYLLUM Chen, 1941 (PTT)
   ferruginascens (Stirton) Giacomini
   recurvirostrum (Hedwig) Chen

BRYOXIPHIUM Mitten, 1869 (BRX)
   norvegicum (Bridel) Mitten
BRYUM Hedwig, 1801 (BRY)
   affine = *Ptychostomum turbinatum*
   algovicum = *Ptychostomum pendulum*
   alpinum = *Imbribryum*
   amblyodon = *Ptychostomum inclinatum*
   argenteum Hedwig
   barnesii = *Gemmabryum*
   biddlecomiae = *Ptychostomum inclinatum*
   bicolor = *Gemmabryum*
   bimum = *Ptychostomum*
   blindii Bruch & Schimper
   caespiticium = *Ptychostomum imbricatulum*
   calobryoides J. R. Spence
   capillare = *Rosulabryum*
   cirrhatum = *Ptychostomum pallescens*
   creberrimum = *Ptychostomum*
   cuspidatum = *Ptychostomum creberrimum*
   cyclophyllum = *Ptychostomum*
   duvalii = *Ptychostomum weigelii*
   filiforme = *B. julaceum*
   flaccidum = *Rosulabryum*
   gemmiparum = *Imbribryum*
   inclinatum = *Ptychostomum*
   intermedium = *Rosulabryum sp.*
   julaceum Schrader *ex* Gärtner, Meyer, &
      Schreber
   klinggraeffii = *Gemmabryum*
   knowltonii = *Ptychostomum*
   kunzei = *Ptychostomum*
   lanatum (P. Beauvois) Bridel
   lisae = *Ptychostomum creberrimum*
   lonchocaulon = *Ptychostomum*
   mildeanum = *Imbribryum*
   muehlenbeckii = *Imbribryum*
   neodamense = *Ptychostomum*
   obconicum = *Rosulabryum sp.*
   pallens = *Ptychostomum*
   pallescens = *Ptychostomum*
   pendulum = *Ptychostomum*
   porsildii = *Haplodontium macrocarpum*
   provinciale = *Rosulabryum canariense*
   pseudotriquetrum = *Ptychostomum*
   roseum = *Rhodobryum*. Err. rep.
   stenotrichum = *Ptychostomum inclinatum*
   subapiculatum = *Gemmabryum*
   subpurpurascens = *Ptychostomum pallens*
   subrotundum = *Ptychostomum pallescens*
   torquescens = *Rosulabryum*
   uliginosum = *Ptychostomum cernuum*
   veronense De Notaris
   weigelii = *Ptychostomum*

wrightii. Err. rep.

BUCKLANDIELLA Roivainen, 1972 (GRM)
affinis (Weber & Mohr) Bednarek-Ochyra & Ochyra
heterosticha. Err. rep.
macounii (Kindberg) Bednarek-Ochyra & Ochyra
sudetica (Funck) Bednarek-Ochyra & Ochyra

BUXBAUMIA Hedwig, 1801 (BXB)
aphylla Hedwig
indusiata. Err. rep.
piperi Best

CALLIERGON (Sullivant *in* A. Gray) Kindberg, 1894 (AMB)
cordifolium (Hedwig) Kindberg
giganteum (Schimper) Kindberg
megalophyllum. Err. rep.
richardsonii (Mitten) Kindberg
sarmentosum = Warnstorfia
stramineum = Straminergon
turgescens = Pseudocalliergon

CALLIERGONELLA Loeske, 1911(HPN)
lindbergii (Mitten) Hedenäs

CAMPTOTHECIUM (BRC)
lutescens. Err. rep.
nitens = Tomentypnum

CAMPYLIADELPHUS (Kindberg) Chopra, 1975 (CMP)
chrysophyllus (Bridel) Kanda
polygamus = Drepanocladus
stellatus = Campylium

CAMPYLIDIUM (CMP)
hispidulum = Campylophyllum
sommerfeltii = Campylophyllum

CAMPYLIUM (Sullivant *in* A. Gray) Mitten, 1869 (CMP)
chrysophyllum = Campyliadelphus
halleri = Campylophyllum
hispidulum. Err. rep.
polygamum = Drepanocladus
protensum (Bridel) Kindberg
stellatum (Hedwig) C. Jensen

CAMPYLOPHYLLUM (Schimper) Fleischer, 1914 (CMP)
halleri (Hedwig) Kanda
sommerfeltii (Myrin) Hedenäs

CAMPYLOPUS Bridel, 1818 (DCR)
frigidus = Paraleucobryum sauteri
hallii = Paraleucobryum enerve
schimperi Milde
subulatus. Err. rep.

CATHARINEA (PLT)
selwynii = Atrichum

CATOSCOPIUM Bridel, 1826 (CTS)
nigritum (Hedwig) Bridel

CERATODON Bridel, 1826 (DTR)
purpureus (Hedwig) Bridel

CHAMBERLAINIA (BRC)
albicans = Brachythecium
campestris = Brachythecium
collina = Brachythecium
digastra = Brachythecium
erythrorrhiza = Brachythecium
glareosa = Brachythecium
leibergii = Brachythecium
oxyclada = Brachythecium
salebrosa = Brachythecium

CIRRIPHYLLUM Grout, 1898 (BRC)
brandegei = C. cirrosum
cirrosum (Schwägrichen. *ex* Schultes) Grout
piliferum. Err. rep.

CLIMACIUM Weber & Mohr, 1804 (CLM)
americanum. Err. rep.
dendroides (Hedwig) Weber & Mohr

CNESTRUM I. Hagen, 1914 (DCR)
schisti (Wahlenberg) I. Hagen

CODRIOPHORUS P. Beauvois, 1822 (GRM)
acicularis (Hedwig) Bednarek-Ochyra & Ochyra
fascicularis (Hedwig) Bednarek-Ochyra & Ochyra

CONARDIA H. Robinson, 1976 (BRC)
compacta (Müller Halle) H. Robinson

COSCINODON Sprengel, 1804 (GRM)
calyptratus (Hooker *ex* Drummond) C. Jensen *in* Kindberg
cribrosus (Hedwig) Spruce
rauii = Jaffueliobryum
wrightii = Jaffueliobryum

CRATONEURON (Sullivant) Spruce, 1867 (AMB)
commutatum = Palustriella
decipiens = Palustriella
falcatum = Palustriella
filicinum (Hedwig) Spruce
williamsii = Palustriella falcata

CROSSIDIUM Juratzka, 1882 (PTT)
aberrans Holzinger & Bartram
griseum = C. squamiferum var. *pottioideum*
squamiferum (Viviani) Juratzka var. **pottioideum** (De Notaris) Mönkemeyer
squamigerum = C. squamiferum

CRUMIA (PTT)
latifolia. Err. rep.

CTENIDIUM (HPN)
molluscum. Err. rep.

CYLINDROTHECIUM (ENT)
concinnum = Entodon

CYNODONTIUM Bruch & Schimper, 1856 (DCR)
  alpestre. Err. rep.
  gracilescens (Weber & Mohr) Schimper
  polycarpon. Doubtful rep.
  schisti = Cnestrum schisti
  strumiferum (Hedwig) Lindberg
DESMATODON (PTT)
  arenaceus = Tortula obtusifolia
  cernuus = Tortula
  coloradense = Tortula obtusifolia
  convolutus = Tortula atrovirens
  guepinii = Tortula
  latifolius = Tortula hoppeana
  laureri = Tortula
  leucostomum = Tortula
  obtusifolius = Tortula
  plinthobius = Tortula
  systylius = Tortula
DICHELYMA Myrin, 1833 (FNT)
  capillaceum. Err. rep.
  falcatum (Hedwig) Myrin
  uncinatum Mitten
DICHODONTIUM Schimper, 1856 (DCR)
  pellucidum (Hedwig) Schimper
DICRANELLA (Müller Halle) Schimper, 1856
  (DCR)
  crispa (Hedwig) Schimper
  schreberiana (Hedwig) Crum & Anderson
  subulata (Hedwig) Schimper
  varia (Hedwig) Schimper
DICRANODONTIUM (DCR)
  denudatum. Err. rep.
  longirostre. Err. rep.
DICRANOWEISIA Lindberg *ex* Milde, 1869 (DCR)
  cirrata. Err. rep.
  crispula (Hedwig) Lindberg *ex* Milde
DICRANUM Hedwig. 1801 (DCR)
  acutifolium (Lindberg &Arnell) C. Jensen *ex*
      Weinmann
  bergeri. Err. rep.
  bonjeanii. Err. rep.
  brevifolium (Lindberg) Lindberg
  elongatum Schleicher *ex* Schwägrichen.
  flagellare Hedwig
  fuscescens. Err. rep.
  gracilescens = Cynodontium alpestre
  groenlandicum Bridel
  montanum Hedwig
  muehlenbeckii Bruch & Schimper
  pellucidum = Dichodontium
  polysetum Swartz
  rhabdocarpum Sullivant
  scoparium Hedwig

  spadiceum. Err. rep.
  strictum = D. tauricum
  tauricum Sapehin
  varium = Dicranella
  virens = Oncophorus
DIDYMODON Hedwig, 1801 (PTT)
  acutus = D. rigidulus
  anserinocapitatus (X. J. Li) Zander
  asperifolius (Mitten) Crum, Steere & Anderson
  australasiae. Err. rep.
  brachyphyllus (Sullivant) Zander
  fallax var. *reflexa = D. ferrugineus*
  ferrugineus (Bescherelle) Hill
  nevadense Zander
  reedii = D. tectorum
  recurvirostris = Bryoerythrophyllum
  revolutus. Err. rep.
  rigidulus Hedwig
  rubellus = Bryoerythrophyllum recurvirostrum
  subandreaeoides (Kindberg) Zander
  tectorum (Müller Halle) Saito
  tophaceus (Bridel) Lisa
  vinealis (Bridel) Zander
DISSODON (SPL)
  froelichianus = Tayloria
  hornschuchii = Tayloria
DISTICHIUM Bruch & Schimper, 1846 (DTR)
  capillaceum (Hedwig) Bruch & Schimper
  inclinatum (Hedwig) Bruch & Schimper
DITRICHUM Hampe, 1867 (DTR)
  flexicaule (Schwägrichen) Hampe
  giganteum = D. gracile
  gracile (Mitten) O. Kuntze
DREPANOCLADUS (Müller Halle) Roth, 1899
  (AMB)
  aduncus (Hedwig) Warnstorf
  capillifolius = D. longifolius
  crassicostatus = D. longifolius
  exannulatus = Warnstorfia
  fluitans = Warnstorfia
  kneiffii = D. aduncus
  longifolius (Mitten) Paris
  polygamus (Bruch & Schimper) Hedenäs
  revolvens = Scorpidium
  sendtneri. Err. rep.
  sordidus (Müller Halle) Hedenäs [to be expected]
  trichophyllus = Warnstorfia
  uncinatus = Sanionia
DRYPTODON (GRM)
  elatior = Grimmia
  pilifer = Grimmia pilifera
ENCALYPTA Hedwig, 1801 (ENC)
  alpina J. E. Smith

ciliata Hedwig
commutata = E. alpina
intermedia. Taxonomic status doubtful
macounii. Err. rep.
procera Bruch
rhaptocarpa Schwägrichen
streptocarpa. Err. rep.
vulgaris Hedwig
ENTODON Müller Halle, 1845 (ENT)
compressus. Err. rep.
concinnus (De Notaris) Paris
orthocarpus = E. concinnus
ENTOSTHODON Schwägrichen, 1823 (FNR)
americanus = Funaria
tucsonii (Bartram) Grout
EUCLADIUM Bruch & Schimper, 1846 (PTT)
verticillatum (Bridel) Bruch & Schimper
EURHYNCHIASTRUM (BRC)
pulchellum = Eurhynchium
EURHYNCHIUM Bruch & Schimper, 1854 (BRC)
diversifolium = E. pulchellum
fallax = E. pulchellum
piliferum = Cirriphyllum. Err. rep.
pulchellum (Hedwig) Bruch & Schimper
serrulatum = Steerecleus
striatum (Hedwig) Schimper
strigosum = E. pulchellum
FABRONIA Raddi, 1808 (FBR)
ciliaris Bridel
pusilla Raddi
wrightii = F. ciliaris
FISSIDENS Hedwig, 1801(FSS)
bryoides Hedwig
crispus Montagne
debilis = F. fontanus
exiguus = F. bryoides
fontanus (B. la Pylaie) Steudel. To be expected.
grandifrons Bridel. To be expected.
julianus = F. fontanus
limbatus = F. crispus
obtusifolius Wilson
osmundoides Hedwig
sublimbatus Grout
viridulus = F. bryoides
FONTINALIS Hedwig, 1801 (FNT)
antipyretica Hedwig
duriaei = F. hypnoides
hypnoides C. J. Hartman
neomexicana Sullivant & Lesquereux
nitida = F. hypnoides
FUNARIA Hedwig, 1801 (FNR)
americana Lindberg ex Sullivant
hygrometrica Hedwig

microstoma Bruch & Schimper
muhlenbergii Turner
GEMMABRYUM J. R. Spence, 2005 (BRY)
barnesii. Err. rep.
bicolor (Dickson) J. R. Spence
gemmilucens. Err. rep.?
klinggraeffii (Schimper *in* Klinggräff) Spence & Ramsay
subapiculatum (Hampe) Spence & Ramsay
valparaisense (Thériot) J. R. Spence
GRIMMIA Hedwig, 1801 (GRM)
affinis = G. longirostris
agassizii = Schistidium
alpestris (Weber & Mohr) Schleicher
alpicola = Schistidium rivulare
anodon Bruch & Schimper
anomala Hampe
apocarpa = Schistidium
apocarpa var. *ambigua = Schistidium ambiguum*
apocarpa var. *conferta = Schistidium confertum*
apocarpa var. *gracilis = Schistidium trichodon*
apocarpa var. *rivularis = Schistidium rivulare*
apocarpa var. *tenerrima = Schistidium* sp., status unclear
arizonae = G. pilifera
atricha = Schistidium atrichum
bernoullii = G. ovalis
brachyodon = G. montana
brandegei = G. plagiopodia
caespiticia (Bridel) Juratzka
californica = G. trichophylla
calyptrata = Coscinodon
cognata = G. elatior
coloradensis = Schistidium atrichum
commutata = G. ovalis
conferta = Schistidium
cribrosa = Coscinodon
crinitoleucophaea Cardot
donniana J. E. Smith
dupretii = Schistidium
elatior Bruch *ex* Balsamo & De Notaris
elongata. Err. rep.
hartmanii var. *anomala = G. anomala*
incurva Schwägrichen.
jamesii = G. montana
laevigata (Bridel) Bridel
leucophaea = G. laevigata
longirostris Hooker
mollis = Hydrogrimmia
montana Bruch & Schimper
moxleyi. Err. rep.
occidentalis = Schistidium
ovalis (Hedwig) Lindberg

ovata = G. longirostris
ovata var. *affinis = G. longirostris*
pilifera P. Beauvois
plagiopodia Hedwig
platyphylla = Schistidium sp., *aff. S. rivulare*
pulvinata (Hedwig) J. E. Smith
raui = Jaffueliobryum
reflexidens = G. sessitana
santaritae = G. pilifera
sessitana De Notaris
subincurva = G. torquata
tenera = Schistidium tenerum
tenerrima = G. sessitana
tenuicaulis = Schistidium tenerum
teretinervis Limpricht
torquata Hornschuch & Greville
trichophylla Greville
wrightii = Jaffueliobryum
GYMNOSTOMUM Nees & Hornschuch, 1823 (PTT)
aeruginosum J. E. Smith
calcareum = G. aeruginosum
recurvirostrum = Hymenostylium
rupestre = G. aeruginosum
GUEMBELIA (GRM)
laevigata = Grimmia
longirostris = Grimmia
teretinervis = Grimmia
HAPLODONTIUM Hampe, 1865 (BRY)
macrocarpum (Hooker *ex* Drummond) J. R. Spence
HEDWIGIA P. Beauvois, 1804 (HDW)
albicans = H. ciliata
ciliata (Hedwig) P. Beauvois
nivalis (Müller Halle) Mitten
HELODIUM Warnstorf, 1905 (HLD)
blandowii (Weber & Mohr) Warnstorf
paludosum. Err. rep.
HENNEDIELLA Paris, 1896 (PTT)
heimii (Hedwig) Zander
HERZOGIELLA (PGT)
turfacea. Err. rep.
HOLMGRENIA (HPN)
chrysea = Orthothecium
diminutiva = Isopterygiopsis pulchella
HOMALOTHECIUM Schimper, 1851 (BRC)
aureum (Lagasca) H. Robinson
nevadense = H. aureum
nitens = Tomentypnum nitens
HOMOMALLIUM (Schimper) Loeske, 1907 (HPN)
adnatum (Hedwig) Brotherus
mexicanum Cardot

HUSNOTIELLA (PTT)
revoluta = Didymodon
HYDROGRIMMIA Loeske (GRM)
H. mollis (Bruch & Schimper) Loeske
HYGROAMBLYSTEGIUM (AMB)
fluviatile = Amblystegium varium
irriguum = A. varium
noterophilum = A. varium
orthocladon =A. varium
tenax = A. varium
HYGROHYPNELLA (AMB)
ochracea = Hygrohypnum
HYGROHYPNUM Lindberg, 1872 (AMB)
alpestre (Hedwig) Loeske
bestii (Renauld & Bryhn *ex* Renauld) Holzinger *ex* Brotherus
cochlearifolium (Venturi *ex* de Notaris) Brotherus
dilatatum = H. duriusculum
duriusculum (De Notaris) Jamieson
eugyrium. Err. rep.
luridum (Hedwig) Jennings
molle (Hedwig) Loeske. Err. rep.
ochraceum (Turner *ex* Wilson) Loeske
palustre = H. luridum
smithii (Swartz) Brotherus
styriacum (Limpricht) Loeske
HYLOCOMIASTRUM Fleischer *ex* Brotherus, 1925 (HYL)
pyrenaicum (Spruce) Fleischer
HYLOCOMIUM Bruch & Schimper, 1852 (HYL)
alaskanum = H. splendens
pyrenaicum = Hylocomiastrum
splendens (Hedwig) Bruch & Schimper
HYMENOSTYLIUM Bridel, 1827 (PTT)
recurvirostrum (Hedwig) Dixon
HYPNUM Hedwig, 1801 (HPN)
abietinum = Abietinella
arcuatum = Calliergonella lindbergii
aduncum = Drepanocladus
aduncum var. *giganteum = Drepanocladus*
brandegei = Cirriphyllum cirrosum
campestre = Brachythecium
collinum = Brachythecium
coloradense = Cirriphyllum cirrosum
commutatum = Palustriella
commutatum var. *falcatum = Palustriella falcata*
cupressiforme Hedwig
curvifolium. Err. rep.
diversifolium = Eurhynchium pulchellum
eugyrium. Err. rep.
fastigiatum. Err. rep.
filicinum = Cratoneuron

fluitans = *Warnstorfia*
giganteum = *Calliergon*
hamulosum. Err. rep.
hispidulum = *Campylium*
imponens. Err. rep.
laetum = *Plagiothecium*
lindbergii = *Calliergonella*
molluscum = *Ctenidium*
nitens = *Tomentypnum*
nitidulum = *Isopterygiopsis pulchella*
orthocladon = *Amblystegium varium*
pallescens (Hedwig) P. Beauvois
paludosum = *Helodium*. Err. rep.
patientiae = *Calliergonella lindbergii*
pratense = *Breidleria*
radicale = *Campylium*. Err. rep.
reptile = *H. pallescens*
revolutum (Mitten) Lindberg
rugosum = *Rhytidium*
sprucei = *Platydictya jungermannioides*
strigosum = *Eurhynchium pulchellum*
subtile = *Platydictya*. Err. rep.
uncinatum = *Sanionia*
vaucheri Lesquereux
IMBRIBRYUM Pederson, 2005 (BRY)
　alpinum (Withering) Pederson
　gemmiparum (De Notaris) J. R. Spence
　mildeanum (Juratzka) J. R. Spence
　muehlenbeckii (Bruch & Schimper) Pedersen
ISOPTERYGIOPSIS Iwatsuki, 1970 (HPN)
　alpicola (Lindberg & H. Arnell) Hedenäs
　pulchella (Hedwig) Iwatsuki
ISOPTERYGIUM (PGT)
pulchellum = *Isopterygiopsis*
ISOTHECIUM (BRC)
stoloniferum. Err. rep.
JAFFUELIOBRYUM Thériot, 1928 (GRM)
　raui (Austin) Thériot
　wrightii (Sullivant) Thériot
KIAERIA I. Hagen, 1915 (DCR)
　starkei (Weber & Mohr) I. Hagen
LEPTOBRYUM (Schimper) Wilson, 1855 (MEE)
　pyriforme (Hedwig) Wilson
LEPTODICTYUM (AMB)
humile = *Amblystegium varium*
kochii = *Amblystegium varium*
riparium = *Amblystegium varium*
trichopodium = *Amblystegium varium*
LEPTODON Mohr 1803 (LPT)
　smithii (Hedwig) Weber & Mohr
LEPTOPTERIGYNANDRUM Müller Halle (PTR)
　austroalpinum Müller Halle

LEPTOTRICHUM (DTR)
glaucescens = *Saelania*
LESCURAEA Bruch & Schimper, 1851 (LSK)
arizonae = *Pseudoleskeella*
frigida = *L. saxicola*
　incurvata (Hedwig) Lawton
　patens (Lindberg) H. Arnell & C. Jensen.
　radicosa (Mitten) Mönkemeyer
　saxicola (Bruch & Schimper) Molendo *ex* Lorentz
LESKEA Hedwig, 1801 (LSK)
cyrtophylla = *Pseudoleskeella tectorum*
nervosa = *Leskeella nervosa*
　polycarpa Hedwig
williamsii var. *filamentosa* = *Lescuraea incurvata*
tectorum = *Pseudoleskeella*
LESKEELLA Loeske, 1903 (LSK)
arizonae = *Pseudoleskeella*
　nervosa (Bridel) Loeske
LIMNOBIUM (AMB)
ochraceum = *Hygrohypnum*
palustre = *Hygrohypnum luridum*
LINDBERGIA Kindberg, 1897 (LSK)
　brachyptera (Mitten) Kindberg
MEESIA Hedwig, 1801 (MEE)
　longiseta Hedwig
　uliginosa Hedwig
MEIOTRICHUM G.L.S. Merrill, 1992 (PLT)
　lyallii (Mitten) G.L.S. Merrill
MERCEYA (PTT)
latifolia = *Crumia*
METANECKERA (NCK)
menziesii. Err. rep.
MIELICHHOFERIA Nees & Hornschuch, 1831 (BRY)
　elongata (Hornschuch ex Hooker) Nees & Hornschuch
macrocarpa = *Haplodontium*
mielichhoferi = *mielichhoferiana*
　mielichhoferiana (Funck) Loeske
porsildii = *Haplodontium macrocarpum*
MNIOBRYUM (BRY)
albicans = *Pohlia wahlenbergii*
wahlenbergii = *Pohlia*
MNIUM Hedwig, 1801 (MNC)
affine var. *rugicum* = *Plagiomnium ellipticum*
affine var. *elatum* = *Rhizomnium magnifolium*
ambiguum = *M. thomsonii*
　arizonicum Amann
　blyttii Bruch & Schimper
ciliare = *Plagiomnium*
cuspidatum = *Plagiomnium*
ellipticum = *Plagiomnium*

hornum Hedwig
Mnium lyocopodioides. Err. Rep.
marginatum (Withering) Bridel *ex* P. Beauvois
medium = Plagiomnium
orthorrhynchum = M. thomsonii
pseudopunctatum = Rhizomnium
punctatum = Rhizomnium sp.
punctatum var. *elatum = Rhizomnium magnifolium*
riparium = M. marginatum
rostratum = Plagiomnium rostratum
rugicum = Plagiomnium ellipticum
saximontanum = M. arizonicum
serratum var. *marginatum = M. marginatum*
spinosum (Voit) Schwägrichen.
spinulosum. Err. rep.
stellare. Err. rep.
thomsonii Schimper
MOLENDOA Lindberg, 1878 (PTT)
sendtneriana (Bruch & Schimper) Limpricht
MYURELLA Bruch & Schimper, 1853 (THL)
julacea (Schwägrichen.) Bruch & Schimper
tenerrima (Bridel) Lindberg
NECKERA Hedwig, 1801 (NCK)
douglasii. Err. rep.
menziesii = Metaneckera
oligocarpa Bruch *in* Ångström
pennata var. *tenera = N. oligocarpa*
NIPHOTRICHUM Bednarek-Ochyra & Ochyra, 2003 (GRM)
canescens (Hedwig) Bednarek-Ochyra & Ochyra
NYHOLMIELLA (ORT)
obtusifolia = Orthotrichum
OCHYRAEA (AMB)
alpestris = Hygrohypnum
cochlearifolia = Hygrohypnum
smithii = Hygrohypnum
OCTODICERAS (FSS)
fontanum = Fissidens fontanus
OLIGOTRICHUM (PLT)
lyallii = Meiotrichum
ONCOPHORUS (Bridel) Bridel, 1826 (DCR)
gracilescens = Cynodontium
polycarpus = Cynodontium
virens (Hedwig) Bridel
wahlenbergii Bridel
OREAS Bridel, 1826 (DCR)
martiana (Hoppe & Hornschuch) Bridel
ORTHODICRANUM (DCR)
montanum = Dicranum
rhabdocarpum = Dicranum
strictum = Dicranum tauricum
tauricum = Dicranum
ORTHOGRIMMIA (GRM)

alpestris = Grimmia
caespiticia = Grimmia
donniana = Grimmia
montana = Grimmia
sessitana = Grimmia
ORTHOTHECIUM Schimper, 1852 (HPN)
chryseum (Schwägrichen *ex* Schultes) Bruch & Schimper
diminutivum = Isopterygiopsis pulchella
strictum Lorentz
ORTHOTRICHUM Hedwig, 1801 (ORT)
affine Schrader *ex* Bridel
alpestre Hornschuch *ex* Bruch & Schimper
anomalum Hedwig
cupulatum Bridel
diaphanum Bridel
elegans. Err. rep.
hainesii = O. laevigatum
hallii Sullivant & Lesquereux *ex* Sullivant
hutchinsiae = Ulota. Err. rep.
jamesianum = O. pellucidum
laevigatum Zetterstedt
lyellii. Err. rep.
macounii = O. rupestre
obtusifolium Bridel
pallens Bruch *ex* Bridel
pellucidum Lindberg
pumilum Swartz
raui = O. laevigatum
rupestre Schleicher *ex* Schwägrichen
schimperi = O. pumilum
schlotthaueri = O. laevigatum
shawii. Err. rep.
speciosum. Err. rep.
tenellum. Err. rep.
texanum = O. rupestre
watsonii = O. alpestre
PALUDELLA Bridel, 1817 (MEE)
squarrosa (Hedwig) Bridel
PALUSTRIELLA Ochyra, 1989 (AMB)
commutata. Err. rep.
decipiens. Err. rep.
falcata (Bridel) Hedenäs
PARALEUCOBRYUM Lindberg *ex* Limpricht) Loeske, 1907 (DCR)
enerve (Thedin *ex* C. J. Hartman) Loeske
longifolium (Hedwig) Loeske
sauteri (Bruch & Schimper) Loeske
PHAROMITRIUM (PTT)
subsessile = Pterygoneurum
PHASCUM (PTT)
cuspidatum = Tortula acaulon
PHILONOTIS Bridel, 1827 (BRT)

americana. Err. rep.

calcarea. Err. rep.

fontana (Hedwig) Bridel

marchica (Hedwig) Bridel

muehlenbergii = P. fontana

tenella var. *coloradensis = P. fontana*

seriata. Err. rep.

tomentella = P. fontana

PHYSCOMITRIUM (Bridel) Bridel, 1827 (FNR)

coloradense = P. hookeri

hookeri Hampe

immersum. Err. rep.

latifolium = P. hookeri

pyriforme (Hedwig) Hampe

turbinatum = P. pyriforme

PLAGIOBRYUM Lindberg, 1863 (BRY)

bimum = Ptychostomum

capillare = Rosulabryum

cyclophyllum = Ptychostomum

demissum (Hoppe & Hornschuch) Lindberg

lisae = Ptychostomum

pallens = Ptychostomum

pallescens = Ptychostomum

pseudotriquetrum = Ptychostomum

uliginosum = Ptychostomum

zieri (Hedwig) Lindberg

PLAGIOMNIUM Koponen, 1968 (MNC)

ciliare. Err. rep.

cuspidatum (Hedwig) Koponen

ellipticum (Bridel) Koponen

longirostrum = P. rostratum

medium (Bruch & Schimper) Koponen

rostratum. Err. rep.

PLAGIOPUS Bridel, 1826 (BRT)

oederi = P. oederiana

oederianus (Swartz) Crum & Anderson

PLAGIOTHECIUM Bruch & Schimper, 1852 (PGT)

cavifolium (Bridel) Iwatsuki

denticulatum (Hedwig) Bruch & Schimper

laetum Bruch & Schimper

pulchellum = Isopterygiopsis

PLATYDICTYA Berkeley, 1863 (HPN)

jungermannioides (Bridel) Crum

PLATYGYRIUM Bruch & Schimper, 1851 (HPN)

repens (Bridel) Bruch & Schimper

PLEUROZIUM Mitten, 1869 (HYL)

schreberi (Bridel) Mitten

POGONATUM P. Beauvois, 1804 (PLT)

alpinum = Polytrichastrum

contortum. Err. rep.

urnigerum (Hedwig) P. Beauvois

POHLIA Hedwig, 1801 (BRY)

albicans = P. wahlenbergii

acuminata = P. elongata

andalusica (Höhnel) Brotherus

bolanderi (Lesquereux) Brotherus

bulbifera (Warnstorf) Warnstorf

camptotrachela (Renauld & Cardot) Brotherus

commutata = P. drummondii

cruda (Hedwig) Lindberg

drummondii (Müller Halle) Andrews

elongata. Err. rep.

elongata var. *greenii = P. greenii*

excelsa: Status unknown. Shaw did not treat this species, described from Colorado

filiformis = Bryum julaceum

greenii Bridel

longicollis (Hedwig) Lindberg [the original spelling!]

ludwigii. Err. rep.

nutans (Hedwig) Lindberg

obtusifolia (Bridel) L. Koch

proligera (Lindberg *ex* Breidler) Lindberg *ex* H. Arnell

tundrae Shaw

wahlenbergii (Weber & Mohr) Andrews

POLYTRICHADELPHUS (PLT)

lyallii = Meiotrichum

POLYTRICHASTRUM G. L. Smith, 1971 (PLT)

alpinum (Hedwig) G. L. Smith

formosum (Hedwig) G. L. Smith

longisetum (Swartz *ex* Bridel) G. L. Smith

lyallii = Meiotrichum

POLYTRICHUM Hedwig, 1801 (PLT)

alpinum = Polytrichastrum

commune Hedwig

gracile = Polytrichastrum longisetum

jensenii Hagen

juniperinum Willdenow *ex* Hedwig

longisetum = Polytrichastrum

piliferum Hedwig

strictum Menzies *ex* Bridel

POTTIA (PTT)

bryoides = Tortula protobryoides

cavifolia = Pterygoneurum ovatum

heimii = Hennediella

latifolia = Stegonia

PSEUDISOTHECIUM (BRC)

stoloniferum. Err. rep.

PSEUDOCALLIERGON (Limpricht) Loeske, 1907 (AMB)

angustifolium Hedenäs

turgescens (T. Jensen) Loeske

PSEUDOCROSSIDIUM R. S. Williams, 1915 (PTT)

replicatum (Taylor) Zander
PSEUDOLESKEA (LSK)
  arizonae = Pseudoleskeella
  atrovirens. Err. rep.
  incurvata = Lescuraea
  oligoclada = Lescuraea incurvata
  pallida = Lescuraea radicosa var. *pallida*
  patens = Lescuraea patens
  radicosa = Lescuraea radicosa
PSEUDOLESKEELLA Kindberg, 1896 (LSK)
  arizonae (Williams) Lawton
  catenulata. Err. rep.
  nervosa = Leskeella
  sibirica (H. Arnell) P. Wilson & Norris
  tectorum (Funck *ex* Bridel) Kindberg *ex*
    Brotherus
PTERIGYNANDRUM Hedwig, 1801 (PTR)
  filiforme Hedwig
PTERYGONEURUM Juratzka, 1882 (PTT)
  ovatum (Hedwig) Dixon
  subsessile (Bridel) Juratzka
PTILIUM De Notaris, 1867 (HPN)
  crista-castrensis (Hedwig) De Notaris
PTYCHOSTOMUM Hornschuch, 1822 (BRY)
  amblyodon = P. inclinatum
  angustifolium = P. imbricatulum
  bimum (Schreber) J. R. Spence
  cernuum Hornschuch
  creberrimum (Taylor) Spence & Ramsay
  cryophilum (Mårtensson) J. R. Spence
  cyclophyllum (Schwägrichen) J. R. Spence
  imbricatulum (Müller Halle) Holyoak &
    Pederson
  inclinatum (Swartz *ex* Bridel) J. R. Spence
  knowltonii (Barnes) J. R. Spence
  kunzei (Hornschuch) J. R. Spence
  lonchocaulon (Müller Halle) J. R. Spence
  pallens (Bridel) J. R. Spence
  pallescens (Schleicher *ex* Schwägrichen) J. R.
    Spence
  pendulum Hornschuch
  pseudotriquetrum (Hedwig) Spence & Ramsay
  turbinatum (Hedwig) J. R. Spence
  weigelii (Sprengel) J. R. Spence
PYLAISIA (HPN)
  intricata = Pylaisiella polyantha
  jamesii = Pylaisiella polyantha
  polyantha = Pylaisiella
PYLAISIELLA Kindberg *ex* Grout, 1896 (HPN)
  polyantha (Hedwig) Grout
PYRAMIDULA (FNR)
  tetragona. Err. rep.
RACOMITRIUM (GRM)

  aciculare = Codriophorus
  canescens = Niphotrichum
  fasciculare = Codriophorus
  heterostichum = Bucklandiella
  heterostichum var. *affine = Bucklandiella affinis*
  heterostichum var. *sudeticum = Bucklandiella sudetica*
  macounii = Bucklandiella
  sudeticum = Bucklandiella
RHABDOWEISIA Bruch & Schimper, 1846 (DCR)
  crispata (Withering) Lindberg
RHEXOPHYLLUM Herzog, 1916 (PTT)
  subnigrum (Mitten) Thériot *ex* Hilpert
RHIZOMNIUM (Brotherus) Koponen, 1968
  (MNC)
  magnifolium (Horikawa) Koponen
  pseudopunctatum (Bruch & Schimper)
    Koponen
RHODOBRYUM (BRY)
  roseum. Err. rep.
RHYNCHOSTEGIELLA (BRC)
  compacta = Conardia
RHYNCHOSTEGIUM (BRC)
  pulchellum = Eurhynchium
  serrulatum = Steerecleus
RHYTIDIADELPHUS (Limpricht) Warnstorf,
  1906 (HYL)
  triquetrus (Hedwig) Warnstorf
RHYTIDIOPSIS Brotherus, 1908 (HYL)
  robusta (Hooker) Brotherus
RHYTIDIUM Kindberg, 1882 (RHY)
  rugosum (Hedwig) Kindberg
ROELLIA Kindberg, 1897 (BRY)
  roellii (Brotherus *ex* Röll) Andrews *ex* Crum
ROSULABRYUM J. R. Spence, 1996 (BRY)
  capillare (Hedwig) J. R. Spence
  flaccidum (Bridel) J. R. Spence
  laevifilum (Syed) Ochyra
  torquescens. Err. rep.
SAELANIA Lindberg, 1878 (DTR)
  glaucescens (Hedwig) Brotherus
SANIONIA Loeske, 1907 (AMB)
  uncinata (Hedwig) Loeske
SARMENTYPNUM (AMB)
  sarmentosum = Warnstorfia
SCHISTIDIUM Bruch & Schimper, 1845 (GRM)
  agassizii Sullivant & Lesquereux
  alpicola = S. rivulare
  ambiguum Sullivant
  apocarpum (Hedwig) Bruch & Schimper
  apocarpum subsp. *canadense = apocarpum*
  atrichum (Müller Halle & Kindberg) W. A.
    Weber
  atrofuscum (Schimper) Limpricht

boreale Poelt
brandegei = S. confertum?
confertum (Funck) Bruch & Schimper
dupretii (Thériot) W. A. Weber
flaccidum (De Notaris) Ochyra
frigidum Blom
occidentale (Lawton) Churchill
papillosum. Err. rep. Misidentifications of *S. pulchrum*
pulchrum Blom
rivulare (Bridel) Podpera
strictum (Turner) Loeske *ex* Mårtensson
tenerum (Zetterstedt) Nyholm
trichodon (Bridel) Poelt
SCIUROHYPNUM (BRC)
nelsonii = Brachythecium
oedipodium = Brachythecium
plumosum = Brachythecium
SCLEROPODIUM (BRC)
obtusifolium. Err. rep.
SCORPIDIUM (Schimper) Limpricht,1899 (AMB)
cossonii (Schimper) Hedenäs
revolvens (Swartz *ex* Anonymo) Rubers
scorpioides (Hedwig) Limpricht
turgescens = Pseudocalliergon
SCOULERIA Hooker, 1828 (SCL)
aquatica Hooker. To be expected.
SELIGERIA Bruch &Schimper, 1846 (SLG)
campylopoda Kindberg *ex* Macoun & Kindberg
donniana (J. E. Smith) Müller Halle
tristichoides Kindberg
SPHAGNUM Linnaeus, 1753 (SPH)
acutifolium. Err. rep.
angustifolium (C. Jensen) C. Jensen
balticum (Russow) C. Jensen
capillaceum. Err. rep.
capillifolium. Err. rep.
contortum K. F. Schultz
cuspidatum. Err. rep.
fimbriatum Wilson *ex* Hooker, f.
fuscum (Schimper) Klinggräff
girgensohnii Russow
lindbergii. Err. rep.
platyphyllum (Braithwaite) Warnstorf
recurvum = S. angustifolium
robustum. Err. rep.
russowii Warnstorf
squarrosum Swartz *ex* Crome
subsecundum. Err. rep.
teres (Schimper) Ångström *ex* C. Hartman
warnstorfii Russow
SPLACHNOBRYUM (SPB)
kieneri = Ptychostomum creberrimum

SPLACHNUM Hedwig, 1801 (SPL)
sphaericum Hedwig
STEERECLEUS H. Robinson, 1987 (BRC)
S. serrulatus (Hedwig) H. Robinson
STEGONIA Venturi, 1883 (PTT)
latifolia (Schwägrichen *ex* Schultes) Venturi *ex* Brotherus
STEREODON (HPN)
complexus = Hypnum cupressiforme
plicatilis = Hypnum revolutum
STRAMINERGON Hedenäs, 1993 (AMB)
stramineum (Bridel) Hedenäs
STROEMIA (ORT)
obtusifolia = Orthotrichum
SWARTZIA (DTR)
montana = Distichium capillaceum
SYNTRICHIA Bridel, 1801(PTT)
bartramii. Err. rep.
cainii (Crum & Anderson) Zander
calcicola Amann
caninervis Mitten
inermis = Tortula
intermedia = S. montana
montana. Err. rep.
norvegica Weber & Mohr
pagorum (Milde) Amann
papillosa (Wilson) Juratzka
papillosissima (Coppey) Loeske
ruralis (Hedwig) Weber & Mohr
sinensis (Müller Halle) Ochyra
subulata var. *inermis = Tortula inermis*
virescens (De Notaris) Ochyra
TAYLORIA Hooker, 1816 (SPL)
acuminata Hornschuch
froelichiana (Hedwig) Mitten *ex* Lindberg
hornschuchii (Greville & Arnott *ex* Arnott) Brotherus
lingulata (Dickson) Lindberg
splachnoides. Err. rep.
TETRAPHIS Hedwig, 1801 (TTR)
pellucida Hedwig
TETRAPLODON (SPL)
urceolatus = T. mnioides
mnioides. Err. rep.
THUIDIUM (THD)
blandowii = Helodium
abietinum = Abietinella
TIMMIA Hedwig, 1801 (TMM)
austriaca Hedwig
bavarica = T. megapolitana
megapolitana Hedwig var. **bavarica** (Hessler) Brassard
norvegica Zetterstedt

TOMENTYPNUM Loeske, 1911 (BRC)
    nitens (Hedwig) Loeske
TORTELLA Limpricht, 1888 (PTT)
    alpicola Dixon
    arctica (H. Arnell) Crundwell & Nyholm
    fragilis (Hooker *ex* Drummond) Limpricht
    rigens. Err. rep.
    tortuosa (Hedwig) Limpricht
TORTULA Hedwig, 1801 (PTT)
    acaulon (L. *ex* Withering) Zander
    atrovirens (J. E. Smith) Lindberg
    bartramii = Syntrichia
    bistratosa = Syntrichia caninervis
    brevipes = T. muralis
    canescens. Err. rep.
    caninervis = Syntrichia
    cernua (Hübener) Lindberg
    eucalyptrata = T. hoppeana
    euryphylla = T. hoppeana
    guepinii (Bruch & Schimper) Brotherus
    hoppeana (Schultz) Ochyra
    inermis (Bridel) Montagne
    intermedia = Syntrichia montana
    latifolia = T. hoppeana
    laureri C. F. Schultz) Lindberg
    leucostoma R. Brown) Hooker & Greville
    mucronifolia Schwägrichen
    muralis Hedwig
    nevadensis (Cardot & Thériot) Zander
    norvegica = Syntrichia
    obtusifolia (Schwägrichen) Mathieu
    pagorum = Syntrichia
    papillosa = Syntrichia
    papillosissima = Syntrichia
    plinthobia (Sullivant & Lesquereux) Austin
    protobryoides Zander
    ruraliformis. Err. rep.
    ruralis = Syntrichia
    subulata. Err. Rep.
    systylia (Schimper) Lindberg
TRICHOSTOMOPSIS (PTT)

    australasiae. Err. rep.
TRICHOSTOMUM Bruch, 1829 (PTT)
    alpinum = Dicranoweisia crispula
    crispulum. Err. rep.
    cylindricum = T. tenuirostre
    perligulatum = T. planifolium
    planifolium (Dixon) Zander
    sweetii = T. planifolium
    tenuirostre (Hooker & Taylor) Lindberg
ULOTA (ORT)
    americana. Err. rep.
    hutchinsiae. Err. rep.
VOITIA Hornschuch, 1818 (SPL)
    nivalis Hornschuch
WARNSTORFIA Loeske, 1907 (AMB)
    exannulata (Bruch & Schimper) Loeske
    fluitans (Hedwig) Loeske
    sarmentosa (Wahlenberg) Hedenäs
    trichophylla. Err. rep.
WEBERA (BRY)
    acuminata = P. elongata
    commutata = Pohlia drummondii
    cruda = Pohlia
    elongata var. *humilis = P. greenii*
    ludwigii. Err. rep.
    nutans = Pohlia nutans
WEISSIA Hedwig, 1801 (PTT)
    andersoniana = W. ligulifolia
    cirrhata = Dicranoweisia. Err. rep.
    condensa (Voit) Lindberg
    controversa Hedwig
    crispula = Dicranoweisia
    euteiches = W. ligulifolia
    glauca = W. condensa
    ligulifolia (Bartram) Grout
    perligulata = Trichostomum planifolium
    viridula = W. controversa
ZIERIA (BRY)
    demissa = Plagiobryum
ZYGODON (ORT)
    lapponicus = Amphidium

CATALOG: LIVERWORTS AND HORNWORTS

This list attempts to account for all names used in publications on the Colorado bryophyte flora, and their present dispositions. Names in bold face are adopted in this book. Names in italic face represent synonyms or erroneous reports (Err. rep.).

ANEURA Dumortier, 1822 (ANR)
  pinguis (L.) Dumortier
ANTHELIA, (Dumortier) Dumortier, 1835 ANH
  juratzkana (Limpricht) Trevisan
ANTHOCEROS (ANT)
  laevis = Phaeoceros
APOMETZGERIA Kuwahara, 1966 (MTZ)
  pubescens (Schrank) Kuwahara
ASTERELLA P. Beauvois, 1895 (AYT)
  gracilis (F. Weber) Underwood
  lindenbergiana (Corda *ex* Nees) Arnell
  ludwigii: erroneous name for *A. gracilis*
ATHALAMIA Falconer, 1851 (CLV)
  A. hyalina (Sommerfelt) Hattori
BARBILOPHOZIA (LPH*)*
  attenuata = Lophozia attenuata
  barbata = Lophozia barbata
  floerkei = Lophozia floerkei
  hatcheri = Lophozia hatcheri
  kunzeana = Lophozia kunzeana
  lycopodioides = Lophozia lycopodioides
BLASIA L., 1753 (BLS)
  pusilla (L.) Micheli
BLEPHAROSTOMA (Dumortier) Dumortier, 1835 (TRC)
  B. trichophyllum (L.) Dumortier
CALYPOGEIA Raddi, 1818 (CLY)
  fissa (L.) Raddi
  integristipula Stephani
  muelleriana (Schiffner) Müller Fribourg
  neesiana (Massalongo & Carestia) Müller Fribourg
  suecica (Arnell & Persson) Müller Fribourg
CARPOBOLUS (NTT)
  orbicularis = Notothylas
CEPHALOZIA (Dumortier) Dumortier, 1835 (CPH)
  ambigua = C. bicuspidata
  bicuspidata (L.) Dumortier
  binsteadii = C. bicuspidata
  pleniceps (Austin) Lindberg
CEPHALOZIELLA (Spruce) Schiffner, 1893 (CZL)
  arctica = C. varians
  byssacea = C. divaricata
  divaricata (Smith) Schiffner
  hampeana (Nees) Schiffner

  rubella (Nees) Douin
  varians (Gottsche) Stephani
CHILOSCYPHUS Corda *in* Opiz, 1829 (GCL)
  fragilis = C. pallescens
  minor = Lophocolea
  pallescens (Ehrhart *ex* Hoffmann) Dumortier
  polyanthus (L.) Corda
  profundus = Lophocolea heterophylla
CLEVEA (CLV)
  hyalina = Athalamia
CONOCEPHALUM Hill, 1773 (CNC)
  conicum. Err. rep.
  salebrosum Szweykowski et al.
FRULLANIA Raddi, 1818 (FRL)
  brittoniae Evans
  inflata Gottsche
  riparia Hampe
GEOCALYX Nees, 1833 (GCL)
  graveolens (Schrader) Nees
GRIMALDIA (AYT)
  fragrans = Mannia
GYMNOCOLEA (Dumortier) Dumortier, 1835 (LPC)
  inflata (Hudson) Dumortier
GYMNOMITRION Corda *in* Opiz, 1829 (GYM)
  concinnatum (Lightfoot) Corda
  corallioides Nees
HAPLOMITRIUM Nees, 1833 (HPL)
  hookeri (J. E. Smith) Nees
HARPANTHUS Nees, 1836 (GCL)
  flotovianus (Nees) Nees [to be expected]
JUNGERMANNIA L., 1753 (JNG)
  atrovirens = J. lanceolata
  cordifolia = J. exsertifolia
  exsertifolia Stephani
  hyalina Lyell
  lanceolata L. emend. Grolle
  leiantha = J. subulata
  obovata Nees
  pumila Withering
  rubra Underwood
  sphaerocarpa Hooker
  subulata Evans
LEIOCOLEA (LPH)
  badensis = Lophozia badensis
  bantriensis = Lophozia bantriensis
  collaris = Lophozia alpestris

heterocolpos = *Lophozia heterocolpos*
muelleri = *Lophozia alpestris*
LEPIDOZIA (Dumortier) Dumortier, 1835 (LPD)
  reptans (L.) Dumortier
LOPHOCOLEA (Dumortier) Dumortier, 1835 (GLC)
  heterophylla (Schrader) Dumortier
  minor Nees
LOPHOZIA (Dumortier) Dumortier, 1835 (LPH)
  alpestris (Schleicher *ex* F. Weber) Evans
  ascendens (Warnstorf) Schuster
  attenuata (Martius) Dumortier
  badensis (Gottsche) Schiffner
  bantriensis (Hooker) Stephani
  barbata (Schmidel *ex* Schreber) Dumortier
  collaris = *L. alpestris*
  excisa (Dickson) Dumortier
  floerkei (Weber & Mohr) Schiffner
  guttulata = *L. longiflora*
  hatcheri (Evans) Stephani
  heterocolpos (Thedin *ex* Hartman) M. Howe
  incisa (Schrader) Dumortier
  kunzeana (Hübener) Evans
  longidens (Lindberg) Macoun
  longiflora (Nees) Schiffner
  lycopodioides (Wallroth) Cogniaux
  muelleri = *L. alpestris*
  obtusa (Lindberg) Evans
  opacifolia = *L. incisa*
  porphyroleuca = *L. longiflora*
  ventricosa (Dickson) Dumortier
  wenzelii (Nees) Stephani
LUNULARIA Adanson, 1753 (LNL)
  cruciata (L.) Dumortier
MANNIA Opiz, 1829 (AYT)
  fragrans (Balbis) Frye & Clark
  pilosa (Hornemann) Frye & Clark
  rupestris (Nees) Frye & Clark
MARCHANTIA L., 1753 (MRC)
  alpestris Nees
  polymorpha L.
MARSUPELLA Dumortier, 1822 (GYM)
  emarginata (Ehrhart) Dumortier
  sphacelata (Giesecke) Dumortier
MASSULA (LPH)
  incisa = *Lophozia*
NARDIA S. F. Gray, 1821 (JNG)
  geoscyphus (De Notaris) Lindberg
  scalaris (Schrader) Dumortier
NOTOTHYLAS Sullivant, (NTT)
  orbicularis (Schweinitz) Sullivant
ORTHOCAULIS (LPH)
  gracilis = *Lophozia attenuata*

kunzeana = *Lophozia*
PELLIA Raddi, 1818 (PLL)
  endiviifolia. Err. rep.
  fabbroniana = *P. endiviifolia*
  epiphylla. Err. rep.
  neesiana (Gottsche) Limpricht
PHAEOCEROS Proskauer, 1951 (ANT)
  laevis (L.) Proskauer
PLAGIOCHASMA Lehmann & Lindenberg, 1832 (AYT)
  rupestre (Forster) Stephani
  wrightii Sullivant
PLAGIOCHILA (Dumortier) Dumortier, 1835 (PLC)
  asplenoides (L.) Dumortier subsp. **porelloides** (Nees) Lindenberg
  asplenoides subsp. **arctica** (Bryhn & Kaalas) Schuster
PLECTOCOLEA (JNG)
  hyalina = *Jungermannia*
  obovata = *Jungermannia*
PORELLA L., 1753 (PRL)
  cordaeana (Hübener) Moore
  platyphylla (L.) Pfeiffer
PREISSIA Corda *in* Opiz, 1829 (MRC)
  quadrata (Scopoli) Nees
RADULA Dumortier, 1822 (RDL)
  complanata (L.) Dumortier
REBOULIA Raddi, 1818 (AYT)
  hemispherica (L.) Raddi
RICCARDIA S. F. Gray, 1821 (ANR)
  multifida (L.) S. F. Gray
  pinguis = *Aneura*
RICCIA L., 1753 (RCC)
  austinii Stephani
  beyrichiana Hampe
  cavernosa Hoffmann emend. Raddi
  crystallina = *R. cavernosa*
  fluitans L.
  frostii Austin
  lescuriana = *R. beyrichiana*
  sorocarpa Bischoff
RICCIOCARPUS Corda *in* Opiz, 1829 (RCC)
  natans (L.) Corda
SACCOBASIS (LPH)
  polita = *Tritomaria*
SCAPANIA (Dumortier) Dumortier, 1835 (SCP)
  brevicaulis Taylor
  curta (Martius) Dumortier
  cuspiduligera (Nees) Müller Fribourg
  degenii = *S. brevicaulis*
  fulfordiae Hong
  glaucocephala (Taylor) Evans

gymnostomophila Kaalaas
hyperborea E. H. Jørgensen
irrigua (Nees) Dumortier
mucronata Buch
paludosa (Müller) Müller Fribourg
subalpina (Nees) Dumortier
uliginosa (Swartz) Dumortier
undulata (L.) Dumortier
SOLENOSTOMA (JNG)
cordifolium = Jungermannia exsertifolia var. *cordifolia*

pumilum = Jungermannia
sphaerocarpa = Jungermannia
TRICHOSTYLIUM (ANR)
pinguis = Aneura
TRITOMARIA Schiffner *ex* Loeske, 1909 (LPH)
exsecta (Schmidel) Schiffner
exsectiformis (Breidler) Loeske
polita (Nees) E. H. Jørgensen
quinquedentata (Hudson) Buch

www.ingramcontent.com/pod-product-compliance
Lightning Source LLC
Chambersburg PA
CBHW080609270326
41928CB00016B/2982